D1479424

The Palgrave Handbook of Bottom-Up Urbanism

Mahyar Arefi · Conrad Kickert
Editors

The Palgrave
Handbook of
Bottom-Up Urbanism

Editors
Mahyar Arefi
Department of Planning and Landscape
    Architecture
University of Texas at Arlington
Arlington, TX, USA

Conrad Kickert
School of Planning
University of Cincinnati
Cincinnati, OH, USA

ISBN 978-3-319-90130-5   ISBN 978-3-319-90131-2   (eBook)
https://doi.org/10.1007/978-3-319-90131-2

Library of Congress Control Number: 2018939742

Cover image: © Adrienne Bresnahan

Printed on acid-free paper

This Palgrave Macmillan imprint is published by the registered company Springer International Publishing AG part of Springer Nature
The registered company address is: Gewerbestrasse 11, 6330 Cham, Switzerland

CONTENTS

Notes on Contributors

Stefan Al is an architect and urban designer based in New York whose design credentials include work on the 2000-tall Canton tower, briefly the world's tallest tower. His research focuses on high-density and transit-oriented cities, urbanization in developing countries, and climate change adaptation.

Nezar AlSayyad is a Professor of Architecture, City Planning, Urban Design, and Urban History at the UC Berkeley College of Environmental Design. He is the Faculty Director of the Center for Arab Societies and Environments Studies (CASES) and is currently serving as a President of the International Association for the Study of Traditional Environments (IASTE) and Editor of its journal *Traditional Dwelling and Settlements Review.* He has authored and edited numerous books in the field of urban studies.

Mahyar Arefi is a Professor of Planning at the University of Texas at Arlington, USA. He received his Ph.D. in Planning from the University of Southern California. In addition to his books, his publications have appeared in edited volumes and peer-reviewed flagship journals, and he is the Co-Editor of Palgrave's *Urban Design International.*

Evangelos Asprogerakas is an Assistant Professor of Spatial Planning in the Department of Spatial Planning and Regional Development at the University of Thessaly, Greece. His research interests include spatial planning and governance, urban network policy and development planning, and urban regeneration.

David Brain is a Professor of Sociology and Environmental Studies at the New College of Florida. His research and teaching interests focus on the connections between place-making, community-building, and civic engagement, and on sociological issues related to the planning and design of good neighborhoods, humane cities, and sustainable development at the regional scale.

Stephen Coyle is an urban planner, architect, developer, and an international expert in sustainable and resilience in development. He cofounded the National Charrette Institute, now part of Michigan State University's School of Planning, Design and Construction, the board of the California Chapter, Congress for the New Urbanism, Ingénieurs sans frontières, Gabon (Engineers Without Borders) and is currently Community Development Director in Woodland, CA.

Rosa Danenberg is a Ph.D. Candidate at KTH Royal Institute of Technology, and freelance urbanist in the field of Urban Planning and Design. Her research focuses on the intersection between buildings, public space, and the migrant economy.

Hank Dittmar is a global sustainability authority and urbanist, advising governments, companies and communities all over the world on making cities and towns more liveable and resilient. He is a former Chief Executive of The Prince's Foundation for the Built Environment in London, Board Chair of the Congress for the New Urbanism, and President and CEO of Reconnecting America.

Gordon C. C. Douglas is an Assistant Professor of Urban and Regional Planning, and Director of Metropolitan Studies at San Jose State University. His research addresses questions of local identity, peoples' relationships to their physical surroundings, and social and spatial inequality in the city.

Kim Dovey is a Professor of Architecture and Urban Design in the faculty of Architecture, Building, and Planning at the University of Melbourne. His research on social issues in architecture and urban design has included investigations of the relations of place to power, urban informality and the politics of public space.

Sujin Eom is an historian of the built environment and holds a Ph.D. in architecture from the University of California, Berkeley. She is currently working on a book manuscript that examines the mobility of architecture and urban form, with a regional focus on East Asia, interrogating how technologies of dividing cities are produced and circulated at specific historical junctures.

Tigran Haas is an Associate Professor of Urban Design and Planning and the Director of the Centre for the Future of Places and Director of the Graduate Program in Urbanism at the School of Architecture and the Built Environment at KTH. His research and teaching focus on contemporary trends and paradigms in urban planning and design, new urbanism, sustainable urbanism, social housing and urban transformations, and city development.

Paul M. Hess is an Associate Professor int the Department of Geography and Planning at the University of Toronto. His research focuses on pedestrian environments and design, streets as public space, transport equity, and suburban form and transformation.

Jeffrey Hou is a Professor of Landscape Architecture and Adjunct Professor in Architecture and Urban Design and Planning at the University of Washington, Seattle. His work focuses on community design, design activism, public space and democracy, and transcultural urbanism.

Douglas S. Kelbaugh FAIA is Professor and former Dean at the University of Michigan's Taubman College of Architecture and Urban Planning. He has taught design at nine schools of architecture in the USA, Europe, Japan, and Australia and is a former principal in Kelbaugh, Calthorpe and Associates in Seattle and in Kelbaugh+Lee in Princeton, New Jersey, both award-winning firms. Among many other writings, he coauthored the national best seller *The Pedestrian Pocket Book*.

Luna Khirfan is an Associate Professor at the School of Planning at the University of Waterloo. Her current research focuses on community climate change adaptation, community engagement, participatory planning, and urban governance, knowledge transfer, and historic preservation.

Conrad Kickert is an Assistant Professor of Urban Design at the University of Cincinnati, USA. He received his Ph.D. in Architecture from the University of Michigan. He has worked as an urban researcher and designer in The Netherlands, the UK, and the USA, specializing in urban morphology.

Petra Kuppinger is a Professor of Anthropology at Monmouth College, Illinois and former President of the Society of Urban National and Transnational Anthropology. Her research focuses on cities and spaces; space and power; Middle Eastern cities and space culture and Islam in Germany.

Anastasia Loukaitou-Sideris is an Associate Provost for Academic Planning at UCLA and a Professor of Urban Planning at the UCLA Luskin School of Public Affairs. Her research focuses on the public environment of the city, its physical representation, aesthetics, social meaning and impact on the urban resident.

Vikas Mehta is an Associate Professor of Urbanism at the University of Cincinnati's School of Planning. He is the Fruth/Gemini Chair and Ohio Eminent Scholar of Urban/Environmental Design and the school's coordinator for the Urban Design Certificate. Dr. Mehta's work focuses on the role of design and planning in creating a more responsive, equitable, stimulating and communicative environment.

Neda Mohsenian-Rad is an Architect and Urban Designer in Atlanta. She graduated from the University of Cincinnati's School of Planning graduate program on the redesign of downtown Cincinnati.

Vinit Mukhija is a Professor and Department Chair of Urban Planning in the Luskin School of Public Affairs at the University of California, Los Angeles (UCLA). His research focuses on informal housing and slums in developing countries and "Third World-like" housing conditions (including colonias, unpermitted trailer parks, and illegal garage apartments) in the USA.

Kristien Ring is an Assistant Professor of Architecture at the University of South Florida's School of Architecture and Community Design. She is the Principal of the interdisciplinary studio AA-PROJECTS, which is engaged in the production of interdisciplinary projects on future-oriented themes in the realm of architecture and urban planning and was the Founding Director of the DAZ German Architecture Center, Berlin.

Konstantinos Serraos is a Professor in Urban Planning and Design at the National Technical University of Athens (NTUA), School of Architecture. He is an architect engineer and city and regional planner.

Scott Shall is an Associate Professor of Architecture and Associate Dean of the College of Architecture and Design at Lawrence Technological University. His work focuses on developing effective socially responsive design and pedagogic practices for architects, planners, and others working within extra-legal settlements.

Quentin Stevens is an Associate Professor of urban design at RMIT University in Melbourne. His research focuses on people's perception and behavior in urban open spaces, with a particular interest in playful appropriations of underused spaces, waterfronts, memorials, and public artworks.

Emily Talen is a Professor of Urbanism at the University of Chicago. Her research is devoted to urban design and urbanism, especially the relationship between the built environment and social equity.

LIST OF FIGURES

LIST OF TABLES

CHAPTER 1

Introduction

Conrad Kickert and Mahyar Arefi

This book delves into a question that has become increasingly prevalent over the past decade: Who shapes our cities? For over a century the answer to this question included a range of usual suspects, dependent on the global context. In the Modern era of the Global North, cities were usually shaped by governments, giving way over the decades to powerful private interests. The inflexibility of this order has prompted some of the most influential urban theorists of the twentieth century such as Jane Jacobs, Jan Gehl, and William Whyte to reappreciate individual agency in the Modern city, and Kevin Lynch to argue for a more democratic urban environment in which citizens can achieve a better 'fit' for their needs (Gehl 1987; Jacobs 1961; Lynch 1984; Whyte 1988). In the Global South cities have been shaped by a far more complex interplay of governmental and non-governmental organizations, politicians, and citizens. Conversely to the perhaps overly ordered environment of Western cities, informal cities in developing countries have often struggled to maintain a physical, social, and economic infrastructure for their populations, prompting concern among global organizations such as the United Nations and the World Health Organization. The status quo between Northern rigidity and Southern informality has increasingly been uprooted from the bottom up, shaking up global distinctions. This book explores the various new ways in which our urban environment is transformed by new stakeholders, challenging the public–private hegemony in Western countries, while bolstering a new urban order in the Global South.

C. Kickert (✉)
School of Planning, University of Cincinnati, Cincinnati, OH, USA

M. Arefi
Department of Planning and Landscape Architecture,
University of Texas at Arlington, Arlington, USA

M. Arefi and C. Kickert (eds.), *The Palgrave Handbook of Bottom-Up Urbanism*, https://doi.org/10.1007/978-3-319-90131-2_1

This book expressly bridges current bottom-up theories and their implementations across the globe. First, it compares the main schools of thought on bottom-up and informal urbanism by introducing a dialogue between key thinkers in the field of urban planning, architecture, sociology, and anthropology. Rather than picking sides in the increasingly crowded landscape of urbanisms, this book surveys their differences and similarities. Furthermore, it compares and contrasts the bottom-up initiatives that fulfill the unmet desire for spontaneity and serendipity in the orderly city with the order and ordering of informality in the overwhelmed city. The new agency of bottom-up urbanism can complement and describe the urban condition across the globe. By connecting the intellectual positions of bottom-up urbanism and the geographical positions of informality, the book provides an unprecedentedly detailed and comprehensive perspective on agency in the contemporary global city.

New Urbanists and New Urbanisms?

The question of agency in cities is as old as their order itself. Commonly accepted as a manifestation of power, the ordering of urban environments has consistently reflected the relations between rulers and citizens. While famous for the order of the gridded castra, Romans were unable to keep their capital city from devolving into a notoriously overcrowded and unsanitary hotbed for dissent. When London burned down in the seventeenth century, institutional structures could not prevent citizens rebuilding to largely the same structure, despite visionary plans and legislative support. Efforts to structure the city were only successful in small doses, from Paris' Place Royale to London's Covent Garden, or on more or less virgin land, such as Rome's restructuring under Pope Sixtus V and Amsterdam's new canal districts in its Golden Age. An inner-urban leap in the scale of urban order involved new and largely undemocratic power structures, prompting Hausmann's restructuring of Paris under Napoleon Bonaparte, and the construction of the Viennese Ringstrasse under Joseph II—arguably the onset of urban modernity (Abrahamse 2011; Berman 1980; Morris 1972; Mumford 1961). This shift prompted the first pushback among urban observers and scholars, such as Charles Baudelaire who described the lingering inequality behind Paris' new spatial order (1869), or Camillo Sitte's reappreciation of the bottom-up 'artistic' order of medieval squares as opposed to Vienna's new technocratic urban order (Sitte 1889). Sitte was one of the first urban observers to discover the "organized complexity" behind the emergence, evolution, and functioning of the urban environment as a self-organizing process, anticipating the human and urban ecologies of Geddes (1915), Park and Burgess (1925), the urban manifestos of Jacobs (1961, 1969), the pattern language of Alexander et al. (1977), and Contextualists like Rowe and Koetter (1984: 128–129; Shane 2004).[1] Meanwhile, the Structuralist movement that started in Europe in the 1960s acknowledged the different dynamics between individual, collective and institutional agency (e.g. Habraken and Teicher 2000). As Chapter 2

---

[1] An excellent critical overview of agency in urban history is provided in Stephen Marshall's *City Design and Evolution* (2009).

by Jeffrey Hou in this volume explains, bottom-up urbanism can almost read as a counter-narrative of our conventional history of Western urban development.

Urban agency has taken a decidedly different shape outside of Europe. In the fever of North American urbanization, the speculative forces that shaped the city commonly considered any spatial ordering beyond the perceived neutrality of the grid as a burden to progress. American urban planning has surprising roots as a bottom-up effort of urban beautification, originating in local art and civic societies, only later taking root as the Progressive start to professional city planning and urban reform (Peterson 1976; Talen 2015). While the urban grid arguably originates in prehistoric India and China, top-down urban order often has colonial underpinnings in the Global South, from the sixteenth-century Laws of the Indies imposed on many Central and South American cities (Morris 1972: 302–306), to Lutyen's New Delhi (Hall 2002: 198–206; Ridley 1998), and Geddes' designs for many other Indian cities (Tyrwhitt 1947). Even in the post-colonial era, many countries continued to look toward the West for urban order, with Le Corbusier, Gropius, and Wright inspiring or building projects in South America and the Middle East, followed by Doxiades and a slew of internationally active architecture–engineering conglomerates. Today, the renderings for new African, Middle Eastern, and Asian middle-class developments are frighteningly difficult to distinguish, as the Generic City is emerging across the globe (Koolhaas and Mau 1995; Murray 2017). Resistance has grown against these top-down or "top-up" visions of urban order, as explained in Chapter 17 by Nezar AlSayyad and Sujin Eom. By and large, these external orders have had relatively little impact on the overwhelmed cities in the Global South, as the majority of urban growth still happens outside formally designed global enclaves in the form of rapidly expanding informal settlements. Rather than responding to local or national authorities, most cities in Latin America, Africa, and Asia are shaped by a growing influx of individual agents responding to global forces of migration and opportunity (Burdett and Sudjic 2007). In the Global South, the formally designed and planned city is the counter-narrative against a domination of informal urban development.

Urban informality has become an increasingly salient topic in contemporary scholarly and professional debates on the city due to the fact that these counter-narratives have grown in importance. The division between formal development in the Global North and informal development in the Global South has started to blur. An increasing number of scholars—such as Margaret Crawford, Jeffrey Hou, Vinit Mukhija, and Anastasia Loukaitou-Sideris—recognize that informality is also part of Western cities, and many of these authors elaborate on this standpoint in this book. Conversely, informal settlements in the Global South are increasingly transforming and formalizing through various strategies and by new agents, often related to global power structures such as the United Nations, the World Bank, and the International Monetary Fund, trends that this book also surveys. Increasingly, bottom-up and top-down trends are becoming intermingled, and new agents are look-ing for a middle ground between individual customization and civic improve-ment. Now more than ever, we are seeing new urbanists shaping the city, leading to new forms of urbanism.

WHY NOW?

The question of urban agency that this book posits is an inevitable result of rapidly changing urban conditions over past decades. One of the trends that has accelerated has been the rise of a pluralist perspective on urbanism. In the contemporary era of "splintering urbanism" (Graham and Marvin 2001), even defining the city itself has become far more difficult, let alone defining those who shape it. Some continue to consider the city a monolith with a dominant elitist narrative, while others see it as an amalgam of multiple nar-ratives and realities. The rise of this pluralist perspective reflects the changing ways in which public and private stakeholders' roles, identities, and resources shape and transform a city over time. Major trends in turn, including globali-zation, regionalism, neoliberalism, and immigration, have significantly ampli-fied these forces of change, opening up new avenues for citizen intervention while closing others. Most importantly, they have changed the landscape of urban agency. Globalization, for example, has had tremendous impacts on the transfer of capital, labor, and knowledge, with a highly ambiguous effect on urban democracy. The fluidity of global capital has given rise to global cities that operate in alliances beyond the nation state (Sassen 1994, 2001), eroding its traditional national role in maintaining urban order. Global citi-zens have become similarly footloose, as immigrants and asylum seekers pur-sue new ways of obtaining a 'right to the city' to either assimilate or intensify ethnic or religious differences (Lefebvre 1968; Harvey 2008). By eroding the stabilizing effect of the welfare state, neoliberal orthodoxy has simultaneously promoted the values of individualism over governmental interventions in the market economy, celebrating the virtues of public participation in the deci-sion-making process while simultaneously limiting the power of citizens to shape an increasingly privatized urban environment. The contested territory of these flows, forces, their forms of expression, and the policies associated with them are in a constant state of flux. The aggregation of these voices and forces, as opposed to a single dominant narrative, reflects an ongoing inter-play of aspirations, regulations, restrictions, intentions, and interventions. How can we examine or offer a taxonomy of urbanisms, including top-down and bottom-up, formal and informal?

As a result of this pluralism, urban planners and scholars have increas-ingly accepted the phenomenon of informality—if not always by choice. For many decades city administrators and public officials have branded whatever lay outside of the urban development mainstream as social deviance, from the formation of informal settlements to an increase in spontaneity in the public space, expressed by, for example, graffiti culture and street vending.

Yet these informal urbanisms have become more difficult to contain and control. The advent of globalization, economic liberalization, a massive influx of immigrants, and—more recently—powerful political demonstrations such as 'occupy Wall Street' and the Arab Spring, which resulted in some countries in regime change, have multiplied both the magnitude and complexity of the way citizens not only inhabit but also shape cities and their experiences. While the informal shaping of cities far precedes any planned urban order (Kostof 1991; Morris 1972), and has arguably even led to the rise of formal planning itself (Talen 2015), bottom-up and top-down urbanism have lived in a seeming state of perpetual opposition.

In scholarly circles, the traditionally negative connotation of informality has turned into an energized and more positive discourse on informality as epitomizing manifestations of spontaneity in the urban realm, especially in the Western world. Arising from De Certeau's (1984) reappreciation of individualist ingenuity to navigate and influence institutionalized environments in *The Practice of Everyday Life* and its urban translation in Crawford's *Everyday Urbanism* in 1999, urban scholars and planners have changed their perspective on the relationship between governments, developers and citizens. Whereas Crawford's concept of familiarization aimed to critique the discrepancy between planned and lived environments, urbanists are increasingly looking to actively bridge this divide. In cities across the Western world, scholars and professionals are challenging the status quo of top-down urbanism, seeking alternative models (Francesca 2014; Franke et al. 2015; Kee et al. 2014). If informality marked the main challenge of yesteryear, nowadays, what can be summarized as bottom-up urbanism covers a plethora of forms ranging from Do-It-Yourself and Tactical, to Guerrilla, Insurgent, Everyday, and Lean Urbanism. Each of these forms of lived experience, while considered an archetype of bottom-up informality, has its unique characteristics.

On the other hand, bottom-up urbanism has increasingly become part of the mainstream toolkit of urban intervention and development. Admittedly, some bottom-up movements do continue to reflect a sense of insurgency, an opposition to institutionalized forces. Movements like 'Occupy Wall Street' or demonstrations in Tahrir Square, Cairo, symbolize dissatisfaction and a desire for political change in public spaces by fostering national support. Flash mobbing, on the other hand, showcases the power of social media in reshaping the public space by an instantaneous mobilization of people surrounding a shared concern. Graffiti goes back further in time and constitutes a way of physical identification and protest in public space, even if these are underutilized or abandoned. Other insurgencies, such as street vending, are more passive, but still aim to challenge existing economic systems. Most of these movements aim to challenge the established neoliberal hegemony over space and power, and do so in public space (Hou 2010). How do these antagonistic bottom-up movements relate to informal urbanism? Is illegality a common characteristic of both terms? Do these trends threaten, destabilize, or challenge the political and formal establishments of power?

These questions are increasingly complicated to answer, especially as new bottom-up movements have emerged that are enabled by the very top-down institutions they were originally so distant from. The spontaneous nature of bottom-up interventions has percolated into the mainstream culture of young urbanites, and governments eager to attract them have begun to take heed (Mould 2014). Furthermore, in an age of urban austerity, relinquishing part of the urban development process has become increasingly attractive for governments (Tonkiss 2013). Governments are becoming interested in informality as an urban driver. As described in the work of Mukhija and Loukaitou-Sideris, informality has begun to seek a formal response (2014), and new movements, such as Tactical Urbanism, expressly look to bridge informality and top-down institutional structures (Lydon et al. 2015). Furthermore, the informality that characterizes urban development in many countries in the Global South is increasingly embraced by their governments as assets for empowerment, and not just as pestiferous expressions of social failure. AlSayyad and Eom explain in Chapter 17 how efforts that could be seen as formalization also contain significant informal characteristics. In many cases, governments in developing countries hardly have a choice in this matter, as globalization simply eclipses their search for order. How can planners effectively operate in rather amorphous atmospheres of the supremacy of globalization? Are formal planning agencies succumbing to more globalized forces in some cases, even unbeknown to them?

As informal, bottom-up and town-down urbanisms increasingly intermix, what is the interface between the temporary and tactical versus the permanent and strategic types of planning practice? As David Brain and many others explain in this book, the binary distinction between individual and institutional agency is fictional and will not lead to democratic improvements of the city. Are we experiencing new paradigms that arise from or grow out of the new social, economic, and political necessities of a globalizing era, in which 'transitory' and 'tactical' are becoming the norm, rather than the 'permanent' or the 'orthodoxy of coming of age'? What is the role of the professional if citizens take urban transformations into their own hands? Does bottom-up urbanism make more formal planning and design irrelevant or obsolete? And who are these citizens who shape their environment, and for whom do they take action?

STRUCTURE

This book aims to address the questions posed by this introduction. By addressing the 'what,' 'where,' 'how,' 'when,' and 'who' dimensions of these questions, a more comprehensive, albeit nuanced, understanding of bottom-up urbanism is gained.

In Part I a range of key thinkers in the field provide various theoretical perspectives on what constitutes bottom-up urbanism, framing it in the tension between individual and collective agency and authority. This section demonstrates that informality certainly not only equals insurgency, and that bottom-up tactics do not preclude top-down strategies. Instead, the relationship between both types of agency is more complex. Some authors position bottom-up action in existing institutional frameworks, as in the case of David Brain's argument for the value of design charrettes in shaping cities. Others see room for bottom-up frameworks that exist alongside current governance, like Hank Dittmar and Douglas Kelbaugh's appeal for embracing informality in Western countries through Lean Urbanism. Jeffrey Hou, Anastasia Loukaitou-Sideris, and Vinit Mukhija argue for informality that is more oppositional to formal state structures, and Scott Shall argues for the rise of insurgent urbanism in a world of receding public powers. The variety of voices in this introductory part helps the reader navigate the existing typology of bottom-up urbanisms. Despite their differences, the authors commonly frame bottom-up urbanism as a growing global phenomenon, even if it is described, received, and embedded differently in cultural and political contexts.

The two following parts explore this difference in further detail, focusing on the vastly different meaning of bottom-up urbanism in Global North and Global South. Whereas Western cities are embracing initiatives that introduce spontaneity and individual expression, and are coming to terms with citizen-led initiatives, and even urban insurgency, developing countries face informality on a vastly different scale and from a different angle. This introduces a seeming dichotomy between the popularization of informality in the Global North and the simultaneous rise of formalization efforts in the Global South. The categorization of bottom-up urbanism along the lines of where it takes place helps to define the responses of public and private stakeholders, although many chapters also describe the exceptions to this rule.

Part II describes the rise and reception of bottom-up initiatives in various cities in the United States and Europe. Transatlantic comparisons are complemented by a focus on the role of identity in shaping the urban experience in Western countries by stakeholders that range from squatter groups and religious minorities to urban families and billionaire foundations, in places as varied as Sweden, Germany, Greece, and Turkey. It demonstrates that there is no singular description that fits bottom-up initiatives in the Global North, as they represent a wide range of goals and perspectives. Several chapters present the curiously similar struggles faced by bottom-up projects navigating Western institutional structures: from a lack of cooperation between initiators and authorities in Sweden and Turkey, as described by Rosa Danenberg and Tigran Haas; to the stigmatization of religious minorities in Germany described by Petra Kuppinger. In other cases, governments have embraced bottom-up initiatives, either by choice, as in the case Kristien Ring's description of families self-commissioning urban housing in Germany, or through a lack of alternatives, as described by Greek urbanization in the age of fiscal crisis, as described by Konstantinos Serraos and Evangelos Asprogerakas.

Part III also focuses on the tension between bottom-up urbanism and established social and political structures, but switches perspective to developing countries. The authors show a remarkably similar struggle between the informal and the formal in the Global South to that in the Global North, even when the proportion of both urbanisms is vastly different. As in most Western countries, a tension exists between individual agency, collective demands, and institutional structures to shape the city. Rather than seeing this struggle as binary, the chapters describe how to bridge the divide between bottom-up and top-down urbanism: Vikas Mehta looks at the system of organized complexity behind the seeming disorder of Indian streets, challenging our notions of disorder and obsolescence as did Jane Jacobs in the 1960s; Stephen Coyle applies the Western concept of Lean Urbanism to African development, mixing models of informality; Mahyar Arefi, Neda Mohsenian-Rad, and Luna Khirfan all describe the range of successful and unsuccessful strategies of enabling informal settlements in the Middle East; and Stefan Al warns us about taking formalization too far in China. All the chapters demonstrate the remarkable self-organizing capacity of the South's pervasive informal settlements, concluding that formalizing structures are not always necessary or effective. Similar to the growing acknowledgment of informality in the West, they find a formal order in the developing world.

The final part introduces critical perspectives on salient questions in bottom-up urbanism. Many authors agree that the question of agency in cities is more than binary, and challenge the notion of a democratic, individual, and informal bottom-up movement that resists an undemocratic, institutional, and formal top-down process. Instead, bottom-up movements can represent highly undemocratic and formal structures, and top-down structures have far more symbiosis with informality than expected—for better and for worse. While Nezar AlSayyad and Sujin Eom demonstrate the, often futile, nature of bottom-up urbanism in the Global South, Paul Hess reminds us that the formal structures of cities strongly influence the lived reality and coping mechanisms of its citizens. Moreover, bottom-up urbanism does not always achieve improvements for citizens, and if it does, it may be for a highly select demographic. Seemingly insurgent bottom-up initiatives often play into the hands of receding Western states, and run the risk of becoming simply less accountable mechanisms of gentrification. As Gordon Douglas describes, one person's placemaking can become another person's placetaking, especially in urban environments where space is at a premium. Quentin Stevens and Kim Dovey add that bottom-up initiatives have "ironically become a new form of top-down strategic planning" as they are coopted by neoliberal regimes, raising questions how to govern this new form of agency.

To demonstrate the broad nature and impacts of bottom-up urbanism, the contributors to this volume consist of experts from various disciplines, including planning and public policy, architecture, anthropology, geography, and landscape architecture. Key thinkers on bottom-up urbanism are deliberately complemented by new voices, many of which are critical of the rise of informality in the Global North and formality in the Global South. By comparing and contrasting various views on bottom-up urbanism, this book aims to provide an overview of urban agency that is independent from established schools of thought. To keep abreast of some of the most exciting and contemporary research on bottom-up urbanism, most of these chapters draw on case studies from around the world. These case studies also demonstrate a kaleidoscope of informality, from impoverished slums to a half-billion-dollar Greek library. The proliferation of research on bottom-up initiatives worldwide attests to the fact that it is indeed pervasive, dynamic, and effective, not only in scope, but also in terms of its political legacies and ramifications. In some cases the impacts are short-lived, but they set the tone for more long-term and ambitious socio-economic and political goals. Bottom-up urbanism has grown from an inevitability in the Global South and a trend in the Global North to a permanent and significant movement shaping global cities. This book provides a critical perspective on this movement as part of a growing canon on agency in cities.

REFERENCES

Abrahamse, J. E. (2011). *De grote uitleg van Amsterdam: stadsontwikkeling in de zeventiende eeuw*. Bussum: Thoth.

Alexander, C., Ishikawa, S., & Silverstein, M. (1977). *A pattern language: Towns, buildings, construction*. New York: Oxford University Press.

Baudelaire, C. (1869). *Le Spleen de Paris - Petits Poemes en prose*. Paris: Editions Nelsson.

Berman, M. (1980). *All that is solid melts into air: The experience of modernity*. New York: Viking Penguin.

Burdett, R., & Sudjic D. (2007). *The endless city: An authoritative and visually rich survey of the contemporary city*. London: Phaidon Press.

Crawford, M., Chase, J., & Kaliski, J. (1999). *Everyday urbanism*. New York: Monacelli Press.

De Certeau, M. (1984). *The practice of everyday life*. Berkeley: University of California Press.

Francesca, F. (2014). *Make shift city, renegotiating the urban commons*. Berlin: Berlin Jovis Verlag.

Franke, S., Niemans, J., & Soeterbroek, F. (2015). *Het nieuwe stadmaken: van gedreven pionieren naar gelijk speelveld*. Haarlem: Trancity Valiz.

Geddes, P. (1915). *Cities in evolution*. London: Williams & Norgate.

Gehl, J. (1987). *Life between buildings: Using public space*. New York: Van Nostrand Reinhold.

Graham, S., & Marvin, S. (2001). *Splintering urbanism: Networked infrastructures, technological mobilities and the urban condition*. London: Routledge.

Habraken, N. J., & Teicher, J. (2000). *The structure of the ordinary: Form and control in the built environment*. Cambridge, MA: MIT Press.

Hall, P. (2002). *Cities of tomorrow: An intellectual history of urban planning and design in the 20th century*. Oxford: Blackwell.

Harvey, D. (2008). The right to the city. *New Left Review, 53*(September–October), 23–40.

Hou, J. (2010). *Insurgent public space: Guerrilla urbanism and the remaking of contemporary cities*. New York: Routledge.

Jacobs, J. (1961). *The death and life of great American cities.* New York: Random House.

Jacobs, J. (1969). *The economy of cities.* New York: Random House.

Kee, T., Miazzo, F., Pineda Revilla, B., Webster, C., Foundation & Trancity. (2014). *We own the city: Enabling community practice in architecture and urban planning.* Amsterdam: TrancityxValiz.

Koolhaas, R., & Mau, B. (1995). *S, M, L, XL.* New York: Monacelli.

Kostof, S. (1991). *The city shaped: Urban patterns and meanings through history.* Boston: Little, Brown.

Lefebvre, H. (1968). *Le droit a la ville: Suivi de Espace et politique.* Paris: Anthropos.

Lydon, M., Garcia, A., & Duany, A. (2015). *Tactical urbanism: Short-term action for long-term change.* Washington, DC: Island Press.

Lynch, K. (1984). *Good city form.* Cambridge: The MIT Press.

Marshall, S. (2009). *Cities design and evolution.* Abingdon and New York: Routledge.

Morris, A. E. J. (1972). *History of urban form, prehistory to the Renaissance.* London: George Godwin Ltd.

Mould, O. (2014). Tactical urbanism: The new vernacular of the creative city. *Geography Compass, 8*(8), 529–539.

Mukhija, V., & Loukaitou-Sideris, A. (2014). *The informal American city: Beyond taco trucks and day labor.* Cambridge: The MIT Press.

Mumford, L. (1961). *The city in history: Its origins, its transformations, and its prospects* (Vol. 67). New York: Harcourt, Brace & World.

Murray, M. J. (2017). *The urbanism of exception: The dynamics of global city building in the twenty-first century.* Cambridge: Cambridge University Press.

Park, R. E., & Burgess, E. W., (Eds.) (1925). *The city* (1st ed.). Chicago, IL: University of Chicago Press.

Peterson, J. A. (1976). The city beautiful movement: Forgotten origins and lost meanings. *Journal of Urban History, 2*(4), 415–434.

Ridley, J. (1998). Edwin Lutyens, New Delhi, and the architecture of imperialism. *The Journal of Imperial and Commonwealth History, 26*(2), 67–83.

Rowe, C., & Koetter, F. (1984). *Collage city.* Cambridge: The MIT Press.

Sassen, S. (1994). *Cities in a world economy.* Thousand Oaks: Pine Forge Press.

Sassen, S. (2001). *The Global City: New York.* London and Tokyo: Princeton University Press.

Shane, D. G. (2004). *Recombinant urbanism: Conceptual modelling in architecture, urban design and city theory.* New York: Wiley.

Sitte, C. (1889). *Der Städte-Bau nach seinen künstlerischen Grundsätzen: ein Beitrag zur Lösung modernster Fragen der Architektur und monumentalen Plastik unter besonderer Beziehung auf Wien* (2e. Aufl. ed.). Wien: C. Graeser & Co.

Talen, E. (2015). Do-it-yourself urbanism: A history. *Journal of Planning History, 14*(2), 135–148.

Tonkiss, F. (2013). Austerity urbanism and the makeshift city. *City, 17*(3), 312–324.

Tyrwhitt, J., (Ed.) (1947). *Patrick Geddes in India.* London: Lund Humphries.

Whyte, W. H. (1988). *City: Rediscovering the center* (1st ed.). New York: Doubleday.

PART I

Agency Versus Authority

CHAPTER 2

Everyday and Bottom-Up: A Counter-Narrative of American City Design

Jeffrey Hou

On a summer evening in Seattle in 2013, hundreds of after-work urbanites and curious tourists poured into the city's downtown for the *Alleypalooza*, an annual festival taking place in a network of alleys extending from Pike Place Market to Pioneer Square, supported by local advocates and businesses. Empty and deserted on a typical day, the alleys were transformed into a place for live performances, beer gardens, games of ping pong and air hockey, and overflow spaces for nearby galleries and restaurants. Back façades of buildings became a canvas for temporary artworks. The smell of urine and garbage was momentarily substituted by that of free hot dogs and lemonade. The scale and dimensions of the alleys made a perfect setting for an improvised social event—a burst of urban life in the otherwise overlooked and decaying urban environment (Fig. 2.1).

A relic of late nineteenth century city building to bring in air and light and to facilitate garbage pick-up and other services, alleys seem to be unlikely candidates for signifying the rebirth of American inner cities after decades of decline. But in Seattle and other cities from coast to coast, alleyways and other leftover urban spaces have increasingly been transformed and adapted for new and innovative uses, often by citizens themselves. In Detroit, abundant vacant lands have been turned into productive urban farms. In Los Angeles, downtown parking lots are now sites for vibrant night markets with food trucks and

J. Hou (✉)
Department of Landscape Architecture,
University of Washington, Seattle, WA, USA

© The Author(s) 2019
M. Arefi and C. Kickert (eds.), *The Palgrave Handbook of Bottom-Up Urbanism*, https://doi.org/10.1007/978-3-319-90131-2_2

Fig. 2.1   Performances in Seattle's Nord Alley. Photograph by Jeffrey Hou

accessory vendors. In San Francisco, the temporary occupation of street parking spaces through the annual Park(ing) Day event has led to the creation of 'Parklets' through the city's Pavement to Parks program (now Groundplay). Similar initiatives have appeared in Chicago, New York City, and Philadelphia, among other cities.

Under different monikers, including pop-up, tactical, and DIY urbanisms, these movements signal not only a growing interest in everyday urban spaces and activities, but also a counter-discourse against the prevailing practice of city planning and design that has long-favored visionary ideas and master plans over incremental, spontaneous, and adaptive responses. But is the current wave of tactical urban responses a new invention in city design? Have American cities and cityscapes always been a product of top-down professional planning? Has the discourse of the everyday always been outside the mainstream in American city design?

Through a brief account of the evolution of city design discourses in the United States, this chapter examines the influence of the everyday on American city design. Specifically, it argues that rather than an overnight sensation, bottom-up everyday urbanism as evoked in these interventions is in fact a recurring theme in the discourses of city design in America.[1]

---

[1]Talen (2015) makes a similar argument in her work by examining the precedents of DIY urbanism in American history. This chapter builds on similar sources but uses the everyday and bottom-up as a focus for examining the thread that continues to this day. This chapter uses the term 'city design' to encompass historical traditions in the design of American cities, and distinguishes it from the discourse of 'urban design' that emerged in the 1950s.

Instead of esoteric diversions, the everyday experiences of urban dwellers and the non-pedigreed spatial practices of ordinary citizens have arguably been a constant source of inspiration and innovation for theories and practices. Rather than frivolous interventions, bottom-up urbanism and the everyday practices of citizens and communities should be recognized as having a central and defining role in the evolution of discourses and practices in American city design.

Retelling the History of American City Design

The concept of urbanity has a peculiar place in the American consciousness. In a nation in which freehold agrarianism is upheld as a core founding philosophy, where cities are often associated with crime, disease, and social disorder, the 'urban' has a historically negative connotation. As such, the goal of planning and design has long been to bring order and hierarchy to the perceived mess of everyday life. It is in this context that a typical narrative of American city design begins—with the emergence of the Garden City Movement, the City Beautiful Movement, and the Urban Parks movements, as responses to the degenerative conditions of cities in the nineteenth century (Hall 1988). In these movements, introducing order and beauty to cities was a priority. The Garden City movement, as introduced to American cities from Britain, would afford residents with "modest income and good tastes ... country air and country life" within striking distance of the existing city.[2] In the City Beautiful movement, designers and architects envisioned a renewal of civic order and pride through building neoclassical landmarks and large-scale civic spaces, inspired by the grand boulevards of European cities (Hall 1988). Through the Urban Parks Movement, large, picturesque, open spaces were conceived to serve as a refuge for residents of dense neighborhoods, to bring nature to the city and to uplift the spirit of working-class urban dwellers. Instead of treating everyday urban life as a source of inspiration, each of these movements sought to remove urban dwellers from the everyday city, a place synonymous with the social ills of society at the turn of the twentieth century.

In this prevailing narrative of American city design, the everyday is seen as antithetical to the notion of a planned and orderly city. In such context, how did the everyday become a significant and even central part of American city design? How can discourses of American city design be revisited through the lens of the everyday and bottom-up? The following presents a counter-narrative of American city design by weaving together generations of discourses and actions that suggest the significance of the everyday and bottom-up in the evolution of theories and practices in the United States. The sequence here follows both a thematic order and a chronological structure that reflects how such an evolution of thoughts is historically situated and contingent upon major events in society.

---

[2]From the Russell Sage Foundation brochure for the Forest Hills Gardens, New York, one of the early Garden City-inspired projects.

City Making Through Voluntary Civic Actions

Starting with the colonial settlements of the early era, city making and town building in America have long been associated with the exercise of power and authority. But top-down power and authority have not been responsible for the totality of towns and cities. In the case of the City Beautiful movement, for instance, planning historian Jon Peterson points out that before the movement became synonymous with architects such as Daniel Burnham and with a bias toward grandiosity, it had existed mostly as a localized municipal art movement, aiming at incremental, piecemeal improvements of city spaces (Peterson 1983). "Civic improvement began as a laymen's cause and flourished initially in small- and medium-sized cities" (Peterson 1983: 48). Unlike the large-scale interventions that have long been associated with the City Beautiful movement, the earlier work was promoted through thousands of civic improvement associations across the country that engaged in vacant lot cultivations, improvements of church exteriors, and landscaping of factory grounds, school yards, and city streets (Peterson 1983).

Unlike the common association of the City Beautiful movement with the boulevards and promenades of the great European capitals, this alternative history of early planning actions points to a different focus, scale, and approach, one that was strongly related to the everyday and bottom-up, vis-à-vis the visionary and top-down. In chronicling the historical precedents of DIY urbanism, Talen (2015: 140) points to yet another movement, the settlement house movement, that contributed to the creation of "parks, playgrounds, baths, and other facilities at the neighborhood level," distinct from the large-scale, civic beautification projects in late 1800s. She further argues that "[t]oday's DIY urbanism can be interpreted as a revival of a civic spirit that was the hallmark of an earlier era" (Talen 2015: 145).

DIY, indeed, has been a hallmark associated with a strong American tradition. Going further back in time, Hester (1999) traces citizen participation in the United States to the long traditions of local self-help, New England town meetings, civil disobedience, and voluntary associations: "For more than a hundred years ... most civil plans were created and improved through voluntary efforts" (Hester 1999: 15). As evident in these early movements, the everyday activities of ordinary citizens were not an antithesis to, but rather partner of, planning and city design, at least before such activities was eventually subsumed by a process of professionalization. Hester (1999) argues that by the late 1800s, the inability of elected officials to deal with increasingly complex urban problems led to a city manager form of local government and the professionalization of civic affairs. As professionals assumed more responsibilities, citizens' ability and power in deciding local affairs was slowly undermined.

REVOLT AND RESEARCH OF THE EVERYDAY

Through the building of central city freeways, master-planned communities, and large-scale infrastructure projects, the professionalization of planning practice in the USA reached a high point in the post-World War era of the 1950s and 1960s. However, the era of top-down planning also planted the seeds for a radical rethink and a return to the everyday. Specifically, the backlash against urban renewal and top-down planning set the stage for the resurgence of the everyday as a potent ingredient for a more reflexive approach to city design.

During the massive turmoil of the American inner city in the 1950s and 1960s, the familiar narrative began with neighborhoods and ghettos being demolished to make way for freeways and large-scale civic projects. A revolt ensued against the top-down planning machine. In *The Death and Life of Great American Cities*, Jacobs (1961) not only attacked the predominant planning practice of the time, but also drew attention to the nuances of everyday life in an inner-city neighborhood with a focus on the activities that took place on the sidewalks and in urban neighborhoods and the networks, reciprocity, and relationships that formed among neighbors. She further identified density, mixed use, and smaller blocks as critical components of successful cities (Jacobs 1961).

Aside from the work of Jane Jacobs, several other important works emerged at the time, reflecting a renewed, shared interest in the everyday. One such was Kevin Lynch's *The Image of the City* (1960). Using cognitive mapping surveys, he examined how ordinary users in three cities—Boston, Jersey City, and Los Angeles—perceived and navigated the urban landscape. Lynch's work brought attention to everyday experiences and perceptions in the city and the importance to its users of legibility and imageability. Unlike professional dogmas and canons passed down through different schools of design, everyday perceptions of users now formed the foundation for a theory of urban forms.

The work of William H. Whyte also came to prominence around this time. His Street Life Project, conducted in the late 1960s and 1970s, documented everyday uses of public spaces in New York City. Using filmmaking and still-motion cameras, the project contributed new methods to urban design observation and analysis (Larice and Macdonald 2013). By focusing on how urban spaces are used by ordinary people in their everyday activities, the project brought to light the key ingredients for producing a socially successful urban space. Whyte argues that these ingredients include food, water, seating, sunlight, shade, and so on. Later published as *The Social Life of Small Urban Space* (1980), Whyte's work made a lasting impact on the theory and practice of urban design and has been promoted by groups such as the Project for Public Space.

Parallel to the interest in the social life of urban spaces, cultural landscape emerged as a field of study. Through the work of J.B. Jackson, D.W. Meinig, and others, geographers and architectural historians began to give attention to the everyday, vernacular aspects of American landscapes, both rural and urban. Meinig (1979: 6) stated that "in its focus upon the vernacular, cultural landscape study is a companion of that form of social history which seeks to understand the routine lives of ordinary people." Groth (1997: 1, italics in original) further defined cultural landscape studies as focusing most on "the history of how people have used *everyday* space—buildings, rooms, streets, fields, or yards—to establish their identity, articulate their social relations, and derive cultural meaning." For cultural landscape scholars, the subjects include everything, "fences and roads and barns, the design of factories and office buildings, the layout of towns and farms and graveyards and parks and houses" (Jackson 1980: 113). While this focus on the everyday and vernacular was, perhaps, marginal to the mainstream planning and design practice of the time, it profoundly influenced the work of Robert Venturi and Denise Scott Brown whose book *Learning from Las Vegas* (1972) had a much wider, albeit debated, impact on the design profession.[3]

Pioneering work from the 1950s to the 1970s, from Jane Jacobs's critique of the rational planning paradigm and Lynch's and Whyte's research, to the explicit focus on the everyday and vernacular by cultural landscape scholars, together set an important theoretical and epistemological foundation for the evolution of American city design discourse in the decades that followed.

PRESERVATION, URBAN NATURE, AND PUBLIC LIFE

As America's inner cities went through a major transformation in the 1950s and 1960s, the issue of historic preservation also came into focus. With greater attention on the destruction of historic landmarks and the urban fabric, historic preservation became a subject of argument and contestation. In *The Power of Place: Urban Landscapes as Public History* (1995), Dolores Hayden recalls the debate in 1975 between architectural critic Ada Louise Huxtable and urban sociologist Herbert J. Gans concerning the preservation of singular architectural landmarks versus public history as represented by ordinary buildings. Making a passionate argument for "the power of ordinary urban landscapes to nurture citizen's public memory" (Hayden 1995: 9), the book represented the culmination of a project that Hayden began in 1984, focusing on uncovering histories of women and minority groups that have not been adequately represented in the historic preservation practice widely adopted by cities since the 1970s. Hayden's focus on the struggles of minority groups contributed to a much more critical discourse and practice of preservation in the 1990s with a more explicitly political and personal application, in addition to a longstanding focus on ordinary and everyday landscapes in the tradition of cultural landscape studies (Groth 1997).

---

[3]For details of this influence and various other historic connections, see Denise Scott Brown's own account (Scott Brown 2003).

Fig. 2.2 Community gardens, such as this in the West Seattle neighborhood, serve as nearby nature for urban dwellers. Photograph by Jeffrey Hou

If the modern American environmental movement was a product of the 1960s and 1970s, the concept of everyday, nearby nature served as the contribution of the 1980s and 1990s. The most prominent contributors to this work were environmental psychologists Rachel and Stephen Kaplan, who conducted research and wrote extensively on the human perception of nature in this period. Specifically, they focused on the role of everyday nature to the well-being of everyday people (Kaplan et al. 1998), including parks and opens spaces, street trees, vacant lots, and backyard gardens. They found that "the healing power of the restorative experience can be experienced in nearby and undramatic natural environment as well" (Kaplan and Kaplan 1990: 243). In *Community Open Space* (1984), Francis, Cashdan and Paxson reached similar conclusions in their research on community gardens as an alternative form of urban open space, as distinct from traditional parks. They found that these everyday sites promote an increased sense of local pride and a sense of ownership, among other benefits (Francis et al. 1984) (Fig. 2.2). In *The Granite Gardens*, Anne Whiston Spirn (1984) brings light to a full spectrum of complex relationships between everyday natural processes and the health and survival of cities. Together, these research outcomes on urban open spaces have enriched our understanding of the critical role of everyday nature in the urban context. They provided the basis for a subsequent development of discourse on health and places as well as the foundation for the reintegration of city and nature through discourses such as landscape urbanism and urban ecological design.

The decentralization of American cities and the urban sprawl of the 1960s and 1970s, coupled with the privatization and fortification of public spaces, led to growing scholarly attention on the loss of public life. In a manifesto crafted in 1980, and then published in 1987 as an article titled "Toward an Urban Design Manifesto," Allan Jacobs and Donald Appleyard responded to what they perceived as the major problems facing cities at the time—including poor living environments, giantism and loss of control, large-scale privatization and the loss of public life, centrifugal fragmentation, destruction of valued places, placelessness, injustice, and rootless professionalism (Jacobs and Appleyard 1987).

Placelessness, a concept put forward by geographer Edward Relph (1976), became a common critique of American urban spaces. In *Variations on a Theme Park*, Michael Sorkin (1992: xiii–xiv) brought together a group of leading scholars and critics to examine the conditions of "generic urbanism," "obsession with security," and "the city of simulation" in the North American cities of the late-twentieth century. With notions and observations on the Analogous City (Boddy 1992), Exopolis (Soja 1992), Panopticon Mall (Davis 1992), City Tableau (Boyer 1992), and many more, contemporary cities are presented as secured, sanitized, and simulated spaces, as vehicles for consumption (shopping) devoid of public life. Through the introduction of skyways and tunnels, for instance, Boddy (1992: 126) notes, "the messy vitality of the metropolitan condition, with its unpredictable intermingling of classes, races, and social and cultural forms is rejected, to be replaced by a filtered, prettified, homogeneous substitute." One might argue that, in a way, this focus on public life had set the scene for the emergence of New Urbanism with its focus on neo-traditional urban forms as the vessel for urban activities and public life. The growing discussion also set the stage for an expanding body of literature that examined the politics and transformations of public space.

Everyday Urbanism and Tactical Cities

The publication of *Everyday Urbanism* in 1999 marked a new beginning for late-twentieth century discourses on city design with the rediscovery and explicit reaffirmation of the bottom-up and the everyday. Instead of dense urban settings in traditional, mixed-use fabrics of older East Coast cities, however, the iconography of late-twentieth century, everyday urbanism was that of parking lots, garages, roadside signs, fast food restaurants, and gas stations—a defiant departure not only in terms of the locations of everyday places, but also their meanings and aesthetics (Fig. 2.3). Taking cues from Louis Wirth (1938) and the work of French theorists Henri Lefebvre, Guy Debord, and Michelle de Certeau on the practice of the everyday, Margaret Crawford (2008: 6) suggested that "the utterly ordinary reveals a fabric of space and time defined by a complex realm of social practices—a conjuncture of accident, desire, and habit." The focus here was not only the spatial form,

Fig. 2.3 Parking lot shrine in East Los Angeles: adaptation of an everyday urban landscape. Photograph by Jeffrey Hou

but also on the realm of lived space and spatial practice. *Everyday Urbanism* suggests "a radical repositioning of the designer, a shifting of power from the professional expert to the ordinary person" (Crawford 2008: 9). In some ways, the arguments and observations presented in *Everyday Urbanism* set the scene for much of what was to follow.

In the new millennium, as capital investment and the creative class returned to the inner city, many American cities have experienced a social and economic resurgence, albeit with the high cost of widespread gentrification and displacement.[4] The refocus on the urban along with the influx of young professionals has brought new vitality and renewed interest to the remaking of cities and urban places, including direct, spontaneous interventions in everyday settings. In Portland, Oregon, the group City Repair painted street intersections to transform the car-dominated urban landscape into a place of community gathering. The activities that began in the late 1990s continued into the new millennium and have since become Village Building Convergence, an annual placemaking event involving coordination among communities, professionals, volunteers, and city agencies. In 2005, art and design group Rebar staged the first Park(ing) event by transforming a street parking space in San Francisco into a temporary green space.

[4]From 2000 to 2010, the growth of urban populations in the USA outpaced that of the entire nation (US Census 2012).

Fig. 2.4 Park(ing) Day, Seattle (2012). Photograph by Jeffrey Hou

The videos and images of the event went viral on the Internet and led to the founding of Park(ing) Day—as an open-source, annual event that takes place in over a hundred cities globally, with different groups developing their designs and staging their own interventions (Bela 2015) (Fig. 2.4). Direct interventions like these present a renewed engagement in the everyday by both citizens and professionals.

The blossoming of direct action in the new millennium was accompanied by a continued expansion of discourses that took cues from ordinary cityscapes. In *Loose Space: Possibility and Diversity in Urban Life*, Karen Franck and Quentin Stevens (2007) took note of how people in urban places engaged in a rich variety of everyday activities not originally intended for those locations. They argued, "Loose spaces give cities life and vitality" (Franck and Stevens 2007: 4). In *Insurgent Public Space*, my own work examines public spaces as created by ordinary citizens and communities, often outside or on the border of the regulatory domain (Hou 2010). The concept of Insurgent Public Space addresses the political significance of those activities ranging from the work of Rebar to everyday acts of Latino residents in East Los Angeles (Merker 2010; Rojas 2010). What began as defiant images of sidewalks and parking lots in the book *Everyday Urbanism*, have become subjects of further research (see Loukaitou-Sideris and Ehrenfeucht 2009; Ben-Joseph 2012). In *The Informal America City* (Mukhija and Loukaitou-Sideris 2014), such focus expands to front yards, accessary units, community gardens, urban farms, garage sale food trucks, and parking markets. All of these efforts have contributed to a greater understanding of the significance of the everyday and the importance of its engagement.

In recent years, recalling the early development of the City Beautiful movement, the proliferation of actions and discourses has been accompanied by a growing professionalization and institutionalization of the bottom-up initiatives, as exemplified in the acceptance of Tactical Urbanism as a legitimate approach to planning by local governments and professional practices, including followers of New Urbanism. As these approaches are increasingly adopted by local governments, professionals, and even private developers eager for quick fixes, many scholars and practitioners have expressed concerns with its ongoing co-optation. In his review of *Tactical Urbanism* (Lydon and Garcia 2015), Campo (2016: 389) believes that the authors who coin the concept "have distanced themselves from the more radical notions of DIY and the quest for more transformative actions." He argues that Tactical Urbanism is "a counter-urbanism well positioned for the neoliberal world" (Campo 2016: 389). Similarly, Mould (2014: 529) argues that despite the movement's origins in "community-led, activist, and unsanctioned, and even subversive activities, [Tactical Urbanism] is becoming (if it is not already so) co-opted by prevailing 'neoliberal development agendas'. ... [Tactical Urbanism] is being divorced from its citizenry and activist ethos and fast becoming the latest iteration of 'cool', creative urban policy language" (Mould 2014: 529–530). Indeed, in a study of DIY urban design cases in the USA, Douglas (2014: 18) found such actions to be more common in newly hip and "gentrifying" neighborhoods than in impoverished inner-city "ghettos." He also found that the vast majority of individuals engaged in the cases would qualify as members of the so-called "creative class" (Douglas 2014).

This latest twist in the plot of the everyday and bottom-up in American city design is important to note, as it signals on one hand the growing influence and acceptance of the discourse, while on the other hand indicating the persistent tension between informal and formal, unplanned and planned, and bottom-up and top-down in the making of American cities and cityscapes. It also exemplifies the entangled realities of everyday planning practices in the context of seemingly inescapable neoliberal governance.

Learning from the Everyday: A Recurring Theme of American City Design

A century of planning discourse is impossible to capture comprehensively in the limited pages of this chapter. This brief account is by no means complete and exhaustive. Rather, it attempts to highlight and illustrate the influence of the everyday and bottom-up in the short history of American city design. From the civic improvement movements in the nineteenth century, to the revolt against large-scale master planning of the 1950s and 1960s and the development of new theories and discourses in the decades that followed, the everyday and bottom-up has been important in, if not central to, the continued evolution of planning discourses and practices in American cities. In many instances, rather than a momentary preoccupation or esoteric diversion, the everyday and bottom-up have been a consistent source of inspiration and innovation in the evolution of professional thought.

Specifically, the recurring focus on the everyday and bottom-up has reminded planners and designers to consider the city as a lived space, a stage and repository of present life and past memories. It has led planners and designers to see beyond the professional lens to consider phenomena and nuances that are important to the life of urban places. Within the realm of professional discourse and practice, specifically the work of William H. Whyte, Kevin Lynch, and Rachel and Stephen Kaplan, the everyday has contributed to the development and application of methods and techniques for analyzing perceptions and activities of ordinary citizens. In the USA, the focus on the everyday and the experience of the ordinary citizen has also served as an important foundation for the development of citizen participation and engagement in the decision-making process concerning the design of cities.

While the everyday and bottom-up have played a vital role in the development of city planning and design practice, it is important to note that not all attempts have fulfilled their promise and potential. While *Learning from Las Vegas* led to a new appreciation of everyday cultural icons and an understanding of meanings and symbolism, the architectural experiment that followed did not depart much from the formal preoccupations of past practices. What resulted was often not much more than an exercise of coded, symbolic expressions in the form of an architectural style. Similarly, while New Urbanism focuses attention on aspects of urban fabric and urban life previously ignored by the machine modernists, its focus on master-planned developments and form-based codes has also little to do with the complexity and agency of the everyday. The challenges facing the recent discourse on Tactical Urbanism represent the latest struggle for the profession to engage the everyday in a meaningful and critical manner. The ephemeral nature of many of the tactical interventions also raises important questions about their lasting impact. Seattle's Alleypalooza, for example, has since ceased to exist, although other forms of temporary urban adaptations have flourished in the city.

Despite these shortfalls and challenges, it is important not to discount the impact of the everyday and bottom-up on the practice of city planning and design in the United States. What we learn from the everyday and bottom-up, in terms of the importance of physical and social fabrics, the complexity and diversity of urban communities, the agency of individuals and social groups, and the politics of placemaking, is what continues to hold professionals and city authorities accountable to citizens, communities, and the public(s). For more than a century, the everyday and bottom-up and the full spectrum of complexity and nuance have served as a mirror for professionals and bureaucrats to reflect on their actions, values, and ideology. They have kept the design and planning of American cities more grounded in the complexity and richness of urban life. By contributing to greater accountability and reflexivity in professional actions, I argue that the everyday and bottom-up have served not only as a source of inspiration and innovation but also as a cornerstone of democracy in American city design.

Acknowledgements An earlier version of this chapter was written in 2013 for a proposed book *Companion to American Urbanism*. Although the book did not materialize in the end, the author wishes to thank the editor Joseph Heathcott for the opportunity to conceptualize this piece. He is also grateful to the editors of this book, Conrad Kickert and Mahyar Arefi, for providing a new home for this work, and to the anonymous reviewers for their critical comments and suggestions.

REFERENCES

Bela, J. (2015). User-generated urbanism and the right to the city. In J. Hou, B. Spencer, T. Way, & K. Yocom (Eds.), *Now urbanism: The future city is here* (pp. 149–164). London and New York: Routledge.

Ben-Joseph, E. (2012). *Rethinking a lot: The design and culture of parking*. Cambridge: The MIT Press.

Boddy, T. (1992). Underground and overhead: Building the analogous city. In M. Sorkin (Ed.), *Variations on a theme park: The new American city and the end of public space* (pp. 123–153). New York: Hill and Wang.

Boyer, C. M. (1992). Cities for sale: Merchandizing history at South Street Seaport. In M. Sorkin (Ed.), *Variations on a theme park: The new American city and the end of public space* (pp. 181–204). New York: Hill and Wang.

Campo, D. (2016). Book review. Tactical urbanism: Short-term action for long-term change. *Journal of Urbanism, 21*(3), 388–390.

Crawford, M. (2008). Introduction. In J. L. Chase, M. Crawford, & J. Kaliski (Eds.), *Everyday urbanism* (expanded ed.). New York: The Monacelli Press.

Davis, M. (1992). Fortress Los Angeles: The militarization of urban space. In M. Sorkin (Ed.), *Variations on a theme park: The new American city and the end of public space* (pp. 154–180). New York: Hill and Wang.

Douglas, G. C. C. (2014). Do-it-yourself urban design: The social practice of informal "improvement" through authorized alteration. *City & Community, 13*(1), 5–25.

Francis, M., Cashdan, L., & Paxson, L. (1984). *Community open spaces: Greening neighborhoods through community action and land conservation*. Washington, DC: Island Press.

Franck, K., & Stevens, Q. (Eds.). (2007). *Loose space: Possibility and diversity in urban life*. London and New York: Routledge.

Groth, P. (1997). Frameworks for cultural landscape study. In P. Groth & T. W. Bressi (Eds.), *Understanding ordinary landscapes*. New Haven: Yale University Press.

Hall, P. (1988). *Cities of tomorrow*. Malden, MA: Blackwell.

Hayden, D. (1995). *The power of place: Urban landscapes as public history*. Cambridge: The MIT Press.

Hester, R. T. (1999). A refrain with a view. *Places, 12*(2), 12–25.

Hou, J. (Ed.). (2010). *Insurgent public space: Guerrilla urbanism and the remaking of contemporary cities*. London and New York: Routledge.

Jackson, J. B. (1980). *The necessity for ruins and other topics*. Amherst: The University of Massachusetts Press.

Jacobs, J. (1961). *The death and life of great American cities*. New York: Vintage Books.

Jacobs, A., & Appleyard, D. (1987). Toward an urban design manifesto. *Journal of the American Planning Association, 53*(1), 112–120.

Kaplan, R., & Kaplan, S. (1990). Restorative experience: The healing power of nearby nature. In M. Francis & R. T. Hester Jr. (Eds.), *The meaning of gardens* (pp. 238–243). Cambridge: The MIT Press.

Kaplan, R., Kaplan, S., & Ryan, R. L. (1998). *With people in mind: Design and management of everyday nature*. Washington, DC: Island Press.

Larice, M., & Macdonald, E. (2013). *The urban design reader* (2nd ed.). London and New York: Routledge.

Loukaitou-Sideris, A., & Ehrenfeucht, R. (2009). *Sidewalks: Conflict and negotiation over public space*. Cambridge: The MIT Press.

Lydon, M., & Garcia, A. (2015). *Tactical urbanism: Short-term action for long-term change*. Washington, DC: Island Press.

Lynch, K. (1960). *The image of the city*. Cambridge: The MIT Press.

Meinig, D. W. (Ed.). (1979). *The interpretation of ordinary landscapes*. New York: Oxford University Press.

Merker, B. (2010). Taking place: Rebar's absurd tactics in generous urbanism. In J. Hou (Ed.), *Insurgent public space: Guerrilla urbanism and the remaking of contemporary cities* (pp. 45–58). London and New York: Routledge.

Mould, O. (2014). Tactical urbanism: The new vernacular of the creative city. *Geography Compass, 8*(8), 529–539.

Mukhijia, V., & Loukaitou-Sideris, A. (Eds.). (2014). *The informal American city: Beyond taco trucks and day labor*. Cambridge: The MIT Press.

Peterson, J. A. (1983). The city beautiful movement: Forgotten origins and lost meanings. In D. A. Krueckeberg (Ed.), *Introduction to planning history in the United States* (pp. 40–57). New Brunswick, NJ: Center for Urban Policy Research.

Relph, E. (1976). *Places and placelessnes*. London: Pion.

Rojas, J. (2010). Latino urbanism in Los Angeles: A model for urban improvisation and reinvention. In J. Hou (Ed.), *Insurgent public space: Guerrilla urbanism and the remaking of contemporary cities* (pp. 36–44). London and New York: Routledge.

Scott Brown, D. (2003). "Learning from Brinck". In C. Wilson & P. Groth (Eds.), *Everyday America: Cultural landscape studies after J. B. Jackson* (pp. 49–61). Berkeley: University of California Press.

Soja, E. W. (1992). Inside exopolis: Scenes from Orange County. In M. Sorkin (Ed.), *Variations on a theme park: The new American city and the end of public space* (pp. 94–122). New York: Hill and Wang.

Sorkin, M. (Ed.). (1992). *Variations on a theme park: The new American city and the end of public space*. New York: Hill and Wang.

Spirn, A. W. (1984). *The granit gardens: Urban nature and human design*. New York: Basic Books.

Talen, E. (2015). Do-it-yourself urbanism: A history. *Journal of Planning History, 14*(2), 135–148.

US Census. (2012). Growth in urban population outpaces rest of nation, census bureau reports, on the Internet at http://www.census.gov/newsroom/releases/archives/2010_census/cb12-50.html. Accessed August 29, 2013.

Venturi, R., Scott Brown, D., & Izenour, S. (1977). *Learning from Las Vegas: The forgotten symbolism of architectural form*. Cambridge: The MIT Press.

Wirth, L. (1938). Urbanism as a way of life. *The American Journal of Sociology, 44*(1), 1–24.

CHAPTER 3

Guerrilla Architecture and Humanitarian Design

Scott Shall

Extra-Legal Practices

Cities of the future will be largely extra-legal.[1] Most will not be planned, will not conform to any building regulations, and will not be built on land that is legally owned. These cities will not be mapped, permitted, or otherwise documented (WHO 2000).[2] Their residents will have no formal access to sewage or waste disposal and only intermittent access to transportation, schools, water, and electricity. Large swaths of these cities will breed disease, be a danger to the occupants, and a significant drain on our planet's civic and environmental resources (WHO 2000).[3]

[1] To understand this assertion, two points must be considered. First, informal settlements represent the largest and fastest growing urban condition on the planet. In fact, according to a 2013 UN-Habitat report, "since 1990, 213 million slum dwellers have been added to the global population." Second, the definition for informal settlement offered by the UN-Habitat Programme, which is arguably the definition most widely applicable, includes only two primary characteristics: informal settlements are illegally located; and they are illegally constructed. It naturally follows that, now and in the future, most settlements will be extra-legal (UN-Habitat 2015: 3; see also WHO 2000).

[2] Politically, the areas occupied by informal settlements are rarely mapped with any detail and the edges left undefined, often purposefully (WHO 2000).

[3] "Living in these settlements often poses significant health risks. Sanitation, food storage facilities and drinking water quality are often poor, with the result that inhabitants are exposed to a wide range of pathogens and houses may act as breeding grounds for insect vectors. Cooking and heating facilities are often basic, with the consequence that levels of excessive exposures to indoor

S. Shall (✉)
College of Architecture and Design, Lawrence Technological University, Southfield, MI, USA

© The Author(s) 2019
M. Arefi and C. Kickert (eds.), *The Palgrave Handbook of Bottom-Up Urbanism*, https://doi.org/10.1007/978-3-319-90131-2_3

Buildings of the future will be built largely of locally harvested scrap. Most structures will be detailed using the conventions of localized practice, based upon the knowledge of whoever is at hand, and the discernment of whoever is most vested in the construction. These structures will not be resistant to earthquakes, landslides, floods, extreme temperatures, or even rainfall. They will be unsafe, unsanitary, and unsustainable (Davis 2006).[4]

In the future, most cities and buildings will be realized without the formal input of civic or private agencies. Politicians, policy-makers, planners, and civic leaders will have limited impact upon them. Architects and engineers will have even less.

This marginalization is not due to a lack of expertise, as the knowledge and skills offered by planners, architects, and engineers could have a profound and positive impact upon these future-cities. Nor is it due to a lack of desire, as both fields have recognized the emergence of these future-cities and have dedicated a great deal of energy attempting to address the related concerns. It is because the academic and professional frameworks that support these fields are based upon patronage, which privileges a pace and structure of interaction that is fundamentally counter to that used to build future-cities.

The pace of interaction favored by architects and engineers is driven by the need for a clearly delineated interaction of predetermined actors with a project, wherein the design develops through the accumulated wisdom of these parties until it is complete and the project constructed. In this model, study drives design, which, in turn, drives construction. Inhabitation is largely external to the process, occurring after it has concluded and bearing limited impact upon it. In contrast, the illegal settlement operates much more fluidly, prioritizing inhabitation, which drives construction. Design, and the study sound design demands, only occurs at the margins, when the conventions of construction demand evolution, such as when new resources emerge, old resources become scarce, or the context of inhabitation changes. This fluidity has allowed the formation and growth of illegal settlements to far outpace the capacity of the governing authorities, who are operating in the much more linear study[then] design[then]construction approach, to support them. The result is the production of these future-cities at an incredible scale, far beyond what can be accommodated using current means of engagement (Thieme and Kovacs 2015).[5]

pollution may occur. Access to health and other services may be limited; overcrowding can contribute to stress, violence and increased problems of drugs and other social problems" (WHO 2000).

[4]To quote author Mike Davis, "the cities of the future, rather than being made out of glass and steel, as envisioned by earlier generations of urbanists, are instead largely constructed out of crude brick, straw, recycled plastic, cement blocks, and scrap wood. Instead of cities of light soaring toward heaven, much of the twenty-first century urban world squats in squalor, surrounded by pollution, excrement and decay" (Davis 2006: 86).

[5]"In the shadows of rapid urbanization and economic liberalization across the world, the formation and demographic growth of informal settlements or 'slums' far outpaces the availability and capacity of urban planning" (Thieme and Kovacs 2015: 1).

Additionally, this structure of interaction compels both architect and engineer to seek out alliances with more powerful actors to have significant influence upon large-scale concerns. This ties those architects and engineers desiring to significantly impact future-cities to those governmental and private entities who are capable of acting as patron and sponsor in this context—a situation that limits the impact of both fields in two ways. First, the highly technocratic leaning of these sponsoring agencies creates a pronounced reliance upon individualized, highly engineered solutions and expert knowledge, with a related bias against the development of hybridized addresses that span multiple departments, donors, or ministries. As a result, municipal water, city-wide waste disposal, affordable housing, and other civic amenities offered by said agencies tend to be technocratic and isolated from the patterns found within future-cities, despite the obvious benefit of overlapping these concerns with each other and with the prevailing tendencies of the settlements they intend to serve (Thieme and Kovacs 2015).[6]

Second, the limited influence held by the residents of illegal settlements within this process can lead to the reallocation of benefits, resulting in public- and state-assisted efforts benefiting urban middle- and upper-class residents more directly than those to whom the aid was originally pledged (Davis 2006).[7] The unfortunate result is the perpetuation of the conditions that have come to typify, and define, life within these future-cities—a situation that, not coincidentally, benefits the landlords of said settlements (Thieme and Kovacs 2015).[8] Correspondingly, Davis argues that the growth of illegal

---

[6] "The practice of approaching services in an individualized, technocratic form highly reliant upon engineering solutions and expert knowledge reflects institutional and management overlaps and incoherencies between sectors that are not required or in the habit of communicating, whether across governmental ministries, departments or donors, and indeed, is valid across the services' spectrum, whether for waste, water, food or energy. ... Approaches to municipal waste tend to be fairly technocratic in provision and analysis, ignoring the overlapping effects of waste on water, sanitation, food and health, with emphasis on the lack of political will and finances for operationalizing an effective waste management system, but one that does not explicitly address these interdependencies. Consequent to a lack of funds and communication strategies or data streaming between government agencies, waste disposal and management options have largely stagnated and failed to evolve to address new needs and waste forms" (Thieme and Kovacs 2015: 8–9).

[7] "Both 'poaching' and fiscal bias, of course, are expressions of the poor majority's lack of political clout throughout most of the Third World; urban democracy is still the exception rather than the rule, especially in Africa. ... A consensus of urban scholars agrees that public- and state-assisted housing in the Third World has primarily benefitted the urban middle class and elites, who expect to pay low taxes while receiving high levels of municipal services." (Davis 2006: 68–69).

[8] "Slums can be, from the landlord's point of view, a lucrative but risky investment. Seeking a return on their investment, landlords have to contend with risks of high tenant turnover, mobility and rent default. The primary motive is purely profit-driven" (Thieme and Kovacs 2015: 4).

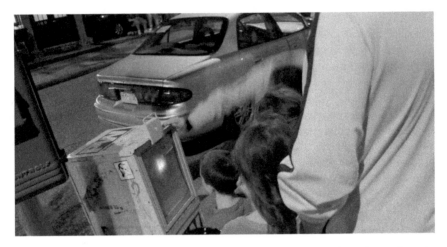

Fig. 3.1 projectionMAIL is a $2 projector made from commandeered postal service boxes. It borrows resources, such as newsbins in off-hours or the shadows generated by park benches, to display image-based artwork within the streetscape, expanding the gallery space and, transitively, the definition of patron, artist, curator, and critic (Image courtesy of the International Design Clinic)

settlements is driven not by the supply of jobs, but by the reproduction of poverty (Davis 2006).[9]

In any case, whether rooted in the inefficiency or the exploitation of the mechanisms used, the inability of governmental and private sector agencies to cope with the housing needs of people within these future-cities is well documented, established by decades of evidence. To meet the challenges and opportunities presented by this emerging world of blue plastic tarps, borrowed land, and commandeered electricity, requires an evolution of practice—one that trades the rigid hierarchies and linear approaches currently deployed for more inclusive and heterarchical terminologies and practices.

For the architect and engineer to contribute well in this place, a sound address requires a shift from the linear, research[then]design[then]construction[then]occupation process for practices that overlap these activities in significant ways.

For, within future-cities …

… site is more about persistent conditions and less about physical geography.

… size, like respect, is earned, not bestowed.

… program is not fixed, but emerges as new users re-imagine, re-inhabitat, or re-construct (Fig. 3.1).

---

[9] "Overcrowded, poorly maintained slum dwellings, meanwhile, are often more profitable per square foot than other types of real-estate investment… speculators are developing the urban periphery at 'monopoly prices' and enormous profits" (Davis 2006: 16).

To impact this world, the architect and engineer must shift from author to instigator.

Their office must move from a place of design, to a place of design, making, inhabitation, measuring and remaking.

Their work must focus less upon the production of constructs, to which others must respond, and more on the production of creative actions that inspire various publics to iteratively realize a sustained address.

In this place, in organization and action, those desiring to offer sound work must act less like an engineer and more like a guerrilla.

STAGE I: DIALOGUE WITH SELF

The guerrilla fighter, who is generally outmanned by a larger, more entrenched, and better-equipped adversary, cannot succeed using traditional techniques or direct confrontation. Instead they must work to redefine the field of battle in a way that places great value upon their assets, while limiting the impact of those held by their adversary (Piao 1977). As the physical field of battle can only shift modestly, due to the limitations of the guerrilla's resources and the nature of the conflict itself, the guerrilla fighter instead prioritizes the gradual, but systematic, redefinition of the people's perception of this field. Through the massive publication of small victories, the publication (and at times instigation) of the injustices of their adversaries, and other acts of self-promotion, the guerrilla leverages the unifying principles of the people to bring them to the cause of the guerrilla or against the cause they oppose (Parenti 1978). Over time, as the guerrilla learns from small, initial engagements, they shift resources from less effective to more impactful techniques. In so doing, they not only clarify their tactics, but also the approach and intent of the movement itself—all of which are designed to support the systematic movement of active support from one position (the cause opposed) to another (the cause of the guerrilla). This is the primary objective of the guerrilla fighter.

Paolo Friere notes the importance of this conversion within humanitarian efforts, labeling it an "issue of indubitable importance" that plays a "fundamental role" in the historic struggle against oppression.[10] At the same time, Friere observes that this shift from the support of one position to another can pose difficulties for the well-intentioned convert, especially as related to the importation of prejudices and a related lack of trust for those they intend to support—a concern not foreign to the guerrilla (Freire 2010).[11]

[10] "Given the preceding context, another issue of indubitable importance arises: the fact that certain members of the oppressor class join the oppressed in their struggle for liberation, thus moving from one pole of the contradiction to the other. Theirs is the fundamental role, and has been so throughout the history of this struggle" (Freire 2010: 60).

[11] "It happens, however, that as they cease to be exploiters or indifferent spectators or simply the heirs of exploitation and move to the side of the exploited, they almost always bring with them the marks of their origin: their prejudices and their deformations, which include a lack of confidence in the people's ability to think, to want, and to know. Accordingly, these adherents to

Unfortunately, for the architect and engineer desiring to assist the residents of illegal settlements, the structure of interaction favored by each profession is an extension of an unjust order, through which powerful actors have historically leveraged both professions to extend their influence and solidify their favored position. To understand the full impact of this arrangement, one must first understand the utility and value of the built environment, and transitively those who help to shape it, to the most influential members of our social system. To explain: If one regards the body as the "irreducible element in our social scheme" and acknowledges that this element must exist in two conditions—space and time—then it naturally follows that all forms of spatial manipulation that impact either of these conditions have great influence.[12] Thus, for Foucault space becomes a "metaphor for a site or container of power which usually constrains but sometimes liberates processes of becoming." The built environment, which helps to craft spatial values and hierarchies, thus becomes a useful ally in shaping the identity of not only individuals, but also of society. Everyone is forced to respond to the values put forward by it; either they accept the values embedded in their environment or they choose to resist by carving out "personal spaces of resistance and freedom."[13] Either way, the "symbolic ordering of space and time" maintains its influential position. The built environment, whether realized at the scale of furniture, building, or city exudes an inevitable influence over those residing therein (Foucault 2002).[14]

If it is "through the dialectical relationship between the body and a structured organization of space and time that common practices and representations are determined," then it follows that those who wish to influence said "practices" and "representations" would seek out a partnership with those professions trained to design and construct this structured organization

the people's cause will constantly run the risk of falling into a type of generosity as malefic as that of the oppressors. The generosity of the oppressors is nourished by an unjust order, which must be maintained in order to justify that generosity. Our converts, on the other hand, truly desire to transform the unjust order; but because of their background they believe that they must be the executors of the transformation. They talk about the people, but they do not trust them; and trusting the people is the indispensible precondition for revolutionary change. A real humanist can be identified more by his trust in the people, which engages him in their struggle, than by a thousand actions in their favor without that trust" (Freire 2010: 60).

[12]"spatial and temporal practices are never neutral in social affairs. They always express some kind of class or other social content, and are more often than not the focus of intense social struggle. That this is so becomes doubly obvious when we consider the ways in which space and time connect with money, and the way that connection becomes even more tightly organized with the development of capitalism" (Harvey 1990: 213).

[13]These "personal spaces of resistance and freedom" are termed "heterotopias" (Foucault 2002: 367–379).

[14]Perhaps this is why Mies Van der Rohe identifies architecture as "the real battleground of the spirit" (Mies van der Rohe 1975: 154).

(Foucault 2002).[15] The course of politics and the built environment are, thus, intertwined—"to break the code of architecture at the scale of the city is to grasp the structure of society" (Hatch 1984: 9). This is true whether one is discussing the legally sanctioned city or the illegal environments that are thereby formed.

From this intertwining of politics and the built environment, the architect and engineer have, over time, developed a deep allegiance to those who wish to influence common practices and representations. The value of the built environment to those wishing to extend or solidify their influence encouraged the engineer, architect, and planner (all of whom have, at times, struggled with questions of professional identity) to develop a livelihood around supporting society's most influential people through the production of cultural and symbolic capital (Stevens 1998). This has created symbiotic relationships between these professions and society's most influential actors (Pecora 1991: 46). Those designing the built environment have access to the highest reaches of society, resulting in opulent commissions, popular respect, and a modicum of professional security (Crawford 1991: 27–45). In trade, the architect, engineer, and planner have developed a deep, albeit sometimes tacit, allegiance to those who can afford their work; the elite citizen, whether religious or commercial, and the state are their patrons and their focus.[16]

The impact of this relationship on both profession and professional is significant, defining the manner in which architects and engineers are educated, organized, deployed, and supported—frameworks that are in direct opposition to, and somewhat responsible for, the conditions faced by those living in illegal settlements. Thus, it is not surprising to note that when creative professionals have attempted to use this professional frame to address the issues faced by those living within these future-cities, the results have been, at best, mixed. Researchers Thieme and Kovacs believe that these attempts to help have actually served to create what they term "malevolent urbanism" (Thieme and Kovacs 2015: 1).[17] Although unfortunate, the architect's and

---

[15] "that in the 18th century one sees the development of reflection upon architecture as a function of the aims and techniques of the government of societies … If one opens a police report of the times—the treatises that are devoted to the techniques of government—one finds that architecture and urbanism occupy a place of considerable importance" (Foucault 2002: 368).

[16] "By distancing themselves from contractors and builders with economic control of the field, they (architects) also effectively repudiated the interests of moderate-income clients. Instead, the profession linked its professional identity to large-scale monumental commissions requiring wealthy patrons. This left architects dependent on the restricted group of clients who could afford to support their ambitions: the hoped for, but only occasionally awarded, patronage of the state (far less active than in Europe), but more often, the backing of large business and corporate interests" (Crawford 1991: 30).

[17] "Particularly when it comes to basic service provision, a form of 'malevolent urbanism' has generated across urban areas in the global South, where unequal access to and use of the city is prevalent. At the same time, a mosaic of actors, sectors, and initiatives seek to address the 'challenges of slums', usually purporting to work *with* local communities, but often misunderstanding how everyday practices and expectations might differ from externally defined development goals and impact measures" (Thieme and Kovacs 2015: 1).

engineer's inability to create a dialogue between formal and informal address is altogether logical. After all, when read through the biases of the frameworks described previously, informal activities, whether they are designed to provide food, shelter, water, or trade, are not something to be studied so that they might become the foundation for a systematically more permanent address. Rather, they are irregular, casual, precarious, and, often, illegal activities—addresses that are best ignored.[18]

This perspective separates the architect and engineer from the residents of illegal settlements, causing the former to retain their identity as state-sponsored actors, even when they are working voluntarily or under the auspices of a community group. Within such an arrangement, the professional can only impose solutions; grassroots initiatives are impossible.[19] Addressing this bias is not simple, nor is it quick. Instead, it requires a process of continual self-examination.[20]

The constancy of this required examination forces the architect and engineer to intertwine the process of self-examination with the process of study, design, construction, and inhabitation and to reconceive of all engagements in the work as a prolonged series of experimental dialogues through which they might learn more about themselves, their current and future collaborators and their shared terrain. In this way, the actions of these actors are similar to those of the guerrilla, prioritizing long-term survival over short-term gain and having as much to do with defining the tactics and beliefs of the movement as they do with their direct consequences. After all, the chief purpose of all actions for both parties is to reveal a shared terrain of future activity and to build a sustainable dialogue between future collaborators. To quote Friere: "In the revolutionary process, the leaders cannot utilize the banking method as an interim measure, justified on grounds of expediency, with the intention of *later* behaving in a genuinely revolutionary fashion. They must be revolutionary— that is to say, dialogical—from the outset" (Freire 2010: 86) (Fig. 3.2).

[18] "The implication has been that informal economic activities and by extension informal provision of goods and services were not only described as irregular, casual and potential precarious, but also outside the remit of state regulation and surveillance. Therefore, as urban slums are characterized by informality in all spheres of life, they become to an extent invisible to the state, especially in terms of public provisions" (Thieme and Kovacs 2015: 18).

[19] "Many upgrading approaches continue to inappropriately import solutions from other places without adapting operations to the local context. They are therefore unable to neither take full advantage of local knowledge nor develop city-wide/'at-scale' responses" (UN-Habitat 2015: 3).

[20] "Those who authentically commit themselves to the people must re-examine themselves constantly. This conversion is so radical as not to allow of ambiguous behavior. To affirm this commitment but to consider oneself the proprietor of revolutionary wisdom—which must then be given (or imposed on) the people—is to retain the old ways. The man or woman who proclaims devotion to the cause yet is unable to enter into communion with the people, whom he or she continues to regard as totally ignorant, is grievously self-deceived. The convert who approaches the people but feels alarm at each step they take, each doubt they express, and each suggestion they offer, and attempts to impose his 'status', remains nostalgic toward his origins" (Freire 2010: 60–61).

Fig. 3.2 Vista Oculta packages art education into small lessons that could be completed in less than five minutes and disseminated on a standard postcard. The authors charted the use, movement and popularity of the cards to collaboratively generate a street-based educational system in La Paz, Bolivia (Image courtesy of the International Design Clinic)

STAGE II: DIALOGUE WITH OTHERS

As the movement's belief system intertwines with the inherent nature of the battlefield, the guerrilla begins to seek out opportunities to test the strength of this relationship and, transitively, their advantage. Unfortunately for the guerrilla fighter, this task is made quite difficult by the fact that these fundamental conditions are typically subject to the ever-shifting nature of the social, economic, or political climate and the stakes of misalignment are significant. As observed by revolutionary Che Guevara, "If the military situation will be difficult at first, the political will be no less ticklish. And if one single military error can liquidate the guerrilla movement, a political error can stop its development for long periods" (Guevara 1977: 208–209). The only way the guerrilla can overcome these difficulties is to develop engagements that reverse the flows of knowledge and cultivate the dialogical relationship described by Friere.[21] UN-Habitat's (2015) report offers a similar

---

[21] Theime and Kovacs also urge those attempting to positively impact the conditions found within slums to reverse the "flows of knowledge and expertise so as to theorize the nexus from the slum, where inhabitants experience everyday relationships to water, food, energy and waste as integrated" (Thieme and Kovacs 2015: 15).

exhortation, encouraging those who attempt to productively engage the residents of illegal settlements to prioritize the creation of platforms to "draw
on the knowledge of stakeholders involved in the improvement of slums"
and "facilitate information and experience exchange as well as peer learning opportunities" (UN-Habitat 2015: 3). In this way, those engaging the
future-city can allow "for meaningful negotiations and encounters between
local communities, local authorities, development agencies and the entrepreneurial sectors" to emerge (Thieme and Kovacs 2015: 15).

Unfortunately, for the reasons already described and the related need to
preserve professional value, the frameworks traditionally deployed by the
architect and engineer are not designed to create peer learning opportunities
nor the reversal of process demanded by authentic dialogue, as evidenced by
the biases found within the tactics oft-used by both fields to realize community-based work. Take, for example, focus groups or community charrettes—
two arrangements used by the creative professional to allow community
members to offer insights into given topics and then propose strategies of
address. Although both meetings have the appearance of an open, inclusive
dialogue, there are several factors that undermine the inclusivity and dialogical nature of both. First, the sponsoring agency—whether a governing official,
private organization, or non-profit—is generally regarded as the unquestioned
authority of the meeting, deciding whether or not to even include either
mechanism in the design process and to what extent the findings uncovered
will impact the work. It is quite difficult to believe that the members of a
committee formed under this hierarchy could possibly feel that their propo-
sitions would be treated equally to those offered by the client, which naturally calls into question the sincerity of the meeting and greatly reduces the
possibility of communicative action or effective dialogue. Second, the people
gathered in this manner will have been recruited using specific forms of advertisement, most of which will have been selected by the sponsoring group or
their agents. Given the difficulties of positioning this advertisement campaign
in a manner that will gain the interest of all groups impacted by the work, it is
highly likely that those gathered will have a preordained bias based upon the
nature of the advertising used to promote the gathering. Third, the parameters of the meeting itself, in terms of time, place, and format have a tendency
to skew participation. For example, holding a meeting at night may welcome
those who work during the day, but will limit the participation of those with
children, night jobs, or extra-curricular responsibilities. Similarly, holding the
meeting in one part of town will bias the proceedings toward people who
have easier access to the space; those with cars, along the bus route, or within
walking distance will be far more likely to attend than those who are located
less conveniently or lack the necessary transportation.

These factors will severely limit the diversity of the group—creating a relatively homogenous gathering that will create groupthink and undermine
the intended dialogue. As noted by James Suroweicki, author of *Wisdom of
Crowds*: "homogeneous groups are great at doing what they do well, but they

become progressively less able to investigate alternatives" (Surowiecki 2005). Radical ideas or unpopular notions are quickly overlooked, regardless of their validity, in favor of those points or beliefs held by the majority. Popularity, not soundness of argument prevails. A false consensus emerges as "the groups' sense of cohesiveness works to turn the appearance into reality, and in doing so helps dissolve whatever doubts members of the group might have" (Surowiecki 2005).[22] Over the course of the meeting, groupthink steels the minds of the participants, closing them from ideas offered by the minority or overlooked by the group as a whole. In so doing, ideological communication has effectively compromised the ability of the group to realize Friere's dialogue.

Even consensus-building, a seemingly inclusive and participatory approach to design, is an inherently flawed aspiration, more often leading to ill-founded conclusions and faulty recommendations than useful insights.[23] The reasons for this extend beyond the intent of the deliberation and to the sociological structure of the debate. First, without thoughtful framing to combat natural tendencies, any discussion or debate will encourage two very harmful group patterns: *information cascade* and *polarization*. Information cascade is a result of the linear process of conversation, in which each insight offered is impacted by that which preceded it. This tendency naturally prioritizes the points raised first, instead of those that are judged to be most prudent through argument or thoughtful consideration, granting the most outspoken participants an exaggerated impact upon the course of the deliberation, and, thus, the conclusion reached. This occurrence is made especially dangerous by the fact that groups tend to polarize through discussion.[24] There are three explanations for this. First, during a deliberation people tend to compare their position to that held by the group. Second, people tend to believe that if lots of people believe a certain thing, they must have a good reason for doing so—a tendency known as 'herding'.[25] Third, within a deliberation, extremists—who tend to

---

[22] "the important thing about groupthink is that it works not so much by censoring dissent as by making dissent seem somehow improbable ... even if at first no consensus exists—only the appearance of one—the groups' sense of cohesiveness works to turn the appearance into reality, and in doing so helps dissolve whatever doubts members of the group might have" (Surowiecki 2005: 37).

[23] "In terms of maximum participation, consensus decision making is the most inclusive" (Cherry 1998: 57).

[24] According Cass Sunstein, who conducted numerous studies on this phenomenon: "As a general rule, discussions tend to move the group as a whole and the individuals within it toward more extreme positions than the ones they entered the discussion with" (Surowiecki 2005: 185).

[25] "Herding" is demonstrated clearly through an experiment by Milgram, Bickman and Berkowitz. In it, the researchers placed a single individual on a street corner, and asked them to look skyward. As others passed, a few stopped to look skyward as well. After a time, they placed five people on the corner looking skyward, which caused four times as many people to gaze skyward. They then placed fifteen skyward-looking people on the corner, resulting in almost half of all passersby following suit. As they continued this progression, more and more people were convinced to stop and look at the sky, until 80% of the passersby ended up so doing by the end of the experiment (Milgram et al. 1969: 79–82).

be more rigid and are generally convinced of their own rightness—tend to have greater influence than moderates. Eventually, due in large part to the first two tendencies, their conviction is transferred to the group, pulling the debate toward extreme positions (Surowiecki 2005: 188).

As group members shift their positions in accordance with the beliefs of the group, they tend to leave behind points and ideas that are unique. This results in consensus-driven groups squelching debate in favor of the familiar and creating tepid solutions that offend no one rather than exciting everyone. Garold Stasser demonstrates this tendency through a simple experiment in which he asked eight people to rank the performance of 32 psychology students. He supplied all participants with two common pieces of information (grades, etc.). He also gave two members two extra pieces of info (i.e. performance in classroom) and one member another two pieces. Stasser found that the ratings of the group were based almost entirely upon the two pieces of shared information. All other pieces of data, despite the fact that they were actually quite telling, were discounted entirely. The reason: in unstructured, free-flow conversations, the information that tends to be discussed the most is that which is shared. Any new or innovative messages are generally either modified to fit old messages or discounted altogether (Stasser 1985: 1467–1478). It is important to note that, at times, this tendency to conform can even lead the group to embrace ideas that are blatantly wrong. In Solomon Asch's famous experiment, he asked nine people to select the longest line on a sheet of paper. The first eight respondents, who were in on the experiment, had been previously instructed to select the wrong line. This caused 70% of the subjects (the final respondent) to select the wrong line at least once and 33% to do so over half the time. Rather than believing their eyes, these respondents believed the group (Asch 1956: 70). One can only imagine the sway of the group when dealing with matters of greater dispute than the length of a line.

To address these patterns, the architect and engineer must first redesign their methods of engagement so as to encourage healthy, inclusive dialogue—a shift that starts with the manner in which participants are gathered. For the reasons cited above, those who intend to develop a truly dialogical process cannot rely upon a single source to determine the correct body of people to invite. Nor can they rely upon mechanisms of advertising for recruitment, or a single time and space for discussion, both of which have biases that will not permit the diversity of participation required in a truly community-centered work. Instead, they must construct methods that will allow wisdom to be collected at a variety of points and times, all of which are located, in time and space, in accordance with whatever facts of the work are known. Whether in the form of smaller, street-side gatherings, large-scale negotiable installations, or text-based events, the creative professional must find ways to collect the wisdom of a wide range of people simultaneously, without prioritizing the views of the majority, the powerful, or the convicted. Done correctly, this will minimize groupthink and cascade thinking, both of which occur when decisions are made sequentially. It is worth noting that in Solomon Asch's experiment, when the scientist instructed just one other respondent to select the correct line, the subject did likewise to an overwhelming degree. Apparently, allowing a single voice of dissent is enough to encourage most people to stay true to their convictions. Just as homogeneity creates pressures toward conformity, diversity contributes to difference, making it easier for everyone to offer their ideas and help to realize healthy dialogue.

Secondly, the architect and engineer must develop practices that allow this diverse body to independently offer their ideas and explore as many alternatives as possible—a process that occurs quite often in the world of business. At the birth of a new technology—the automobile, the television, the Internet—there is generally a boom in the number of businesses that grow around the promise therein offered. More businesses than can possibly succeed vie for supremacy, each attempting to offer the best product to the consumer and make the case for their existence. Over time, the customer, through their purchases, judges some ideas to be better than others. Businesses respond to these trends, causing shifts in investment, until a much smaller set of products have each found a niche within the market. Similar processes can be found within the manner in which people help to establish a betting line or bees locate honey.[26] In each case, the process allows for the generation of lots of losers, which are quickly recognized as such and killed off. Compare this to the process used by the community-based designer, who attempts to form groups which debate, using only abstractions of the idea (drawings, arguments, etc.), and then decide upon a single course of action. It is not surprising that the ideas that result too often fail to produce meaningful change (Fig. 3.3).

Properly installed, these new patterns of working will shift the creative professional's chief responsibility from that of expert, who receives all knowledge and then dispenses it to the group, to that of facilitator, who makes specific knowledge globally accessible and then designs frameworks through which the public might determine the best course of action. It is important to note, that the public is not designing in this scenario, which would naturally compel the people to translate their wisdom into practices generally outside their expertise. Instead, the public is offering insight into their practices, supported

---

[26] Rather than sit in the hive and discuss the alternative locations for nectar, gradually choosing a prudent course of action, bees send all members of the hive out in every direction. Once the scouts find a nectar source, they return to the hive and perform a waggle dance, the intensity of which is based upon the excellence of the supply. This dance attracts a corresponding number of scouts, which follow the bee to the source. They then return to the hive and perform a similar dance, until the entire hive has effectively divided itself to harvest the most nectar (few bees tending the smaller sources, more tending the larger sources). Although seemingly inefficient, this method is generally quite productive: if a nectar source exists within two kilometers of the hive, bees will find it over half the time. The bees, like the business market, succeed because they allow everyone to operate independently, in accordance with their own wisdom (Surowiecki 2005).

◀ **Fig. 3.3** To design a school system with the migrant workers living on construction sites in India, the International Design Clinic initiated various small-scale creative actions to reveal the perspective of all parties who might one day be a part of this system. From this, projects such as a $2 water filter and foldable school-scape developed—ideas that are not obviously related to education, but emerged through the process as fundamental to the exercise of any educational activities within this specific community (Image courtesy of the International Design Clinic)

by the creative expertise of the architect and engineer. Far from an abdication of their role as design experts, this repositioning extends it, allowing their expertise to more capably support the expressed wisdom of others at all stages of work within a much more fluid and inclusive process. Their efforts will, in some regards, mirror those of Google's, establishing a framework whereby a myriad independent sources each offers a small bit of knowledge, which is then aggregated to determine the most appropriate result for any given search. Their work will be developed in a manner similar to the process used to develop Linux, which was created by providing an open-source code and allowing anyone with even a small bit of knowledge to contribute their specific knowledge to the global application. To quote author and Linux advocate Eric Raymond: "Given enough eyeballs, all bugs are shallow" (Surowiecki 2005: 72–73). Given enough eyeballs, perhaps so are some of the issues that typify life within illegal settlements.

STAGE III: DIALOGUE WITH WORK

Having established a continually developing connection between themselves and the opportunities offered by their field of battle, the guerrilla movement shifts its attention to encouraging others to take on a progressively more active role within the struggle. Starting with simple tasks that require minimal skill and risk, such as providing information or holding weaponry, and gradually moving to more complex operations, the guerrilla systematically equips the populace to assume vital responsibilities within the movement.[27] If not subverted, this increased investment will eventually result in the populace leading the entire movement.[28]

In some respects, the very nature of illegal settlement has already forced residents to assume such a position, compelling them to collaboratively develop solutions to pressing needs, such as access to food, shelter, security,

[27] "The duty of every revolutionary is to make a revolution" (Marighella 1977: 3).

[28] "The guerrilla movement, in its growth period, reaches a point where its capacity for action covers a specific region for which there is a surplus of men and an overconcentration in the zone. The bee swarming begins when one of its leaders, an outstanding guerrilla, moves to another region and repeats the chain of developments of guerrilla warfare, subject, of course, to a central command" (Guevara 1977: 210).

education, and water. The resulting environment becomes a repository of experimental work based upon indigenous resources and a base of support for future developments. However, it is important to note that this role is generally more limited than popularly thought, prompting researchers Kavita Datta and Gareth Jones to argue that the "praxis of the poor" is a smokescreen by which governing officials distance themselves from their responsibilities to help. They also argue that much of what is construed as self-help within these future-cities—building one's own home, cleverly creating ways to collect water and dispose of waste, crafting business opportunities—is actually executed with the paid support of artisans and skilled labor. This raises the cost of such endeavors substantially, while simultaneously limiting the ability of residents to address their own concerns.[29]

Within this context, the role of the architect and engineer is not to impose creative solutions nor is it to provide a framework for the development of such solutions. It is to provide a framework for dialogue whereby the people are not only able to offer creative addresses to persistent, local problems, but are trained and equipped to carry on this work indefinitely. One way to accomplish this is for the architect, engineer, or planner to initiate the authentic dialogue described earlier using only existing resources and supports and then to slowly and systematically push the conversation until these means are proven insufficient. In so doing, the designer leverages existing resources, as well as their necessary expansion through small, simple, and inexpensive actions, to establish a dialogue with the people—a dialogue that will allow all parties to more clearly understand the capacity of both the settlement and its individual residents. As the capacity of the people develops and the work gathers conviction and support, the architect and engineer can help shift the work to larger, costlier and more complex embodiments. However, it is important to note that, at each stage of development, it is the responsibility of the architect and engineer not only to appropriately limit the scale, cost, and complexity of the works, but to collaboratively develop evaluative techniques that will equip the residents to do so in the future. For, unlike practices based upon the production of symbolic or cultural capital, in this arena, size, like respect, must be earned, not bestowed. So must cost and complexity, as each will limit the ability to those surrounding the work to develop, expand, or maintain it.[30]

---

[29] "As the research of Kavita Datta and Gareth Jones has shown, the loss of economy of scale in housing construction dictates either very high unit prices for construction materials (purchased in small quantities from nearby retailers) or the substitutions of secondhand, poor quality materials. Datta and Jones argue, moreover, that 'self-housing' is partly a myth: 'Most self-help is actually constructed with the paid assistance of artisans, and for specialist tasks, skilled labour.'" (Davis 2006: 72).

[30] UN-Habitat recognizes the importance of this by noting that "participation is often most effective when initiated at the neighborhood level through individual or community projects which are relatively limited in scale and developed progressively" (UN-Habitat 2012).

Fig. 3.4 fencePOCKET uses reclaimed tarp to create usable space within the thickening of the assumed line between public and private realms caused by the horizontal distortion of chainlink fencing. The size and shape of spaces thereby created are determined by the specific nature of the deviation of the fencing, the programmed use is determined by nearby residents and passersby (Image courtesy of the International Design Clinic)

To encourage this slow, progressive, and systematic development, the architect and engineer must not only limit the scale, cost, and complexity of the offered work, but also any restrictions on use. Minimizing the preprogramming of work to those activities already proven necessary will enable others to impart unanticipated definitions into other aspects of the work. Those surrounding it, become able to re-imagine, re-inhabit, or re-construct it, based upon both current and emerging needs and desires. This flexibility of function will allow the nature of existing activity within the informal settlement to gradually embed itself into the work, rather than attempting to force these natures into a premeditated use, as determined by the creative professional. Works attempting to provide water can become places for commerce, social interaction, and waste treatment. Educational systems can move from a centralized system to a distributed, mobile one. Aspects of the work given such latitude effectively function as heterotopias, which Foucault defines as "those singular spaces to be found in some given social spaces whose functions are different or even the opposite of others" (Foucault 2002: 376). They are also reminiscent of the "freespaces" which author Lebbeus Woods advocated for in Radical Reconstruction (Woods 1997) (Fig. 3.4).[31]

[31] "Design can be a means of controlling human behavior, and of maintaining this control in the future. The architect is a functionary in a chain of command whose most important task (from the standpoint of social institutions) is to label otherwise abstract and 'meaningless' spaces with 'functions' that are actually instructions to people as to how they must behave at a particular place and time. The network of designed spaces, the city, is an intricate behavioral plan prescribing social interactions of every kind, prescribing therefore the thoughts and, if possible, the feelings of individuals." (Woods 1997: 23).

Through the provision of heterotopic freespaces, the architect and engineer intensify the ability of the built environment to act as a social organizer, so that its inherent influence might support the cultivation of authentic dialogue, the accumulation of wisdom, and, inevitably, its own evolution. More importantly, by shifting the manner of determination for use (as well as location, size, complexity, and cost) the creative professional provides the latitude necessary for the work to align with the social practices of the people, rather than the inverse. This allows the work, albeit in a very rudimentary and humble state, to have a liberating function. For, although the influence of the value structures promoted by the built environment is powerful, the power of said constructs to control, liberate, or oppress is subject to the practices of those residing within it—a reality Foucault spoke of in his interview with Paul Rabinow.[32]

By encouraging the people to leverage the improvement of the built environment in order to create the partnerships necessary to address immediate need, the architect and engineer shift the measure of value held by the built environment from symbolic and economic capital to social and cultural.[33] Their work, designed to instigate participation through its redefinition, becomes an instrument through which the residents might mobilize to address land challenges, urban planning, management, and governance issues.[34]

Finally, the architect and engineer must design this dialogical process in a manner that permits measured failure. After all, if the size, cost, complexity, location, and program of the work are truly up for discussion, then the option to stop the conversation, at any point, must also be allowed. At times this call to stop a specific aspect of the growing dialogue, manifest in the people ignoring or destroying the created work, will occur at the beginning of the conversation; at times it will occur near its end. Either way, the voice of those living within the community must be respected and the destruction of the

---

[32] [MF] "If one were to find a place, and perhaps there are some, where liberty is effectively exercised, one would find that this is not owing to the order of objects, but, once again, owing to the practice of liberty. Which is not to say that, after all, one may as well leave people in slums, thinking that they can simply exercise their rights there.

[PR] Meaning that architecture in itself cannot resolve social problems?

[MF] I think that it can and does produce positive effects when the liberating intentions of the architect coincide with the real practice of people in the exercise of their freedom" (Foucault 2002).

[33] As noted by UN-Habitat: "Physical upgrading of slums with street networks and improved access to municipal basic services through augmentation of physical infrastructure has proven to make formidable positive social and economic changes in many cities. Socially, upgraded slums improve the physical living conditions, improve the general well-being of communities, strengthen local social and cultural capital networks, the livelihood generation opportunities, quality of life, and access to services and opportunities in towns and cities" (UN-Habitat 2012).

[34] "what ties the rural and urban slum experience in relation to the nexus are the prevalence of social networks and social capital as the dominant albeit informal platform for self-organizing and provisioning that determine *how* things get done" (Thieme and Kovacs 2015: 11).

work permitted. For it is through such failure that knowledge is generated. And, provided that the creative professional did not permit the work to prematurely exceed, in cost, complexity, or determination, the conviction held by its authors, the knowledge gained will more than offset its cost.

STAGE IV: DIALOGUE WITH PRACTICE

In the eyes of the guerrilla, the final stage of engagement is simply an open-ended expansion of the third stage. That is, by systematically repeating the cycle of increased engagement, the guerrilla strengthens their position. At the same time, with each successful campaign, the guerrilla strengthens the viability of their alternative and diminishes the perceived risk of involvement. This increases the number of people involved, enabling the movement to take on larger and more decisive encounters, which, in turn, leads to great viability and enrollment.[35] In this way, by designing experiences that leverage an ever-growing number of people to become vested in their movement, the guerrilla increases the size and capabilities of their army, until they tip the balance of power in their favor (Fig. 3.5).[36]

Although this tipping point has yet to occur within future-cities, there are a few truly dialogical works that have acquired enough visibility to overcome their informal roots and attained enough status to be viewed as an opportunity for improvement by private or municipal concerns. According to researchers Theime and Kovacs, examples of productive collaborations between the work of residents and the support of external concerns include instances of "UN-Habitat water connectivity, public toilet construction by local MPs of a constituency running for re-election, or the growing design-for-development initiatives promoting off-grid solar." Although each of these works contains troubling political implications and unintended, problematic effects, they do represent small proofs that the overlap of bottom-up demands and innovative, externally supported solutions is possible (Thieme and Kovacs 2015: 20–21). It seems reasonable to conclude that architects, engineers, and planners who choose to better align themselves and their processes to the nature of future-cities can build on this foundation and realize even more sustainable and dialogical works. The promise of GIS, which has the potential to democratize data by enabling individuals to access

---

[35] "Revolution that does not constantly become more profound is a regressive revolution" (Guevara 1977: 204).

[36] "Guerrilla warfare or a war of liberation will, in general, have three stages: the first, a strategic defense, in which a small hunted force bites the enemy; it is not protected for passive defense in a small circle, but its defense consists in limited attacks which it can carry out. After this a state of equilibrium is reached in which the possibilities of action of the enemy and the guerrilla unit are stabilized; and later the final moment of overrunning the repressive army that will lead to the taking of great cities, to the great decisive encounters, to the total annihilation of the enemy" (Guevara 1977: 210).

Fig. 3.5 PARK-IN-A-CART borrows the architecture and rituals of vending culture to provide parks within the unplanned community of El Alto, Bolivia. The need for this work emerged through research on a completely different project, as did mobile-MAKERSPACE—a project that provides educational and vocational opportunities for the residents in working wood, metal, and fabric (Image courtesy of the International Design Clinic)

information previously reserved for better-funded entities, the proliferation of Internet-based structures for knowledge and commerce, and the myriad examples of successful leaderless organizations, would seem to provide additional support for this assertion and shift.[37]

However, for the architect and engineer to have any role in this development depends upon their willingness to reframe their perspective, practice, and, over time, position as designers. Else their contribution to the cities of the future will be as negligible as it is today.

REFERENCES

Asch, S. E. (1956). Studies of independence and conformity. *Psychology Monographs, 70*, 1–70.

Brafman, O., & Beckstrom, R. (2006). *The starfish and the spider: The unstoppable power of leaderless organizations*. Toronto: Portfolio.

Cherry, E. (1998). *Programming for design*. New York: Wiley.

[37]For more information on leaderless organizations, reference *The Starfish and the Spider* by Ori Brafman and Rod A. Beckstrom (2006).

Crawford, M. (1991). Can architects be socially responsible? In D. Ghirardo (Ed.), *Out of site: A social criticism of architecture* (pp. 27–45). Seattle, WA: Bay Press.

Davis, M. (2006). *Planet of slums.* New York: Verso.

Foucault, M. (2002). Space, knowledge, power, interview with Paul Rabinow. In N. Leach (Ed.), *Rethinking architecture* (pp. 367–379). New York: Routledge.

Freire, P. (2010). *Pedagogy of the oppressed.* New York: Continuum International Publishing Group.

Guevara, C. (1977). Guerrilla warfare—A method. In W. Laquer (Ed.), *The guerrilla reader.* New York: New American Library.

Harvey, D. (1990). *The condition of postmodernity.* Cambridge: Blackwell.

Hatch, R. (Ed.). (1984). *The scope of social architecture.* New York: Van Nostrand Reinhold Company.

Marighella, C. (1977). Mini-manual of guerrilla warfare. In W. Laquer (Ed.), *The guerrilla reader.* New York: New American Library.

Mies van der Rohe, L. (1975). Technology and architecture. In U. Conrads (Ed.), *Programmes and manifestoes on 20th century architecture.* Cambridge: MIT Press.

Milgram, S., Bickman, L., & Berkowitz, L. (1969). Note on the drawing power of crowds of different sizes. *Journal of Personality and Social Psychology, 13,* 79–82.

Parenti, M. (1978). *Power and the powerless.* New York: St. Martin's Press.

Pecora, V. (1991). Towers of Babel. In D. Ghirardo (Ed.), *Out of site: A social criticism of architecture.* Seattle, WA: Bay Press.

Piao, L. (1977). Encircling the cities of the world. In W. Laquer (Ed.), *The guerrilla reader.* New York: New American Library.

Stasser, G. (1985). Pooling of unshared information in group decision making: Biased information sampling during discussion. *Journal of Personality and Social Psychology, 48,* 1467–1478.

Stevens, G. (1998). *The favored circle: The social foundations of architectural distinction.* Cambridge: MIT Press.

Surowiecki, J. (2005). *The wisdom of crowds.* New York: Anchor Books.

Thieme, T., & Kovacs E. (2015). Services and slums: Rethinking infrastructure and provisioning across Nexus. *Economic and Social Research Council.* Retrieved from http://www.thenexusnetwork.org/wp-content/uploads/2014/08/Thiemeand-Kovacs_ServicesandSlumsNexusThinkpiece2015.pdf.

UN-Habitat. (2012). *Housing and slum upgrading.* Retrieved from https://unhabitat.org/urban-themes/housing-slum-upgrading.

UN-Habitat. (2015). Streets as public spaces and drivers of urban prosperity. *Habitat III: Issue Papers, 22—Informal Settlements.* Retrieved from https://unhabitat.org/wp-content/uploads/2015/04/Habitat-III-Issue-Paper-22_Informal-Settlements-2.0.pdf.

Woods, L. (1997). *Radical reconstruction.* New York: Princeton Architectural Press.

World Health Organization. (2000). *Informal settlement report.* Retrieved from http://www.who.int/ceh/indicators/informalsettlements.pdf.

CHAPTER 4

The Practice of Urbanism: Civic Engagement and Collaboration by Design

David Brain

This chapter explores implications of design-centered collaborative planning as a mode of engagement that has the potential to introduce bottom-up processes into the heart of the field of action typically associated with top-down interventions led by professional practitioners. In particular, it explores lessons from the practice of the charrette, as it has been reinvented by the New Urbanist movement since the 1990s.[1]

The occasion of a volume on bottom-up urbanism points to a sea change in the discourse on urbanism: a shift from urbanism considered as an objective condition of human settlements to urbanism as a practice rooted in social processes that transcend self-conscious purposes. This shift has been manifested, for example, in the idea of an everyday urbanism comprised of small, self-initiated accommodations that people create in their built environment (Crawford 2008; Kirschenblatt-Gimblatt 2008). Exploration of everyday urbanism is rooted in appreciation of the vernacular expressions and activities that reflect cultural and class circumstances, and the informal actions and commonplace arrangements that constitute everyday experiences

---

[1] Explication of the logic of the charrette process and its efficacy is based on the author's research on the New Urbanist movement, involving observation of charrettes, and interviews with participants and practitioners. What began in 1998 as participant observation and interviews of New Urbanist practitioners became focused more generally on varieties of public process in planning. The original research was funded by a fellowship from the National Endowment of the Humanities, and a grant from the Graham Foundation.

D. Brain (✉)
New College of Florida, Sarasota, FL, USA

© The Author(s) 2019
M. Arefi and C. Kickert (eds.), *The Palgrave Handbook of Bottom-Up Urbanism*, https://doi.org/10.1007/978-3-319-90131-2_4

of urban life. At the same time, advocates for everyday urbanism still locate a practice for designers and planners as facilitators of ongoing urban processes, responsive to the needs and concerns embodied in the everyday (Kaliski 2008).

This reflects what may be an unavoidable contradiction as we attempt to understand urbanism in terms of bottom-up processes and still take practical responsibility for the intentional improvement of cities and neighborhoods. There is a certain tension between the role of professional expertise, inevitably embedded in institutions that sustain its authority, efficacy, and market viability, and urbanism that reflects the quality and character resulting from the cumulative effects of self-organizing processes. It is the tension between rational intervention grounded in technical knowledge and a recognition that the object of intervention cannot be fully reduced to technical rationality—a point eloquently made by Jane Jacobs (1961: 432), in her formulation of the city as a problem of "organized complexity." In a manner that is often elided by Jacobs' emphasis on urban sociability, this complexity resides partly in the interaction between the emergent qualities of urban life and collective efforts to apply directive and regulatory intelligence to the process. Between urbanism as social process and the technical expertise of planners there is the domain of the political, in the broadest sense of the capacity for intentional collective action and self-conscious governance. This domain of regularized power, and political action in general, is often lost in the dichotomous view of top-down and bottom-up processes.

The New Urbanism, as a movement, has helped to bring this problem into focus, as well as opening up possibilities for addressing it. Although it emerged as a professional reform movement rather than a direct return to the everyday, over the last three decades the movement has focused on anchoring a discipline of urban design in local experience, observation, and common sense. This effort has been part of reconfiguring the relationship between public consultation and the professional division of labor associated with planning, and it runs counter to the technocratic reduction of planning decisions to specialized technical knowledge controlled by experts, a buttress for the development regime associated with conventional suburban development patterns. New Urbanism has formulated its principles grounded in traditional town building (Congress for the New Urbanism 1999), and on a practice of organizing public engagement around design-centered collaboration. The idea of the charrette was borrowed from a tradition in architectural education related to training architects for intense design work within a compressed time frame (Lennertz and Lutzenheiser 2014). In the context of New Urbanist projects, it has been reworked as a tactic for breaking down obstacles to problem solving that result from professional specialization, and as a strategy for engaging communities in developing design solutions and building consensus around those solutions.

Like its offshoots of Tactical Urbanism (Lydon and Garcia 2015) and Lean Urbanism (Dittmar and Kelbaugh, Chapter 5 in this volume), New Urbanism's focus on charrettes allows for the creation of alternative forms of agency and modes of action, operating either in the gaps left by powerful institutions or across the grain of those institutional structures. All three of these urbanisms share a suspicion of professional authority and bureaucratic rationality, and are advocating the opposite of planning, allowing space for ad hoc solutions that might typically be precluded by the rational order sought by technical planning. They beg two questions, as a result: what is the role of the design professional, exactly? And, how does the professional's practical discipline engage the native processes of urbanism in a manner that is both respectful and constructive? The New Urbanist movement has illuminated a particular challenge implied in a normative theory of urbanism: How do we understand and engage urbanism not as a social ecology, a political economy, or a spatial form, but as a field of human action over time, as a critical and potentially transformative practice operating at the interface between the self-organizing qualities of bottom-up processes and top-down interventions by powerful actors and institutions? This chapter responds to these questions by focusing on the problem of agency at this crucial interface, and the particular way the New Urbanist charrette reconstitutes agency at the point where planning meets place-making.

PLACE-MAKING AND THE PROBLEM OF AGENCY

It can be a little misleading to talk about bottom-up urbanism, at least insofar as it might imply a simple dichotomy between top-down and bottom-up action. Bottom-up action could qualify as a *market*, an aggregation of individual choices, contrasting with *hierarchies* as formally constituted chains of command (Williamson 1983). Markets have become our primary image of an organic and emergent process, the domain of individual freedom, whereas hierarchies are associated with the ideal type of legal-rational authority, procedural fairness, and technocratic efficacy. This distinction is falsely binary, as Williamson demonstrates that markets and hierarchies are complementary forms of control, the dynamism and fluidity of market-type structures often complementing the reliability and rigidity of command hierarchies.

Furthermore, those who invoke bottom-up processes generally seem interested in something beyond individualistic liberty, or what Joseph Schumpeter called the "perennial gale of creative destruction" of the free market (1942: 84). Instead, scholars focus on the emergent orderliness of self-organizing processes and look for ways in which they might be the foundation for intentional action. Under this heading we might include everything from a "pattern language" (Alexander et al. 1977) for intentional design to the social capital that enables effective democratic governance (Putnam 2000). In contrast to classical liberal or pluralist political theories, contemporary democratic theorists emphasize that effective democratic governance entails deliberation and decision making dependent on the formation of a public realm of common interests that are more than simply an aggregation of individual interests. Sandel refers to this as the "formative politics" associated with the republican ideal (Sandel 1996: 6). The idea of a republic is that citizens, through their participation in a certain kind of community, articulate a shared sense of a common good that would not have been apparent to them otherwise—a civic sensibility that orients action toward a stake in a common world.

The distinction between top down and bottom up, then, seems to be the distinction between authoritative action—power and agency constituted by political and institutional structures—and collective action that involves forms of distributed agency rooted in rights and opportunities. Bottom-up action might or might not aggregate from the unselfconscious to the more purposeful organization of these processes that constitutes the distinctive domain of civic engagement.

It may be appealing to imagine that a wholly bottom-up process could be wholly individual, progressive, and transformational, without requiring the work associated with the politics of formulating and implementing collective goals. However, if we are interested in purposeful change that reflects conscious and clearly articulated intentions, if we are interested in the purposeful and principled defense of shared values, we need to gather up the energy of bottom-up processes and bring them into focus as a form of collective agency. This need not be in the form of top-down political authority, but it cannot be simply set in opposition to the structures of authority and governance. The essence of a civic sensibility lies in action oriented by the recognition that each party benefits from the cumulative effects of engagement in a common world. For example, the long-term efficacy, transformative impact, and democratic legitimacy of tactical urban interventions depend on translating its more organic forms of collective agency into the institutional structures necessary to sustain a process of collective decision making.

The concept of place-making suggests a bridge between bottom-up and top-down urbanism, as it is situated between organic processes that define places and the design work that gives them intentional form, although this ambiguity can raise questions of authenticity. Place-making encompasses the full range of urban actions along the dimensions of the top-down/bottom-up distinction. This flexibility has prompted its adoption by a variety of professions and even authoritative institutions, potentially compromising its association with bottom-up action. As place-making has been appropriated within the scope of professional expertise, there has been a tendency to obscure or reify the structural conditions that define the relationship between professionals and the bottom-up social processes they attempt to engage.

Nevertheless, critical examination of the concept of place can help to illuminate the problem of urban agency. From a sociological perspective, places manifest intentions that link individual action to collective outcomes, and also embody the relationship between those things we take as given and those things that we understand as the result of meaningful choices. From this point of view, urbanism is (at least at one level) a manifestation of subjective intentions, and an enactment of social relationships that operate both in the processes of *making* places and in the processes of *occupying* places.

The construction of place is an opportunity for social life to be objectified in material form. There is often, of course, much at stake in this process of making that gives social life its material aspect: the structuring of social practice (Bourdieu 1990, 1992); the reproduction of structures of inequality (Massey and Denton 1997; Logan and Molotch 1987) and the experience of social connectedness (Sampson 2013).

A public space that is designed by authoritative action in service to institutionalized power is not just a stage on which social life is played out. It is itself an expression of a relationship of power, sometimes in symbolic form and sometimes in practice. The awe-inspiring piazza of St. Peter in Rome is one kind of direct expression of power that is also an opportunity to experience that power. In Lefebvre's terms, it reflects spatial practice, representational space, and (in its formal intentions) a representation of space (1991). Mukerji has examined the ways in which the power of the French monarchy was not only represented but also given practical reality in the construction of the gardens at Versailles (Mukerji 1997). In a different way, a four-lane arterial roadway, lined with big-box retail set back behind large fields of parking and fast food restaurants on out-parcels, is an alienated and obscured, but nonetheless real and effective, manifestation of power and authority through technical disciplines. This place-making by engineers and designers generally implies a "program of action" (Latour 1992: 226) and it is as much an example of the way place-making embodies social relationships as the spaces that involve overt symbolic representation. The common arterial, by appearing less as a purposeful statement and more as a product of technical standards and requirements, instantiates and reinforces a profound disconnection between everyday social action and the authoritative practices of the institutional arrangements ostensibly created to accommodate social life. Against the background of the proliferation of "non-places" such as this (Augé 1995), places appear as moments of meaning in an environment otherwise overdetermined by technical circumstances rather than humane purpose. A deeper practice of place-making requires going beyond the usual, relatively narrow focus on the organic character of bubbles of meaningful activity.

A NEW URBANISM

The New Urbanist movement evolved as a pragmatic critique of conventional suburbia, its professional underpinnings and its associated development regime. However, its most significant contribution may reside in its effort to revive the very idea of urbanism as a normative ideal (Kelbaugh 2007). The critique of sprawl brought into focus its damaging economic, social, and environmental consequences, the evident irrationalities of a highly rationalized professional, regulatory, and financial system, and the ways in which planning practice might be reoriented to concerns for the quality of life and a sense of community (Katz 1994; Duany et al. 2000). Key to New Urbanism is a focus on urban form that treats buildings typologically rather than architecturally or even functionally, and regards urban form itself as something that only emerges over time. This move, in itself, has a decentering effect on the role of the designer, who is now compelled to see the formal intentions in any form-giving practice as articulated in a complex relationship with historical and circumstantial contingencies that transcend the designer's role, and that cannot (and should not) be reduced to technical rationality or the expression of a single authorial vision. This is the implication of the neotraditional approach advocated by the New Urbanists, and the revival of interest in the thinking of writers such as Sitte (1979 [1889]), who sought to learn from medieval cities.

Critics of the New Urbanism have noted that relatively small-scale greenfield projects, such as Seaside and Celebration, seemed to have little to do with the scale, density, and grit commonly associated with urbanism. However, this obscures New Urbanism's normative emphasis on form and process, not just scale and density. Attention to the public realm is manifested, for example, in the use of form-based codes to establish a practical framework in which the private realm defines and characterizes a common world of sidewalks, streets, and squares (Duany and Brain 2005; Parolek et al. 2008). Importantly, this is an outcome of a process that has a temporal character, and an order that emerges from rules that enact a shared responsibility for sustaining the quality of a built environment.[2] These rules establish a typological language and syntax, but impose relatively loose restrictions on particular architectural expressions. Consequently, both a lively complexity and a largely implicit orderliness can emerge from the diverse contributions of many individuals, responsive to contingencies and changing circumstances. As a paradigmatic example, Seaside manifests an appreciation of urbanism as a social practice of creating and sustaining places over time, of managing change, and of conserving that sense of identity, coherence, and continuity.

This understanding of the temporal dimensions of urbanism is expressed at a theoretical level in the writing of Leon Krier (1998), Alexander et al. (1977) and Christopher Alexander (1987), both of whom have influenced New Urbanist thinking. Rather than impose relatively simple formal diagrams and rigid geometries on the city, both Krier and Alexander regard urban form in terms of the way living human settlements have to change and develop over time, not by being remade out of whole cloth but by the way each move can both complete a prior project (a "whole" in Alexander's sense), and move toward creating a new "center," a larger whole (Alexander 1987).

In contrast to approaches to urban design that privilege the designer's formal intentions, this normative conception of urbanism regards urban form as a

---

[2] Seaside's early form-based code is a single sheet of diagrams that focus on building typology and the way buildings define a public realm. The effect of the code is to guide each new contributor to the town as they realize whatever individual aspiration they bring to it. It outlines a few simple responsibilities to the common world that anyone choosing to build a house in the community is expected to take on. See discussion of the evolution of the plan and the Seaside code, by Andrés Duany and others, pp. 165–207 in Thadani (2013).

complex expression of social life, as social life in the making (Brain 1994, 2008), taking on properties very similar to the way social practice is both ordered and contingently accomplished (Bourdieu 1992). As these processes play out over time, the orderliness and consistency of urban form provide a foundation for the continuity that supports the way a community's history is embodied and cultivated in place. This conception of urbanism as a practice shifts the agency associated with the production of urban space downwards, giving a certain privilege to bottom-up processes but also highlighting the importance of linking them to the formative aspirations of a community. Purposeful action and intentional interventions are located at an intermediate level, between the organic processes of atomized individual action and a vision of order that is more a matter of facilitative leadership than top-down imposition. Regulation is shifted from the focus on prohibition and procedure, with the associated tendency to encourage gaming, to constituting a practice that enables and facilitates cooperative outcomes that are generated rather than designed (Salingaros et al. 2006).

COLLABORATIVE DESIGN AS POLITICS BY OTHER MEANS

Beyond recognizing urbanism as a temporal and emergent phenomenon, New Urbanism also recognizes the value of bottom-up agency in its professional planning and design projects through the charrette. In contemporary New Urbanist practice, charrettes balance technical expertise, democratic political processes, and the emergence of new forms of place-making practice. Charrettes contain the potential to constitute a mediating practice, a hybrid of social, technical, and political action that is democratically inclusive, technically proficient, capacity-building, and oriented toward the incremental and adaptive achievement of complex goals.

This potential goes well beyond public participation or consultation of the sort that commonly appears as a manifestation of the mandate for "maximum feasible participation" that came out of the experience of the 1960s urban uprisings (Halpern 1995). One of the legacies of that period has been the expectation that every planning decision, large or small, should now involve 'public input,' predicated on the assumption that by allowing opportunities for citizens to speak at meetings, the result would be better outcomes, legitimacy for the process, and higher levels of satisfaction with both the results and the decision makers (Hibbing and Theiss-Morse 2002). Research in political science, however, has suggested that the opposite result is typical; the more people participate in public processes, the more they tend to question the legitimacy of the decisions and the integrity of the decision makers, the more frustrated and disillusioned they become, and the less likely it is that decisions are regarded as satisfactory outcomes (Hibbing and Theiss-Morse 2002). This research suggests that popular thinking about democracy reflects not only unrealistic expectations regarding the benefits of participation, as such, but that insufficient attention is paid to the character and quality of the processes in which people are expected to participate.

Fortunately, public participation has matured, as urbanists have learned from the 1960s legacy of adversity to community organizing and empowerment, and it now aligns with recent practices of "asset-based community development" that look to mobilize community capacity from within associational networks rather than seek solutions from powerful institutions (Kretzmann and McKnight 1993). Theorists of democratic governance have focused on the underlying social foundations of democratic practice, arguing that robust democracy requires more than an aggregation of votes under the principle of majority rule (Putnam 1993). For example, Putnam has called attention to the importance of networks of regular association that produce relations of trust and reciprocity—that is, social capital—as a foundation for civic engagement.

This phrase—civic engagement—has become commonplace among community activists as well as planners, but not always with careful reflection on what it implies. Engagement suggests that one is not just *involved* but *engaged* in ways that manifest a practical capacity for collective action and sustain a sense of "collective efficacy" (Sampson and Raudenbush 1999; Sampson 2013). One does not just turn out to vote or come to public meetings, but one is embedded in active networks of social action, implying opportunities, obligations, and capacities. Engagement becomes civic to the extent that it entails commitment to common interests (Sirianni and Friedland 2001). In urbanism, this translates to a commitment to the urban commons as a social space that is not only shared and collaboratively produced, but that also becomes an asset because it is shared, and that embodies and helps to sustain social capital.[3] Whereas theorists like Putnam have focused on how the resilience of communities can depend on the quality and quantity of associational life, new practices of bottom-up urbanism demonstrate how community building can also manifest and enact community relationships by producing and sustaining places in the built environment. Place-making becomes not only the medium and the reason for community connection, but also the means of actuating intentional community action and civic responsibility.

There is always a certain contradiction in professional discourse on bottom-up urbanism, as professionals are forced to rethink the relationship of their authority to the native processes of the fields in which they operate. Design-centered collaborative processes, from the New Urbanist version of the charrette to the "user centered" design of the "design thinking" movement (Brown 2009), have been efforts to find a standpoint for professional expertise that can be both disciplined and open to bottom-up and emergent processes. Although this has not generally been the explicit intention, the charrette has been reinvented as a tool that allows for the listening necessary

---

[3]This conception of the urban commons goes beyond merely the idea of the "right to the city" as manifested in the occupation of public space (Kohn 2016), to the idea of the public realm as constituted by the qualities of connection through a shared commitment to a common environment.

to cast problems in terms citizens understand, and that facilitates the shared learning necessary to articulate choices that make sense to non-experts while still benefiting from the knowledge and experience of technical experts from multiple disciplines. Although commonly regarded only in terms of their problem-solving and consensus-building capabilities within conventional planning, charrettes could be integrated with everyday, tactical, and lean approaches to reshaping our cities, in a manner that also reshapes our politics (Brain 2008). Collaborative processes not only offer opportunities for practical engagement, but also offer the public a direct experience of collective efficacy as an intentional product. Charrettes can provide the outlines of a distinctively twenty-first-century practice of urbanism in which professional expertise can productively contribute to planning and design processes, respond to complex problems, challenge problematic institutional power structures, and build democratic capacity.

THE CHARRETTE AS URBANISM IN ACTION: FROM PARTICIPATORY PLANNING TO CIVIC ENGAGEMENT

While there are perhaps as many variants of the charrette as there are practitioners, there are key elements of the reinvention of the charrette as a design-centered collaborative planning process that highlight its potential as a methodology relevant to the idea of bottom-up urbanism. The National Charrette Institute, for example, has defined the charrette as a facilitated collaboration that takes place over the course of multiple days, between a multidisciplinary team of experts, and between the charrette team and community stakeholders (Lennertz and Lutzenhiser 2014). Charrettes are distinguished from other workshops by "their intense, collaborative nature and by their holistic approach, focused on a feasible solution" (Lennertz and Lutzenhiser 2014: 3). In other words, they are distinguished as a collaborative design process that moves from vision to practical action.

The charrette remains fundamentally a design process, aimed at responding to complex challenges through an appropriately complex collaboration. In that collaborative process, specialists are encouraged to think as generalists (i.e., holistically) as they define the problem at hand and test ideas in an iterative process of interactions with diverse stakeholders. In these interactions, the process takes advantage of what sociologists call situated knowledge.[4] This integrates the generalized knowledge and technical expertise of a multidisciplinary team with the context-specific knowledge of stakeholders, both individually and collectively, in the pragmatic and political setting of the project.

---

[4]The concept of "situated knowledge" was introduced into science and technology studies by Donna Harraway. Haraway, D. (1988). Situated Knowledges: The Science Question in Feminism and the Privilege of Partial Perspective. *Feminist Studies, 14*(3), 575–599.

The efficacy of charrette methodology depends on three key elements in its process: the quality of the pre-charrette engagement with stakeholders; management of the flow of the collaborative design process; and, finally, the ability to structure the results as a set of actionable projects.[5] After engaging with stakeholders in the preparation phase, charrettes establish a point of departure with a major, public, hands-on workshop, establishing the key elements of a vision—goals, principles, problems, constraints, opportunities. The charrette team then begins to explore alternative schemes that illuminate key choices, trade-offs, conflicting priorities, and underlying values, helping to clarify the political and value-relevant choices embedded in many seemingly technical issues. These schemes are developed in ongoing collaboration with stakeholders, including those with particular technical interests, ensuring both the feasibility of the choices and the clear representation of what might be at stake. These alternatives are presented in a second major public meeting for stakeholders to understand the complexity of constraints and opportunities, and the necessity of making significant choices that might serve conflicting values and interests. The outcome of this meeting is then the basis for moving forward with the development of a "preferred plan synthesis," representing data, expert contributions, and the insights and interests of stakeholders (Lennertz and Lutzenhiser 2014: 103). This synthesis occurs through several feedback loops, in which the charrette team consults with stakeholders in formal public meetings, in scheduled meetings with particular groups, and in informal meetings that are open to walk-ins. As the stakeholders (and the shared universe of relevant facts) push back in response to the exploration of alternatives, solutions adapt to become more informed and politically robust, as they visibly respond to stakeholders' concerns. The process is open in multiple senses: offering opportunities for stakeholders to weigh in; offering opportunities to follow the reasoning that leads to solutions; and establishing a process in which decisions make sense for reasons that are substantively grounded in stakeholder concerns.

One of the primary reasons this process can be so effective as public engagement is this cycle of listening, designing, presenting, and listening again. The heart of the charrette is a process of constructing a shared

---

[5] For the purposes of this chapter, the NCI model provides a schematic framework of something like a typical charrette process, from pre-charrette to charrette, from charrette to implementation. However, it is recognized that the NCI model is an ideal-typical construction, defined in large part as part of an educational curriculum intended to help improve the practice of charrettes. It is rare for charrettes to follow the NCI model fully, and this is not to imply that the NCI model is by any means the only way to accomplish this sort of design-centered collaboration. It is used here to illuminate what are arguably key aspects of the process. The conclusions drawn here are derived from the author's observations of charrettes and analysis of the logic of the process in terms of its sociological properties.

narrative about the origin and meaning of the solutions as they emerge from a web of choices. This narrative is articulated in terms provided by stakeholders, in a way that makes sense within a universe of knowledge that integrates different perspectives, and is presented in a clear sequence, even if the charrette team introduces other principles, precedents, and technical considerations.

This narrative logic enables participants to grasp the reasoning involved in interconnected but heterogeneous decisions that weave together the various dimensions of urbanism, without reducing the process to an overly simplified hierarchy of choices (Mehaffey and Alexander 2016). The charrette narrative can: clarify the relative salience of factors in the decision making; illuminate the reasoning behind key trade-offs; and link action with desired consequences in the plan. Its shared language enables participants not only to grasp planning as merely functional problem solving, but also to account for the formal intentions associated with urban design as a meaningful and appropriate expression of place.

In many ways, the charrette process mirrors the complexity of the conception of urbanism it has come to serve. Its workflow is structured to ensure that the formulation of a plan is the cooperative work of a multidisciplinary team of experts and that it manifests the cumulative impact of engagement with stakeholders throughout the process. This impact plays out over time, enabling stakeholders to engage with ideas in their early formation, to consider alternatives as responsive to their own observations and experiences, and to watch detailed decisions emerge over the course of the charrette.

The charrette process is place-making in action, as it constructs a built environment with both identity and history. Even in the compressed time frame of a charrette, the complexity of its proposed solutions resists the temptation of reducing results to a simple diagram or *parti*.

If the charrette process is executed in a genuinely open fashion, the resulting plan can carry the traces of its own emergence from a sequence of adaptive decisions, retaining explicit awareness of conflicts and contradictions rather than erasing them or masking them behind a false unity. The plan is also resilient as it can be adapted to future changes, in contrast with the less adaptable geometries of a highly rationalized plan that has been tightly connected to a specific set of functions or circumstances. Instead, the framing capacity of a charrette's common narrative will help to sustain consistent commitment to its vision, even as practical circumstances might shift. Furthermore, its solutions become more robust because they embrace the complexities of the urban environment and simultaneously build relationships in a community, heightening the potential for lasting impact on its capacity for democratic self-governance as the plan is implemented and adapted to changing circumstances. The charrette builds communities by linking the order-constructing, form-giving, and rationalizing toolkit of design with the social exchanges of a process that is both facilitated and allowed to sustain a relative openness.

Success in achieving this complex balance depends on attention to five essential functions, representing the potential of the charrette as transcending the falsely dichotomous view of bottom-up/top-down processes.

1. *Shared learning.* The charrette allows for the construction of hybrid, contextually relevant, knowledge by linking a multidisciplinary team with knowledgeable local stakeholders. Participants work toward a shared definition of the problems, a shared sense of the costs and benefits of different responses as they are integrated into an overall scheme.

2. *Vision and design.* Agreement is too often built at a high level of generality and abstraction, under an assumption that this is the only level at which people can find common ground. However, the devil is in the details, and people will often respond to a general proposal by imagining a host of worst-case scenarios that might result from any general decision. In a design-centered process, there is an iterative movement from big ideas and high-level concepts to the details of a proposal and back again. Collaborative design tests not only the ideas themselves but also the different ways of manifesting those ideas in action, leading to a more robust and practical vision.

3. *Building a shared narrative.* The ability to understand the plan and sustain a commitment to its implementation depends on grasping its logic: What are the conditions and assumptions? What are the key problems to be solved and goals to be accomplished? What is the reasoning associated with key decisions and significant trade-offs? The transparency of the charrette process builds legitimacy for the plan because it enables people to *hear* and *tell* a story that makes sense to them. A narrative that can be appropriated by diverse stakeholders retains the complexity of multiple perspectives, rather than simply reducing them to a singular vision. In the absence of this kind of transparency and discursive engagement, people tend to indulge in the logical error of attributing decisions they do not understand to hidden agendas, corruption, or malfeasance.

4. *Consensus building.* The charrette process encompasses a number of techniques commonly described in the literature on conflict resolution and consensus building. For example: establishing a clear process and agreeing on the participants' roles; engaging in collaborative fact-finding and group problem solving; reaching agreement through a series of small steps; and holding participants to their commitments (Susskind and Cruikshank 2006). The distinctive difference is that these techniques are organized around a design process that moves from abstract ideas to concrete consequences of complex choices. Recent literature on design thinking and user-centered design emphasizes the importance of rapid prototyping as a way to test ideas iteratively in collaboration with users (Brown 2009), working toward improvement of the final

outcome. The charrette process uses a similar technique to ensure a shared understanding of the strengths and defensible limitations of the outcome, ultimately increasing its public support.

5. *Democratic legitimation.* Public processes are often assumed to gain legitimacy only by the representativeness of those who participate. While it remains important to ensure that the process is as inclusive as possible, and that all points of view are represented, legitimacy hinges on the quality of the process beyond the quantity of participants. The legitimacy of a planning outcome rests on the openness and transparency of the process that created it—manifested in charrette practice by repeated demonstrations of listening and significant indications of responsiveness to diverse concerns. The dynamism and creativity of the process resides in the fruitful exploration of conflicting perspectives, each recognized and validated in its own terms and subsequently addressed. The real power of the charrette process often depends on its ability to illuminate, clarify, and encompass significant differences.

CONCLUSION

In charrette practice, we can see a form of what French sociologist Michel Callon has called a "hybrid forum," a discursive practice that allows non-specialists to engage effectively with technical issues, and technical experts to think outside their specialized knowledge to connect meaningfully with something that might be called common sense and everyday practice (Callon et al. 2001). In contrast with conventional processes of public participation, the charrette process opens up the domain of technocratic rationality to the complexity, ambiguity, and lively disorderliness of cities and neighborhoods as lived experience and distributed practices—as something we do together, rather than something that must be engineered for us. Although the charrette has typically been put in service to build support for top-down solutions, charrette-based plans commonly define the space for alternatives. Its techniques of design-based collaboration show the potential for an urbanism that is grounded in alternative modes of collective agency. Although the charrette's current embeddedness in professional practices and institutional settings tends to favor consensus building as a path to achieve closure on urban challenges, the charrette process can set up an interplay between stabilizing consensus and the ability to hold open spaces of possibility for diverse perspectives (see Dittmar and Kelbaugh, Chapter 5 in this volume).

Furthermore, charrette practice represents a potentially profound shift from participation as an ideal or obligation to it as a practice of engagement that cultivates and facilitates the bottom-up processes of urbanism, while translating them into clearly articulated common intentions, explicit practical agreements, and well-defined goals. This translation enables the transparency, critical reflection, and purposeful deliberation necessary to sustain effective democratic governance—aspects of political practice that sometimes get lost as a result of an over-emphasis on bottom-up processes. At the same time, bottom-up urbanism is maturing. Tactical Urbanism has developed a modality of direct action that is now being institutionalized from critical interventions to sanctioned events, and the Lean Urbanism initiative aims to open up a broader space for small-scale, incremental, and locally instigated urban development projects that require neither intensive public or private investment nor compliance with unnecessarily burdensome bureaucratic procedures. Many other initiatives build networks of local actors who can take advantage of crowdsourced intelligence and resources, and build new community capacity, new community relationships, and new forms of bottom-up agency.

These tendencies suggest the development of fields of practice that transcend the deeply institutionalized configuration of professional/technical expertise, bureaucratic regulation, and the democratic politics of Sandel's "procedural republic." As bottom-up urbanism establishes new forms of agency in the making of places, collaborative processes organized around design and design thinking could be a way to frame these modalities of action within a clearly articulated and shared sense of common purpose, in connection with a strategic vision. If bottom-up urbanism has to do with organic, informal processes and top-down urbanism has to do with intentionally enacted, formally institutionalized, and authoritative actions, the charrette process provides a methodology for transcending this dichotomy. Taken together, the result is a self-conscious practice of urbanism that might be our best hope for addressing the growth of urban issues ranging from economic development and social justice to sustainability.

References

Alexander, C. (1987). *A new theory of urban design*. Oxford: Oxford University Press.
Alexander, C., Ishikawa, S., Silverstein, M., Jacobson, M., Fiksdahl-King, I., & Angel, S. (1977). *Pattern language: Towns, Buildings, Construction*. NY: Oxford University Press.
Augé, M. (1995). *Non-places: Introduction to an anthropology of supermodernity*. New York: Verso.
Bourdieu, P. (1990). Social space and symbolic power. In *In other words: Essays towards a reflexive sociology* (pp. 123–139). Palo Alto: Stanford University Press.
Bourdieu, P. (1992). *The logic of practice*. Stanford: Stanford University Press.
Brain, D. (1994). Cultural production as 'society in the making'. In D. Crane (Ed.), *The sociology of culture: Emerging perspectives* (pp. 191–220). Cambridge, MA: Blackwell.
Brain, D. (2008). Beyond the neighborhood: New urbanism as civic renewal. In T. Haas (Ed.), *New urbanism and beyond: Designing cities for the future* (pp. 249–254). New York: Rizzoli.
Brown, T. (2009). *Change by design: How design thinking transforms organizations and inspires innovation*. New York: HarperCollins.

Callon, M., Lascoumes, P., & Barthe, Y. (2001). *Acting in an uncertain world: An essay in technical democracy.* Cambridge: MIT Press.

Congress for the New Urbanism. (1999). *The charter of the new urbanism.* New York: McGraw-Hill.

Crawford, M. (2008). The current state of everyday urbanism. In J. Chase, M. Crawford, & J. Kaliski (Eds.), *Everyday urbanism* (pp. 13–15). New York: Monacelli Press.

Duany, A., & Brain, D. (2005). Regulating as if humans matter: The transect and post-suburban planning. In E. Ben-Joseph & T. S. Szold (Eds.), *Regulating place: Standards and the shaping of urban America* (pp. 293–332). New York: Routledge.

Duany, A., Plater-Zyberk, E., & Speck, J. (2000). *Suburban nation: The rise of sprawl and the decline of the American dream.* New York: North Point Press.

Halpern, R. (1995). *Rebuilding the inner city: A history of neighborhood initiatives to address poverty in the United States.* New York: Columbia University Press.

Hibbing, J. R., & Theiss-Morse, E. (2002). *Stealth democracy: Americans' belief about how government should work.* Cambridge: Cambridge University Press.

Jacobs, J. (1961). *The death and life of great American cities.* New York: Random House.

Kaliski, J. (2008). The present city and the practice of city design. In J. Chase, M. Crawford, & J. Kaliski (Eds.), *Everyday urbanism* (pp. 88–109). New York: Monacelli Press.

Katz, P. (1994). *The new urbanism: Toward an architecture of community.* New York: McGraw-Hill.

Kelbaugh, D. 2007. Toward an integrated paradigm: Further thoughts on the three urbanisms. *Places, 19*(2). Permalink: http://escholarship.org/uc/item/25d4w94az.

Kirschenblatt-Gimblatt, B. (2008). Performing the city: Reflections on the urban vernacular. In J. Chase, M. Crawford, & J. Kaliski (Eds.), *Everyday urbanism* (pp. 19–21). New York: Monacelli Press.

Kohn, M. (2016). *The death and life of the urban commonwealth.* New York: Oxford University Press.

Kretzmann, J. P., & McKnight, J. L. (1993). *Building communities from the inside out: A path toward finding and mobilizing a community's assets.* Chicago: ACTA Publications.

Krier, L. (1998). *Architecture: Choice or fate.* Windsor, UK: Andreas Papadakis Publisher.

Latour, B. (1992). Where are the missing masses? The sociology of a few mundane artifacts. In W. Bijker & J. Law (Eds.), *Shaping technology, building society: Studies in sociotechnical change* (pp. 225–258). Cambridge: MIT Press.

Lennertz, W., & Lutzenheiser, A. (2014). *The charrette handbook: The essential guide to design-based public involvement.* Chicago: APA Planner's Press.

Levebvre, H. (1991). *The social production of space.* Malden: Blackwell.

Logan, J., & Molotch, H. (1987). *Urban fortunes: The political economy of place.* Berkeley: University of California Press.

Lydon, M., & Garcia, A. (2015). *Tactical urbanism: Short-term action for long-term change.* Washington, DC: Island Press.

Massey, D., & Denton, N. (1997). *American apartheid: Segregation and the making of the underclass.* Cambridge: Harvard University Press.

Mehaffey, M., & Alexander, C. (2016). *A city is not a tree: 50th anniversary edition.* Portland, OR: Sustasis Press.

Mukerji, C. (1997). *Territorial ambitions and the Gardens of Versailles.* Cambridge: Cambridge University Press.

Parolek, D., Parolek, K., & Crawford, P. C. (2008). *Form-based codes: A guide for planners, urban designers, municipalities, and developers.* Hoboken: Wiley.

Putnam, R. D. (1993). *Making democracy work: Civic traditions in modern Italy.* Princeton: Princeton University Press.

Putnam, R. D. (2000). *Bowling alone: The collapse and revival of American community.* New York: Simon & Schuster.

Salingaros, N., Brain, D., Duany, A., Mehaffy, M., & Philibert-Petit, E. (2006). Social housing in Latin America. In *The future of cities.* Solingen: Umbau Verlag.

Sampson, R. J. (2013). *The great American city: Chicago and the enduring neighborhood effect.* Chicago: University of Chicago Press.

Sampson, R. J., & Raudenbush, S. W. (1999, November). Systematic social observation of public spaces: A new look at disorder in urban neighborhoods. *American Journal of Sociology, 105*(3), 603–651.

Sandel, M. (1996). *Democracy's discontent: America in search of a public philosophy.* Cambridge: Harvard University Press.

Schumpeter, J. (1942). *Capitalism, socialism and democracy.* New York: Harper.

Sirianni, C., & Friedland, L. (2001). Civic innovation and American politics. In *Civic innovation in America: Community empowerment, public policy and the movement for civic renewal* (pp. 1–34). Berkeley: University of California Press.

Sitte, C. (1979). *The art of building cities: City building according to its artistic fundamentals.* New York: Hyperion Press. (Originally published in 1889.)

Susskind, L. E., & Cruikshank, J. (2006). *Breaking Robert's rules: The new way to run your meeting, build consensus, and get results.* Oxford: Oxford University Press.

Thadani, D. A. (2013). *Visions of seaside.* New York: Rizzoli.

Williamson, O. M. (1983). *Markets and hierarchies: Analysis and antitrust implications.* New York: Free Press.

CHAPTER 5

Lean Urbanism Is About Making Small Possible

Hank Dittmar and Douglas S. Kelbaugh

It has become commonplace knowledge that the rapid growth of cities in the twenty-first century has made it the *urban century* (Brown et al. 2009; Heynen 2014), but that does not make population numbers any less staggering. Humans evolved from being hunter-gatherers to adopting an agrarian lifestyle over tens of thousands of years, and the preponderance of the world's population remained in rural areas through the latter part of the twentieth century. But the shift toward city living has accelerated to the point where, according to the United Nations Department of Economic and Social Affairs, "54 percent of the world's population [is] residing in urban areas in 2014. In 1950, 30 percent of the world's population was urban, and by 2050, 66 percent of the world's population is projected to be urban" (Nations 2014).

Sadly, we lost Hank Dittmar in April 2018. He was one of the very first to articulate the environmental paradox of cities, as well as to think about public space and urbanism as more important than individual buildings. He fervently believed that built capital is the physical stage upon which humans live out their lives. His presence, wisdom and insight will be missed.

H. Dittmar (✉)
formerly of The Prince's Foundation, London, UK

D. S. Kelbaugh
University of Michigan, Ann Arbor, MI, USA

© The Author(s) 2019
M. Arefi and C. Kickert (eds.), *The Palgrave Handbook of Bottom-Up Urbanism*, https://doi.org/10.1007/978-3-319-90131-2_5

URBANIZATION SEEN AS A PRODUCTION AND A STORAGE PROBLEM

Contemporary writers describe today's urban challenge as largely a problem of production, of building cities fast enough for the people tumbling into them. This implies that it is also primarily an issue of storage, viewing cities as a container large enough to accommodate the incoming masses. Otherwise, experts warn, we will have an explosion of slums that will precipitate a human health crisis, from epidemics to premature morbidity and mortality.

Urban writers such as Edward Glaeser extol the benefits of dense, high-rise cities. Arguing in his influential book *Triumph of the City*, Glaeser posits a future of rational beings commuting short distances from residential tower to office tower. Glaeser reverses the recent economists' trend of suburban apologia, exemplified by Peter Gordon and Harry Richardson, and promotes a future in which historic buildings and traditional neighborhoods must give way to a Singaporean city of towers produced by market forces. He argues for an even higher density than the traditional European city of row houses and mid-rise apartment buildings. Glaeser wants to loosen planning regulations to recast the city as a high density, high rise, environment (2011).

This conventional answer, both in planning theory and development practice as an adaptation of the modernist idea of towers in the park, consists of high-rise blocks knocked up as quickly as possible. In practice, what is often built is the high-rise version of separated land use, auto-oriented suburbia promoted through planning diagrams and traffic regulations imported by engineering and design firms from the US, the UK, Canada, and increasingly Singapore. These opportunistic tower blocks appear mind-numbingly similar in their detail and banal in their endless search for new shapes and materials. Derived from Le Corbusier's Villa Radieuse, these towers in the park often end up as towers in the parking lot, surrounded and isolated by roadways, making the inhabitants dependent on motor vehicles for mobility.

Modernist architecture and planning ideas promoted all over the developing world use an iconography and language, eerily redolent of the 1960s. The tower building type is produced in two distinct forms: as a plot based sprawl by speculative builders, often without an overall plan and without supporting infrastructure; and as an isolated tower model, on the super block, at a large scale. This high-rise sprawl is the prevalent model in China, India, Africa, and Latin America.

Similar to urban renewal in the West, where towers displaced long-standing neighborhoods, these buildings bring with them social hardship and upheaval, make it difficult to develop new social relationships, and create clear environmental and cultural problems. This reductionist approach ignores social and economic sustainability and development. This chapter argues for the incorporation of the informal and the small scale into our urban toolkit, in both developed and developing cities.

The Persistence of the Informal

Slums are often seen by government as an embarrassment, as a failure of society to provide for the poor, and as a challenge to the orderly provision of housing and jobs. The first impulse in many developing countries has been to demolish slums, to repossess the land for urban development, and transfer it to well connected, conventionally financed developers along with land titles. Squatters often successfully resist this type of urban removal.

The problems with slums have been well documented elsewhere, but they bear repetition: crime, poor sanitation, lack of potable water, lack of security of tenure, lack of public space, and overcrowded, flimsy dwellings. Despite significant progress in improving living conditions, "around one quarter of the world's urban population continues to live in slums. Since 1990, 213 million slum dwellers have been added to the global population. In Africa, over half of the urban population (61.7%) lives in slums, and by 2050, Africa's urban dwellers are projected to have increased from 400 million to 1.2 billion" (Habitat 2015b).

Resistance to these conditions and to relocation policies led to a slum dwellers movement, wherein people living in shacks and slums advocate for secure tenure, an official postal address, infrastructure and sanitation, and organize themselves to improve living conditions. The exemplar for this approach is the Mumbai slum of Dharavi, where Jockin Arputham, founder of the Slum and Shack Dwellers movement, organized the community's women to provide toilets and worked to resist the forced removal of people from the city's largest slum (Perur 2014).

Slums persist—despite government action to destroy them or to relocate their residents—because of the existence and resilience of social and economic networks. The continuation of loose and close ties and affiliations, and the lack of alternative places to live and work, supports resistance to relocation because it disrupts these useful networks (Doshi 2013). Like the settlement houses in Chicago slums or their Victorian equivalents in England, slums are places with formal and informal social services that impoverished urban immigrants use and rely on.

Moreover, slums often replicate the urban patterns of village, town, and city life across the planet, and the lively, mixed-use streets that predated the superblock and the wide arterial road. There are useful patterns for the future in both the social and physical organization of slums, as HRH The Prince of Wales said in a 2009 speech at the conference Globalization from the Bottom Up: "Indeed, whenever I have visited informal settlements such as, for example, Dharavi in Mumbai in 2003, I find an underlying, intuitive 'grammar of design'—that subconsciously produces somewhere that is walkable, mixed-use and adapted to local climate and materials, which is totally absent from the faceless slab blocks that are still being built around the world to 'warehouse' the poor" (2009). To retain both the social and physical characteristics of resilient urban forms land tenure must be enabled, services provided and the gradual improvement of housing, infrastructure, and economy empowered.

Within and outside of informal settlements, a large part of the world participates in the informal economy by working, trading, or living without necessarily asking permission, paying taxes, or following regulations. Although estimates of the scale of the informal economy vary widely, it remains enormous. According to UN Habitat, "In many developing countries, informal employment comprises more than half of non-agricultural employment. In low-income countries, informal employment makes up 70–95 percent of total employment (including agriculture) and is found mainly in the informal sector" (Habitat 2015a: 3).

The informal sector, or 'gray' economy, ranges from the tiny catering business making meals from a home kitchen or the carpenter working for cash, to underground artists living and working in spaces such as Oakland's Ghost Ship, where a fire claimed 39 lives in late 2016, and to oligarchs and gangsters laundering cash through real estate in cities like London, Vancouver, and Sydney. The bulk are small businesses skirting the law because the cost or hassle of compliance is perceived as too great. The gray economy is often untaxed, unregulated, and invisible, representing a route out of poverty for many, or a disaster waiting to happen because of unsanitary kitchens, pit toilets, structurally deficient buildings, and fire safety problems.

Art is a key element in many Western informal economies. The role of artists in stimulating city economies through job creation and property uplift is well understood, and the Ghost Ship fire sparked both a national crackdown on unsanctioned artist spaces around the United States and a dialogue with artists about bringing them into compliance. New York's pioneering Loft Laws permitted the use of former industrial spaces by artists, prompting task forces and other efforts in a number of cities to bridge the gap between regulators and artists (Board 1982; Luis 2016). Similarly, the rise of pop-up restaurants and food trucks as a way to beat the high regulatory and capital cost of opening a restaurant is welcomed by many city governments.

At the same time, the recent emergence of a movement called Tactical Urbanism, in which people seek to make their immediate environments more livable, and exercise a degree of local control over their neighborhoods and streets. Unsanctioned or semi-sanctioned tactical interventions include: making streets safer and more walkable; installing tiny parks and plazas and planting new trees and guerilla gardens; and introducing pop-up activities in vacant storefronts or parking spaces. These and other temporary interventions in both public and private spaces serve as a tangible demonstration of possibility and as the catalyst for more permanent transformation. Tactical urbanism has spawned 'how to' books and projects from Dallas to Mumbai (Finn 2014; Lydon et al. 2015). The growing popularity of this movement indicates a common interest in stimulating the informal economy and in the city-building process.

THINK LOCALLY, ACT LOCALLY

While Tactical Urbanism has its roots in the United States, another indicator of the move toward more recognition of the smaller scale and the community sector, compared to the master planned settlement, is the advent of localism in England, a popular movement and a recognized part of the planning process. The passage of the Localism Act in 2011, instituted a set of tools for English neighborhoods and communities, including neighborhood planning and neighborhood development orders, the identification and protection of community assets, and the right to build for communities (Commons 2011; Government 2011).

Localism has aimed to create a positive framework for growth by bringing decisions about neighborhood character and form closer to the community. For the first time, local authorities are required to follow neighborhood plans adopted by neighborhood forums by referendum, so long as they passed through a formidable set of hoops. These plans indicate where and how growth should occur, what community assets need to be preserved, and can define neighborhood character through design guidance and policy. The implementation of localism reveals a continuing conflict between the ideal of delegation and empowerment with the habit of centralism and regulation. It also reveals the real appetite for making small-scale development and community building easier.

Over 2000 neighborhood plans are underway in the UK and 238 referenda have been held to approve neighborhood plans, all of which have passed with overwhelming support. The average 'yes' vote is 88% and the average turnout is 33%, quite high for a local election (Government 2016). However, the gap between the number of plans underway and the number that have successfully reached a vote is an indicator that the process is difficult and burdensome. In urban areas, neighborhoods must first be designated, a process than can take up to a year. Neighborhood forums are organized from the bottom up, and limited funding is available. Last but not least, many local authorities have supported localism, others have obstructed the movement. This can present a major obstacle, as neighborhood plans must be reviewed both by the local authority and by a planning inspector before a referendum.

Clearly, localism has been treated not as a simpler, more responsive way to improve communities and promote enterprise and business activity, but as another layer of government. The requirements are as cumbersome as those imposed on higher layers of authority and they are often duplicative (Dittmar 2014).

How do we plan for the new way people will use the city? Not by abandoning the long-term vision and the stability of governance, but by reintroducing shorter time frames and smaller scales, and recasting incentives and rules to advantage the short term, the incremental, and the smaller scale, rather than disadvantage them as in the present situation. *What is needed is a middle scale between the temporary pop-up and the visionary, long-term master plan.*

Lean, as in Fit for Purpose

A lean approach to building and revitalization attacks the sclerosis that affects planning and review processes. The lean shift helps entrepreneurs, artists and crafts people, small business people, and community groups with great ideas, including those without deep pockets, contribute to their communities, to put people to work, and to make nicer places. It slims and relaxes codes and plans, challenges thresholds for regulation, and finds workarounds and patches to the glitches, snags, and snafus that seem to plague small and innovative projects more than they do larger developments.

Lean Urbanism is open ended and open access, reduces the barriers for ordinary people to extend their home, improve their business, or develop community assets. It is open-source, providing tools and platforms that all can build upon. Lean Urbanism works on infill and suburban repair, at the scale of the building or the street, developing code-light zones, pilot projects, pre-approved plans, pre-approved architects, and building types that fall beneath thresholds for costly lifts or expensive equipment, such as a four-unit building with a ground floor accessible unit.

A latent demand for more agile urbanism has induced a group of architects, developers, activists, and urbanists in the United States to come together to undertake the Project for Lean Urbanism.[1] This project has produced white papers, tools for lighter codes, leaner infrastructure, and lean building types, and has prompted pilot projects in Savannah, Georgia, Lafayette, Louisiana, and St. Paul in Minnesota, Chattanooga in Tennessee, and Detroit in Michigan. The effort has been given critical grant support by the Knight Foundation and the Kresge Foundation, and match funding has been provided for pilot projects by local governments and affected business communities.

Lean Urbanism is a response to the requirements, complexities, and costs of regulation that disproportionately burden small-scale developers, builders, and entrepreneurs, who cannot afford the lawyers and technical consultants that large developers regularly hire. The tools, techniques, and architectural plans will be released freely, for all to use. These open-source and municipally provided tools will allow more people to participate in the building of their homes, businesses, and communities.

From Gray to Pink

At the core of a lean project is the idea of targeting small-scale and incremental interventions in existing cities and suburbs. Each can identify, unlock, and leverage hidden or underutilized assets: built assets like vacant buildings, or land or transport capacity; financial assets like social capital or crowd sourcing; or community assets like housing associations, knowledge clusters, or artist quarters; or natural assets.

---

[1] For more information and case studies, see www.leanurbanism.org.

The locus for the implementation of Lean Strategies is the pink zone, an area in which the red tape is lightened. This zone identifies an area where codes are relaxed, new protocols are pre-negotiated, and experiments are conducted, all with the goal of removing impediments to economic development and community building.

A pink zone—similar to an innovation district, another recent planning initiative—spurs revitalization, engages populations that have been left out of the development process, stimulates sustainable economic activity and asset-building through incubator and maker spaces, and addresses both physical assets and community assets. The two designations are different in that innovation districts tend to focus exclusively on jobs, while pink zones mainly address housing. Innovation districts primarily encourage new technologies, commercial innovation, and high-growth businesses (Katz and Wagner 2014), while pink zones encourage: community-supportive enterprises, such as Main Street businesses; start-up makers and platforms, such as shared work spaces, kitchens, community hubs, and markets.

Pink zones can vary in size, intensity and composition. Most consist of infill redevelopment, one or two buildings at a time, whether filling vacant lots, refurbishing dilapidated buildings, or selectively demolishing and redeveloping underutilized parcels. The ideal neighborhood for Lean Urbanism to infill is typically in marginal condition but not hopelessly abandoned, as Lean Urbanism is conceived to operate without extensive subsidy or grant programs, and within its own layers of regulation and complexity. The typical project might be the four- to five-unit walk-up building, the live-above-storefront, a market building, a kitchen incubator, a live-work building, or the densification of a neighborhood with accessory units or alley/garage flats. The lean developer can be a small builder, a local business wishing to expand its property, a local not-for-profit or community association, or a property owner or group of property owners. Lean projects might set the stage for further development but in a way that protects current residents and property owners, who can control the process from the outset.

Pink zones can comprise larger scale projects, with available parcels big enough for medium-scale developers or local institutions to develop ten or more new housing units at a time. While these projects do not align with the small-scale nature of Lean Urbanism, they do apply the concept of lightened regulation or pre-approval to areas designated by local governments, such as districts near public transit. A university might want to build specialized housing for students, faculty or staff, or senior housing might be sponsored by a church, synagogue, mosque or temple. Still other projects might be more experimental, introducing new building types and/or more daring designs, or they might be projects that are primarily self-built by residents. All these types of development are encouraged by and benefit from relaxed codes and regulations (Fig. 5.1).

Fig. 5.1   Renovation of old factory for housing (Image courtesy of DPZ Partners)

Tools for Pink Zones

A pink zone first requires an assessment of the impediments and assets in a community, and is implemented with a carefully chosen kit of tools. These tools can stimulate revitalization and change a 'gray' economy into a healthier and more above-board pink one, encouraging artists, underground businesses, and the like to move to the surface through simpler regulations, technical assistance, and market-creating platforms in pink zones. The tools include:

- Lean Scan: A tool for identifying barriers, finding latent opportunities and leveraging underused assets in a way that unlocks synergies between built, financial, social, and natural resources. Unlike a typical charrette or public-engagement process, the outreach component of the Scan finds those with the will, talent, and energy to improve the urban environment (Dittmar 2015).
- Lean Regulation: Reviewing codes, regulations, and permit requirements to identify existing thresholds for small projects, below which review is not required or code provisions are not triggered. Where regulations allow for interpretation, establishing policies that reduce burdens for small actors, small projects, and small businesses. And finally, delegating authority to a lower but competent level that is closer to the project or business.

- Reduced Fees: Where fees for permits, inspection, environmental impact, and so on, are determined to be impediments, they should be reduced, eliminated or thresholds set which do not unfairly burden small-scale projects.
- Lean Codes: Simplified codes that enable small-scale development and business, and that demystify and simplify requirements; thresholds which reduce requirements for occasional use, small-scale enterprise, and small buildings, with mandatory turnaround times.
- Lean Reuse and Renovation: Identifying and removing barriers to bringing abandoned or underused buildings back to productive life, avoiding the trap of spending more money and resources to meet building codes than the rehabbed value of the building.
- Live-Work: Define provisional changes to existing codes that can facilitate the building and rehab of live-work units as a flexible, low-cost way to provide connected housing and work spaces.
- Lean Building Types: Pre-approved building plans for new construction, and rules that allow certain building modifications and changes of use to be carried out without an application.
- Infrastructure and Public Realm: Identify lean alternatives to enhance pedestrian and cycle use, calm traffic, and improve the public realm within a pink zone.
- Lean Development: Recruiting and training small developers with the 'developer-in-a-box' tool, local small developer boot camps, shared learning about workarounds and thresholds.
- Lean Finance: 'De-risking' investment in pink zones for both community lenders and equity investors (including crowd sourcing and pools of investment capital from local business people and professionals such as doctors, dentists and accountants).
- Platforms: Structures that go beyond the single project by pooling resources and supporting emerging markets within a pink zone. They can take many forms, including physical buildings, online portals, open-source information, incubation and local organizations both for-profit and non-profit.
- Lean Green: small-scale and local green techniques, often low tech and passive in nature, that address energy consumption, pollution, and climate change, including both mitigation and adaptation.

The Seven Principles of Lean, Energy-Efficient Buildings: An Example of Open-Source Lean Urbanism

The open-source nature of a Lean Urbanism project allows for proponents to provide their own knowledge and interpretations. Lean can refer to the process of urban development, regulation, and organization, and it can also refer to the outcome of this process at the urban and architectural scale. For example, a lean building is meant to leave a small energy, carbon, and ecological footprint. These guidelines are included as an example of a tool in the Lean Urbanism toolbox.

Energy-efficient buildings engage both energy and ecology. They deal with *material* resources—both organic and inorganic—and with *energy* consumed by buildings for their heating, lighting, cooling, and running equipment, and for maintenance. As for building materials, there are some radical new building products being experimented with: superglazing, interactive surfaces, self-healing materials, bio-composites, 3D fabrics, engineered living materials, bone-like and eggshell-like structural materials … with surely more to come. There are two basic ways to measure energy: by *quantity* (production of BTUs, watts, tons of fuel, etc.); and by *quality* (its efficiency as a fuel and its ability to do work, as well as its waste, including the 'externalities' produced such as CO_2E and other by-products). The principles listed below primarily describe ways to reduce the *quantity* of energy used.

There has been significant progress over recent decades in cutting energy use in buildings, but there is a long way to go to fully and wisely steward our energy resources, and to mitigate and adapt to climate change and Urban Heat Islands. We want simple and understandable ways to do more with less energy, and to cut overall energy use. Lean design is all the more essential as we run out of low-hanging fruit and have to reach higher to conserve and generate energy. The lowest fruit is reducing the energy *demand* of buildings. It is self-evident that a BTU or watt saved—a negawatt—is far cheaper than one produced, even from 'free' energy sources like wind and solar, because of capital and maintenance costs (Lovins and Browning 1992).

Defensive strategies—primarily insulation and weather-stripping—usually have the quickest economic payback. Reused and recycled building materials and components are preferable to new ones, especially synthetic and high-tech materials whose manufacture requires high energy and chemical inputs. Downsizing buildings as much as possible is always a good defense. Urban buildings that share walls, floors, and roofs require less heating and air conditioning per person, as well as less building material and maintenance. Even better is renovating or retrofitting existing buildings. The best option can be to not renovate or build, getting by with existing facilities. Indeed, "the greenest building is the one already built" (Elefante 2012).

Offensive strategies include renewable energy, typically solar, wind, and geothermal. If done at the community level, hydro, biogas, and biomass can also be cost-effective. The sun is an intensive light source, and its most common and effective architectural use is *illuminating* building interiors with daylight. Passive solar *heating* systems—in which the building itself collects, stores, and distributes heat—is another use of the sun's generous supply of energy. Passive systems are more architecturally integrated than active solar systems, whose panels are add-ons to a building and require fluids, pumps, thermostats, and racks. Passive systems—such as attached greenhouses, direct gain, and Trombe Walls—are proven and cost-effective ways to heat space and domestic hot water in many climates. Although typically less architecturally integrated, photovoltaic panels or cells (PVs) and solar hot water collectors can be added to the building, typically on the roof, where they can be arrayed and tilted to face and even track the sun. Importantly, passive solar and PVs emit no GHGs (greenhouse gases) or other pollutants (except in their manufacture and construction).

Energy codes, regulations, and red tape have steadily increased over the decades. Both mandatory and voluntary, they have significantly reduced energy consumption but have added to the bureaucratic delay and hassle of design and construction. Lean codes and a simpler, more locally tailored, and holistic regulatory regime are essential, even if initially limited to certain constituencies or zones within a city or region.

The metrics used should also be appropriate. Small buildings are typically skin-load dominated, which means their thermal performance is driven by energy flows through their walls, windows, and roofs. Their performance is usually intuitive and can often be predicted by simple rules of thumb and charts. However, large, multistory buildings dominated by internal loads, such as lights, equipment, people, etc. typically need manual or computer calculations during the design process. If energy consumption is calculated, the best overall metric is watts or BTUs per square foot, per year, per occupant, which is roughly equivalent to carbon footprint per capita, and includes people in the equation. It is the best metric to adopt if we are to ultimately achieve a sense of fair share in energy consumption worldwide.

There are many design techniques and building technologies that can significantly reduce the use of energy and production of waste heat and GHGs. Below is a list of seven basic architectural design strategies published by co-author Kelbaugh forty years ago, and recently updated as an open-source tool for the Lean Urbanism project. These principles have been tested and proven over centuries of use, the world of passive solar systems has not changed much over time, with the exception of new glazing, insulation, and building materials. The principles address buildings—especially smaller and new buildings—across all climate zones in the USA, although locales that are extremely cold, hot, or humid need further elaboration. They also apply to similar climate zones throughout the world, with some of the principles reversed in the southern hemisphere (Fig. 5.2).

These principles are offered as a short but complete, irreducible list for designers, builders, and the public. When further detail for actual, site-specific applications is needed, local expert advice is required.

To consume less energy and produce less greenhouse gases, buildings should:

1. *Be built with local, low-energy materials and methods, and designed no bigger or more extravagant than needed.*
   Building materials that are in their natural or near-natural state are preferable to processed ones that have additional energy and chemical

Fig. 5.2 Example of housing with PV roof panels in the Vauban neighborhood of Freiberg, Germany (Image courtesy of Douglas Kelbaugh)

inputs. Locally sourced wood, stone, brick, and glass are less energy-intensive than aluminum, plastic, concrete, and steel (unless they are reused or recycled). Materials that are salvaged, non-toxic, renewable, and bio-degradable are also superior, in both environmental and energy terms.

2. *Have an envelope capable of isolating or buffering it from heat, cold, and humidity, consistent with the climate zone.*

In temperate and cold climates, walls, roofs, and floors should be heavily insulated, with double-glazed windows, with minimal air infiltration and cold bridges. Insulation has one of the highest paybacks of any investment, especially in smaller buildings with a high skin-to-volume ratio. Careful attention should be given to vapor barriers, which belong on the warm side of insulation. Glazing on all faces should be covered at night with interior or exterior, air-tight, movable insulation, such as insulating curtains, shutters, and panels.

Light colored roofs and walls reflect unwanted solar heat gain, but their primary benefit is reducing local heat islands. Shade from trees and other vegetation is often essential. Larger, multi-floor buildings have a more favorable ratio of volume to skin than smaller buildings and are thermally more energy-effective per occupant. Engineering studies have shown that defensive strategies are usually the most cost-effective. It is worth noting that, according to McKinsey & Company, wall and roof insulation have the fastest payback of *any* energy-saving technology on the planet (Enkvist et al. 2007)!

3. *Be oriented to take advantage of local climate, face south (in the northern hemisphere) if possible, with sufficient glazing to passively collect solar gain if there is a heating season, and have appropriate shading of south and west glass when solar gain is unwanted.*

   In temperate and cold climates, if at all possible, buildings should have the largest face oriented to the south, preferably within 20 degrees. Buildings and rows of buildings should be stretched east to west to maximize solar exposure. The south face(s) should be generously glazed, as each square foot of vertical double-glazing gains more heat than it loses over the course of most heating seasons. East, west and north faces perform better with minimal window area, preferably no more than needed for daylight, views, and natural ventilation.

   At appropriate times of the day and year, glazing should be shaded with exterior, fixed or movable, shading devices, without compromising appropriate levels of daylight. In most climates, west-facing glass is particularly susceptible to visual glare and overheating, and exterior fixed or movable vertical louvers, shutters, trellises, and living walls can shade glass from the hot sun. There are similar benefits from *interior* window blinds and shades, which are lower in cost, but devices installed inside of the glass are thermally less effective. It is always better to intercept the sun's radiation, especially direct sunlight, before its heat is trapped by glass (just as it is better to reflect solar radiation with clouds and snow, before it hits the ground and makes a similar conversion in the atmospheric greenhouse that drives climate change).

   South-sloping roofs, steeper in latitudes further from the equator, can be devoted to PV solar panels and to thermo-siphoning solar hot water systems with flat-plate or evacuated tube collectors. (Climates with freezing temperatures must be drained down at night, or use freeze-resistant fluids and heat exchangers with the potable water.) These systems can also be deployed in tilted arrays on flat roofs.

4. *Have sufficient mass to store solar gain and to act as a thermal flywheel, radiating warmth in the heating season and absorbing it in the cooling season.*

   Where heating is needed, there should be enough thermal mass inside the thermal envelope to carry any excess heat gain from the day into the night or next day. Heavy, dark-colored masonry with *direct* exposure to the sun is among the best thermal storage/flywheel devices, especially if it is also part of the building's structural system (for example a Trombe Wall, where the mass, which is between the occupants and glazing, holds up part of the roof). Dark-colored water barrels can be useful for thermal storage, especially in attached greenhouses and sunspaces. Multi-zone buildings should redistribute excess solar gain or internally generated heat to other zones that need heat.

Where cooling is needed during the day, there should be enough thermal mass inside or outside the thermal envelope to delay the arrival of the afternoon heat wave until cooler night air can cool and ventilate the building. Interior thermal mass can also help smooth exterior temperature swings and can work effectively together with natural ventilation.

5. *When cooling is needed, be open to and induce natural ventilation, have low albedo roofs, and shade with vegetation, fixed and movable devices.*

Cross-ventilation and chimney-effect, vertical ventilation should be used in hot, humid climates as needed to cool building spaces and building mass, especially at night when outdoor air is cooler. Night sky radiation and evaporative cooling are effective in hot, dry climates with clear skies. Towers can help induce natural ventilation in the cooling season by taking advantage of the chimney effect.

6. *Be readily adapted, renovated, and repaired over time, with materials and components reused at the end of their useful lives.*

Usually the building foundation, structure, and shell should be built to last a century or more in order to shelter different users and needs over time. Buildings should be constructed of materials that are reusable or recyclable. Movable, short-lived, and personalized building components should be flexible on a daily and seasonal basis, as well as adaptable over the years. Buildings should recycle gray water for flushing of toilets and irrigation of plants. Occupants should recycle inorganic waste, compost organic waste, and minimize water consumption.

7. *Employ the first six principles preferably in connected, multistory urban buildings and in ways that are site-specific, context-sensitive, and that do not prevent other buildings from employing them.*

Urban areas have several advantages for reducing the energy footprint of buildings. First, urban buildings have a smaller exterior surface area per occupant, because their party walls and shared floors reduce the need for heat in the heating season and for air conditioning in the cooling season. Urban buildings also contain less embodied energy and require less energy to construct and maintain per occupant. Buildings that are 100% solar or zero-energy are often less cost-effective than an equivalent investment in multiple buildings that are 60–80% solar, as reaching 100% efficiency can exponentially raise costs on the curve of diminishing returns. This economy is especially important when affordability is paramount, as are other economies, such as building heights from three to five stories, which can make stairways a reasonable alternative to elevators.

Daylight and solar access for neighboring buildings should be blocked as little as possible. Buildings can slope or step down to the north to cast shorter shadows.

Secondly, urban areas also provide the benefit of economies of scale in saving energy. Wherever possible, buildings should connect to district heating, cooling, and electricity generation systems, taking advantage of scalar efficiencies. District infrastructure and systems—such as integrated building energy management, renewable energy, energy storage, water recycling, and on-site wastewater treatment—actively reduce energy and climate impacts.

Finally, urban areas reduce the dependence of residents on energy-intensive transportation, especially personal automobiles and other motorized vehicles. If the building is not located in an area that has ample amenities and services within easy walking or biking distance, or if it is not well served by transit, dependence on automobiles can offset the energy and climate benefits promoted by these principles.

These seven principles of energy-efficient buildings will increasingly prevail as energy prices rise, and more and more good exemplars are built in different climates. New economic, political and climatic realities will make Lean Urbanism and sustainable, resilient architecture more compelling and widely deployed.

To conclude, the small and local nature of Lean Urbanism is not antithetical to the widespread change that we need to meet massive problems like concentrated poverty, neighborhood decline, urban heat islands, and climate change mitigation and adaptation. Making small possible does not mean Lean Urbanism cannot scale up, creating local and national platforms that support a large number of incremental improvements. The persistence and prevalence of both informal settlements and the informal economy demonstrate the contagious nature of seemingly small decisions.

Of course, big projects have their place, but their pace and scale of change can be too fast and/or too fast for local communities, as well as too costly and physically disruptive. The institutions and developers that can take on large-scale projects are often not community or individually owned. Lean Urbanism may thus be a tool for a simpler and more cost-effective scale of revitalization, and one that is aimed at ensuring community benefit and control of the process—allowing revitalization through investment in the social capital of a community as well as the built capital. By shifting from a paradigm that sees urbanization as a problem of storing people in buildings—often near informal settlements that are now being seen more as assets to improve than as slums to be razed—to one that sees urbanization as an opportunity for co-creation, we can balance the manageable and the revolutionary. This sweet spot is the genius of Lean Urbanism.

References

Board, N. Y. C. L. (1982). Multiple Dwelling Law article 7-C *Loft Law*. New York.

Brown, L. J., Dixon, D., & Gillham, O. (2009). *Urban design for an urban century: Placemaking for people*. Hoboken, NJ: Wiley.

Commons, H. o. (2011). *The Localism Act*. London: Stationary Office.

Dittmar, H. (2014). Localism in England—Lessons for lean urbanism. *Lean urbanism case studies*. Retrieved from https://leanurbanism.org/publications/localism-in-england-lessons-for-lean-urbanism/.

Dittmar, H. (2015). The lean scan—Activating community assets. *Lean urbanism case studies*. Retrieved from http://leanurbanism.org/publications/the-lean-scan-activating-community-assets/.

Doshi, S. (2013). The politics of the evicted: Redevelopment, subjectivity, and difference in Mumbai's slum frontier. *Antipode, 45*(4), 844–865.

Elefante, C. (2012). The greenest building is… one that is already built. *Forum Journal, 27*(1), 62–72.

Enkvist, P., Nauclér, T., & Rosander, J. (2007). A cost curve for greenhouse gas reduction. *McKinsey Quarterly, 1*, 34.

Finn, D. (2014). DIY urbanism: Implications for cities. *Journal of Urbanism: International Research on Placemaking and Urban Sustainability, 7*(4), 381–398. https://doi.org/10.1080/17549175.2014.891149.

Glaeser, E. L. (2011). *Triumph of the city: How our greatest invention makes us richer, smarter, greener, healthier, and happier*. New York: Penguin Press.

Government, D. f. C. a. L. (2011). *Plain English guide to the Localism Act*. London.

Government, D. f. C. a. L. (2016). *Notes on neighbourhood planning*. London.

Habitat, U. (2015a). *Habitat III issue paper 14 informal sector*. New York, NY: United Nations.

Habitat, U. (2015b). *Habitat III issue paper 22 informal settlements*. New York, NY: United Nations.

Heynen, N. (2014). Urban political ecology I: The urban century. *Progress in Human Geography, 38*(4), 598–604.

Katz, B., & Wagner, J. (2014). *The rise of innovation districts: A new geography of innovation in America*. Washington, DC: The Brookings Institution.

Lovins, A. B., & Browning, W. D. (1992). Negawatts for buildings. *Urban Land, 51*(7), 26–29.

Luis, A. (2016). In the aftermath of Oakland's tragedy, how museums can better serve local arts and DIY venues. *Smithsonian Magazine*. Retrieved from http://www.smithsonianmag.com/smithsonian-institution/aftermath-oaklands-tragedy-how-museums-can-better-serve-local-arts-and-diy-venues-180961418/#KSEk-TGVjuw9sl7IC.99.

Lydon, M., Garcia, A., & Duany, A. (2015). *Tactical urbanism: Short-term action for long-term change*. Washington, DC: Island Press.

Nations, U. (2014). *World urbanization prospects: The 2014 revision, highlights*. Department of Economic and Social Affairs. Population Division, United Nations.

Perur, S. (2014, June 12). Jockin Arputham: From slum dweller to Nobel Peace Prize nominee. *The Guardian*.

Wales, H. T. P. o. (2009). *A speech by HRH: The Prince of Wales at the Prince's Foundation for the Built Environment conference 2009 titled globalization from the bottom up*. Retrieved from https://www.princeofwales.gov.uk/media/speeches/speech-hrh-the-prince-of-wales-the-princes-foundation-the-built-environment.

CHAPTER 6

Informal Urbanism and the American City

Anastasia Loukaitou-Sideris and Vinit Mukhija

Living in Los Angeles, we are in the midst of what many consider the informal revolution in the United States. Food trucks serving customers, front yard garage sales, street vending, and day laboring are some of the more common informal activities that we encounter traversing the city. Indeed, there is substantial evidence that informality is not confined to the Global South, or even the less affluent immigrant neighborhoods in cities like Los Angeles, but rather that it is an integral and growing part of many cities of the Global North (Mukhija and Loukaitou-Sideris 2014). This is in part due to globalization and immigration; in part an economic response to underemployment or lack of opportunities for formal employment; and in part because of both deregulation and the inadequacy of regulations to address the complexity and heterogeneity of contemporary cities in the USA and other countries of the Global North.

Unregulated activities, which, like previous scholars (Castells and Portes 1989), we refer to as informal activities, are widespread and varied. Most are neither criminal in nature, nor limited to instances of economic survival. On the contrary, informal activities are pervasive and spread across different social groups, diverse urban settings, and geographic regions of the Global North. While formal and informal activities may at times conflict with each other, they often overlap or depend on one another.

Urban informality is clearly enabled by and reflected in the built environment. Indeed, an informal urbanism has emerged in US cities that finds its spatial expression in the myriad landscapes of everyday life that host informal activities—the sidewalks and street corners, front lawns, garage apartments,

A. Loukaitou-Sideris (✉) · V. Mukhija
UCLA Luskin School of Public Affairs, University of California,
Los Angeles, CA, USA

© The Author(s) 2019
M. Arefi and C. Kickert (eds.), *The Palgrave Handbook of Bottom-Up Urbanism*, https://doi.org/10.1007/978-3-319-90131-2_6

parking lots, and community gardens. These formal landscapes are not purposely designed to serve informal activities; they are claimed, adapted, and transformed by the bodies, props, and impromptu or deliberate activities of the quotidian city. A conventional view of informal urbanism sees it as a production of urbanization that is independent from the formal city and its networks. We disagree with this duality. We call for a richer and more complex view of informal urbanism, in which formal and informal actors and settings overlap and often sustain each other.

While informal urbanism has become an increasingly visible and relevant part of the city for a number of social groups, it is severely understudied and often misunderstood. Urban designers and planners usually see informal activities and spaces through the lenses of prejudice and duality. Some see informal urbanism at best as marginal and disorganized settings and activities that should be ignored, and at worst as unlawful activities that should be stopped and prosecuted. At the same time, many progressive urban designers and planners also shy away from informal urbanism, fearing that the attention of formal institutions and actors may destroy the independence and vibrancy of informal spaces and activities. We find these perspectives incomplete and ill-conceived.

An area of professional practice where the idea of informality has struck a chord with urban designers is the growing enthusiasm for temporary installations in the built environment. These 'pop-ups' are inspired by the perceived vibrancy and ephemerality of informal activities. While such interventions may break the strict order and monotony of many US cities, we are troubled by the glib superficiality in the rising fervor for pop-ups. A serious urban design response should not only be aesthetic in nature but must also include a comprehensive understanding of the important economic and equity issues and impacts underlying informal urbanism.

In this chapter, we question the commonly-held but narrow conceptions of informal urbanism as spaces of marginality, irrelevance, and ephemerality that are clearly distinguished from the regulated and formal city. We argue that urban design's scope and focus as a discipline and a professional field should expand to include the spaces of informality, and we suggest that formal and informal urbanism are better understood as linked, overlapping, and mutually contingent. Therefore, following a literature review about the concept of informality and its spaces, we develop a framework for planners and urban designers to more meaningfully understand and address informal urbanism.

From Informal Sector in the Global South to Informal Urbanism in the Global North: A Literature Review

Keith Hart (1973), a British anthropologist, and the International Labour Organization (1972) were the first to use the term *informal sector* to describe unaccounted for employment opportunities in Africa. Hart contrasted the informal sector with the regulated, rational, planned, and organized formal sector. He emphasized the productive value of the informal sector and questioned the feasibility and desirability of conventional policies to shift employment from the informal to the formal sector. He also noted that the informal sector provided the poor with necessary opportunities for economic survival.

Patricia and Louis Ferman (1973), contemporaries of Hart, described similar activities in the USA and called them the "irregular economy." Like Hart's description of unaccounted and unenumerated activities, the Fermans pointed to unmeasured and unmonitored economic services in inner-city neighborhoods. They related the presence of the irregular economy to race and prejudice and argued that because of racial discrimination, inner-city residents were denied access to typical services, such as plumbing, electrical repairs, taxis, or licenses to provide these services. Consequently, both consumers and suppliers of such services resorted to the irregular economy. Like Hart, the Fermans claimed that while activities in the irregular economy were generally illegal, they were rarely criminal (Ferman and Ferman 1973).

As scholars became more familiar with informality, they challenged the logic of economic dualism inherent in its conceptualization (Bromley 1978; Moser 1978; Rakowski 1994). It is now generally accepted in the academic literature that although informal economic activities are unregulated by governments, informal and formal activities are often interlinked, integrated in the larger economy, and overlapping (Peattie 1987; Portes et al. 1989; Portes and Sassen-Koob 1987; Sanyal 1988). Researchers have also shown that the informal economy in the Global South is linked through trade and globalization with the advanced economies of the Global North (Beneria and Feldman 1992).

While the concept of an informal economy is still more readily associated with the Global South, several scholars have described its importance in the Global North, particularly in the lives of minority and immigrant groups (Hondagneu-Sotelo 2001; Gowan 2009; Morales 2010; Mukhija and Mason 2015; Valenzuela 2003; Venkatesh 2009; Ward 1999). Scholars argue that informal activities in the Global North are linked to the deepening of global capitalism, the abundant supply of cheap labor, the weakening of enforcement of regulations, and employers' interest in avoiding regulations (Bernhardt et al. 2008; Portes et al. 1989; Portes and Sassen-Koob 1987; Roy 2005; Sassen 1991; Sassen-Koob 1989; Valenzuela et al. 2006). Scholars have also noted contradictions in the neoliberal state that contribute to the increase of informal activities. On the one hand, there is a weakening of the welfare state and a decline in social welfare spending and services for the poor (Hopper 2003). On the other hand, the state is asserting itself through more rigid regulations in the public sphere and stricter public space ordinances (Loukaitou-Sideris and Ehrenfeucht 2009). These simultaneous trends play an important role in generating and criminalizing informality.

Urban Design and Informality

Some planning and urban design scholars have also started exploring the spatial aspects and effects of informality (Chase et al. 2008). This literature traces its roots to urban design scholarship critical of the overwhelming emphasis on order in planning the built environment. Key scholars from the 1960s and 1970s, including Jane Jacobs (1961), Reyner Banham et al. (1969), Richard Sennett (2008, original 1970), and Robert Venturi, Denise Scott Brown, and Steven Izenour (1972), argued for less emphasis on order, control, and conformity in the design of cities. They were critical of orthodox planning and design practices, and explicitly and implicitly advocated for more unplanned, informal, and spontaneous activities in cities. Their arguments are echoed in the work of subsequent scholars, such as Elizabeth Wilson (1991), Margaret Crawford (1995), and Simon Sadler (1998), who called for a change in the conventional approach through which cities are understood and planned.

In a related vein, many urban design scholars, writing during the 2000s, have explicitly focused on the spatial contours of informal activities (Loukaitou-Sideris and Mukhija 2016; Mukhija and Loukaitou-Sideris 2015), including front yard and garage sales (Crawford 2008), food vending in public spaces (Hou 2010), and unpermitted skateboarding in abandoned urban districts (Campo 2002, 2013). Thus, Quentin Stevens (2007) has called for an urban design that accommodates playful and impromptu uses of space, and Karen Franck and Quentin Stevens (2007: 3) have noted the merits of allowing for "the unexpected, the unregulated, the spontaneous, and the risky." Our argument builds on the work of previous scholarship to suggest that conventional city planning and urban design regulations fail to meet the complex needs of society. We, however, are not calling for a *laissez-faire* city, but for different kinds of regulations, interventions, and changing attitudes in urban design. As Richard Sennett has argued, "What needs to happen is a change in the peculiar institutions of affluent city life, in order to create new forms of complexity and new forms of diverse experience" (2008: 81–82). In what follows, we explain how we envision urban design's recognition of informal urbanism.

Addressing Informal Urbanism Through Design

Informal activities are inherently spatial; individuals and groups need to occupy space to perform them. Urban designers, however, are mostly absent from this social arena (Ogbu 2012). Instead, people are left to their own means and devices, or tactics, in adapting, adjusting, and making space for informal activities (Kamel 2014). Can urban designers change the conventional ways in which cities are planned to better understand and respond to urban informality and its spatial settings? We suggest that a spatial lens on informality can offer a guiding framework covering four distinct areas of urban design: scope, context, process, and practice. We explain this framework in more detail below, and in particular we elaborate on the dimension and questions of urban design practice.

Scope

Responding to informal urbanism requires recognizing its everyday presence as a first step and including the concept in the vocabulary of urban designers by expanding the field's scope to encompass the ordinary and residual spaces of daily life in diverse and underprivileged neighborhoods. In addition to focusing on civic settings and public spaces in central and upscale neighborhoods, urban designers should also concentrate on the dilapidated settings of many informal activities with the goal of making them more comfortable, livable, and humane. Central to the profession of urban design should be questions such as: How should an urban sidewalk be redesigned to better accommodate the needs of street vendors and day laborers? How can the single-family lot be reimagined to increase housing supply, accommodate second units, and enhance affordability? How can street and public space design enhance the everyday life of residents of unregulated subdivisions (or *colonias,* as they are often called in US federal policies)? And so on.

A survey of recent urban design plans in US and Canadian cities found that they were dominated by artistic concerns of beauty (Linovski and Loukaitou-Sideris 2013). However, to respond to informal urbanism, urban design professionals must go beyond their narrow focus and preoccupation with aesthetics and help uncover unconventional and invisible practices to accommodate the informal. Michael Rios (2014), for example, argues that while progressive urban designers cannot avoid questions of aesthetics, they need to be proactively involved in redefining the conception of aesthetics to support and accept informality and its perceived disorder and chaos. The street vendors of New York City are a good example of a city's acceptance and accommodation of urban informality in its daily fabric (Fig. 6.1).

Context

Because informal urbanism is embedded in the larger economic and social structures, urban designers must situate it in the economic and political context of marginality before deciding on a design, action, or policy. The distribution of informal activities varies across urban spaces, as does their impact on different neighborhoods. Some neighborhoods may appreciate informal urbanism, others may tolerate it, yet others disparage it, depending on context and circumstance. For example, garage or yard sales, second units, street vending, or urban agriculture may be unwanted and disliked in some neighborhoods but welcomed and appreciated in others. The social context of informal urbanism is likely to be particularly complex with multiple stakeholders and with different interests and attitudes. Consequently, informal urbanism presents the potential for conflict among different groups. Responding to informal urbanism may require recognizing and understanding its particular characteristics, including its socio-spatial circumstances, cultural specificities, power dynamics, and the economic interests of different actors, before deciding on a design action or strategy (Ogbu 2012).

Fig. 6.1   Annual Italian street fair, Manhattan (*Source* Anastasia Loukaitou-Sideris)

Process

Because informality has traditionally been outside the scope and purview of urban design practice, urban designers have no real experience or expertise in responding to its spaces and activities. This makes it even more important that there is intense and meaningful participation in the design process by actors of the informal city. However, as Jeff Hou (2011: 33) has argued: "Engaging multicultural groups in urban design presents another profound challenge. The established practice of participation, focusing on a narrow set of procedures, is ill prepared to deal with the multicultural complexity of contemporary cities." Including informal actors in the design process may require a radical rethinking of conventional participatory practices. We may need to emphasize more informal participatory processes, including neighborhood events, walking tours, and informal conversations and meals or potlucks, to overcome the limitations of typical formal participatory processes. These informal methods of soliciting participation can help to "produce new meanings, relationships, and collective actions that allow community actors to overcome formal barriers and social and cultural differences" (Hou and Kinoshita 2007: 310).

Practice

Whether urban designers directly intend or not, their interventions and ideas affect informality. Therefore, how design intervenes to accommodate the activities and settings of informal urbanism is important. Urban designers need to use creativity to accommodate the interests of different stakeholders by finding a middle ground to blur or soften the boundaries between formal and informal urbanism, and actively envision new spatial forms, or adapt existing ones, to accommodate informal activities. We elaborate with examples below.

SUGGESTIONS FOR A CREATIVE AND DELIBERATE PRACTICE OF URBAN DESIGN

First, the design and creation of a supportive public infrastructure—or a public commons to support the informal city—may contribute substantially toward lessening the hardships incurred by those participating in informal activities. It would also help to address and minimize the negative externalities and perceived nuisances produced by informality. The public provision of such common infrastructure is not inexpensive, but its absence may be costlier. For example, designing public rest stops with showers and toilets for homeless denizens not only offers relief to those without shelter, it also keeps municipal sidewalks clean from human excrement. Similarly, connecting *colonias* and other infrastructure-poor settlements to municipal water systems would not only lift a huge economic and health burden for residents of these neighborhoods, but also protect public health objectives by reducing the risk of diseases caused by contaminated water.

Along these lines, allowing citizens to convert empty lots to temporary sites for gardening and urban agriculture may not just help in partly addressing their needs for inexpensive fresh fruits and vegetables, but also utilize vacant sites and make the public realm more pleasant and walkable. These gardening and farming activities can also help in mobilizing and organizing community members (Fig. 6.2). Similarly, building worker centers would not only benefit day laborers by providing them with safe and hassle-free spaces to congregate and get training or other resources, but would also benefit potential employers and lessen conflicts between neighborhood residents and day laborers. Thus, creative urban design can help facilitate the cohabitation or coexistence of different social and economic groups by finding ways to better integrate and accommodate non-traditional housing types, higher densities, and diverse land uses in contemporary cities, allowing for a variety of formal and informal activities in shared urban public spaces.

Second, when formal and informal activities coexist, skillful urban design may help to reduce the potential of conflict between them. Conflicts or social

Fig. 6.2 Guerilla gardening (*Source* http://greenbookpages.com/blog/291217/ guerrilla-gardening/)

disagreements are frequently generated by a lack of appropriate or available space that forces informal activities to occupy territory used by formal activities. Conflicts can also be generated when informal activities, particularly the economic activities of underprivileged groups, are perceived as creating nuisances for more privileged groups. Creatively identifying an underutilized space that can host such informal activities, and enhancing this space through design, may not only resolve conflicts but even create a public amenity. One of the best examples of this comes from the city of Portland, where planners and urban designers found creative ways to formalize and accommodate vending by food carts and trucks. They developed food cart 'pods' on vacant lots and formerly desolate neighborhood spaces, thus not only enabling and actively supporting informal vendors but also enhancing the city's vibrancy and pedestrian-friendly public realm (Browne et al. 2014).

Third, where formal and informal actors and their activities are rigidly separated, creative urban design can help to soften the edges or separations between different kinds of spaces and can build bridges, thresholds, and middle grounds. We agree with Liz Ogbu (2012: 576) that "Boundaries is a concept relevant to any discussion of the informal sector.... [Good projects] blur these boundaries." Thresholds, common ground, softer edges and boundaries in the form of zones of transit, such as small public parks, where informal and formal activities can coexist, may offer links to connect activities and actors spatially and socially.

Fourth, although we are critical of pop-ups and do-it-yourself (DIY) installations because they are often superficially conceived, we are in favor of urban designers thinking temporally. As Gordon Douglas (2013: 2) notes "unauthorized urban space interventions that challenge the regulated uses of particular urban spaces" may be valuable. Although he finds that most DIY projects take place in newly gentrifying neighborhoods by young middle-class newcomers, such projects may help accommodate informal activities and create support for them. DIY self-provisioning is also common in Detroit, where residents routinely and informally take care of and paint murals on abandoned homes in the hope of attracting new neighbors and buyers (Kinder 2016).

Thinking temporally may also allow for formal and informal activities to occupy the same place over different periods of time. Jaime Lerner (2014) uses the example of open-air night-markets to show how formal and informal urbanism can reconcile with transformative urban effects. Such interventions, or "pinpricks" as Lerner calls them, can play a strategic and constructive role in learning from and responding to informal urbanism.

Lastly, because informal activities are distributed differently across urban space and have disparate impacts on different neighborhoods, a spatially decentralized decision-making approach may be more appropriate than citywide or region-based umbrella policies. Thus, a citywide ban on sidewalk vending may not make sense in some neighborhoods, where vendors provide important retail services. Similarly, garage sales, second units on single-family-zoned lots (Fig. 6.3), or urban agriculture may be highly appreciated and welcomed in some neighborhoods yet disliked and disparaged in others. While citywide regulations are important for matters pertaining to health and safety, other issues relating to where, when, and how informal activities can take place may sometimes be better resolved at neighborhood level, by neighborhood councils, or other groups of locally-based stakeholders.

This privileging of residents, however, can also be problematic. It has the potential to perpetuate exclusionary tendencies and strengthen existing asymmetrical power dynamics. For example, it is unlikely that most neighborhoods

Fig. 6.3 A second unit or accessory dwelling unit (ADU) in Santa Cruz, California (*Source* Vinit Mukhija)

would enthusiastically agree to create welcoming and livable spaces for the homeless. In other words, decentralized and neighborhood-level policies may work well in cases where neighborhood residents have a clear gain from informal activities, but it may not work that well in other cases. Decentralization cannot be a panacea for urban design decision-making in the informal city.

CONCLUSION

In contemporary cities of the USA and other countries of the Global North, informal urbanism is common and has proliferated due to increasing economic inequality, ongoing cutbacks in social welfare spending, and increasing social and cultural diversity. Divisions between the formal and the informal, the legal and the illegal, the public and the private have been at times hardened and exacerbated by design, through the creation of sharply demarcated borders, abrupt discontinuities, and walled, fenced, or gated boundaries (Loukaitou-Sideris 2012). The premise of this chapter is that now more than ever urban designers can no longer ignore the informal city. Since urban designers do not work in a professional and political vacuum, they should play a key role in understanding, advocating for, and creatively responding to the different facets of informal urbanism. Moreover, we suggest that urban designers need to question the conventional duality of informal and formal urbanism and conceive of them as linked, overlapping, and often mutually sustaining. This attitude or approach can also open up avenues and opportunities for urban designers to respond to urban informality.

Not all informal activities can or should be accommodated and supported by urban designers. Some informal activities can be harmful, dangerous, criminal, or illegal, and need to be impeded. But being outside the realm of design, regulation, and government oversight may also lead to increased vulnerability and exploitation, to unsanitary and unhealthy conditions for those undertaking the informal activity or consuming its products, and to nuisances for their neighbors. But while urban design interventions cannot solve poverty and inequality on their own, they should not contribute to worsening these conditions or stigmatizing underprivileged groups further. In fact, we hope and expect that socially conscious urban design strives for and achieves more than simply doing no harm. Cities should employ design action to reduce conflict among groups, and encourage more inclusive rather than exclusive spaces, overlapping rather than segregated uses, and interlinked rather than disconnected urban spaces and landscapes. Twenty-first-century urban design strategies need to promote human interaction and propinquity by emphasizing and reinforcing the public commons and by creating a supportive infrastructure for different groups.

But to accomplish all these normative goals, we need to think of urban design practice differently. We need to rethink and expand the scope and content of urban design practice as well as pedagogy. More specifically, in this chapter, we have suggested that urban informality is both defined by the built environment and helps to define it, and is an integral part of our cities. It should be an active area of focus for urban design practice. We thus shared a four-part framework for appreciating and addressing informal urbanism. First, we emphasized the need for urban design to eschew its narrow fascination with the grand spaces of the city and develop a wider lens or scope that includes the spaces of ordinary and everyday life in neighborhoods. Second, we stressed the need for urban design to also incorporate a broader context for analysis, which, in addition to a spatial or morphological analysis of neighborhoods, also maps how economic realities and political power shape our cities and situates critical practice within this reality. Third, we highlighted the need for urban design to engage with conventionally ignored, non-traditional stakeholders by engaging in dialogue with community members and appreciating the value of their local knowledge. Fourth, we suggested that urban designers and planners explore creative problem-solving in practice through design proposals that involve negotiation, accommodation, hybridity, unconventional configurations, and attention to small spaces that are often ignored. In sum, we suggest that a four-part framework of scope, context, process, and practice can help guide urban designers to respond to urban informality.

In the 1980s, Nobel prize winning economist Amartya Sen pioneered the capability approach, arguing that social welfare policies should not only focus on the resources available to people but also on how to enhance the freedoms and opportunities available to them to improve their lives and widen their choices (Sen 1985; Robeyns 2006). But our contemporary cities do not have the spatial arrangements that can help participants in urban informality to strengthen and increase their capabilities. We believe that the spatial perspective of urban designers can help to document and comprehend informal urbanism. Such a perspective lends itself to a sophisticated ability to respond to the needs and challenges of informality. Urban designers and planners can offer creative spatial solutions to help accommodate, integrate, or allow informal settings to coexist with, and even contribute to, their formal counterparts. Urban designers and planners, particularly of the next generation, need to move from neglect of urban informality to action, and create a more inclusive urban commons that enhances the capabilities of its participants.

It is our assertion that the informal city cannot be ignored. Because of its overwhelming presence and occupation of the spaces of everyday life, as well as its interactions and overlaps with the spaces, settings, and institutions of the formal city, we see informal urbanism as an integral and inherent part of US cities, and we consider it increasingly relevant for a growing number of social groups that are affected by it in both positive and negative ways. Therefore, we suggest that the question for urban designers is not whether they should acknowledge or ignore urban informality, but rather how they should respond to it. We also suggest that urban designers must respond in creative and deliberate ways that help to reduce conflict and to enhance human capabilities in and across informal and formal urbanism.

References

Banham, R., Barker, P., Hall, P., & Price, C. (1969). Non-plan: An experiment in freedom. *New Society, 13*(338), 435–443.

Beneria, L., & Feldman, S. (Eds.). (1992). *Unequal burden: Economic crises, persistent poverty, and women's work*. Boulder, CO: Westview Press.

Bernhardt, A., Boushey, H., Dresser, L., & Tilly, C. (2008). *The gloves-off economy: Workplace standards at the bottom of America's labor market*. Champaign, IL: Labor and Employment Relations Association.

Bromley, R. (1978). Introduction—The urban informal sector: Why is it worth discussing? *World Development, 2*(9/10), 1034–1035.

Browne, G., Dominie, W., & Mayerson, K. (2014). "Keep your wheels on": Mediating informality in the food cart industry. In V. Mukhija & A. Loukaitou-Sideris (Eds.), *The informal American city: Beyond taco trucks and day labor*. Cambridge, MA: MIT Press.

Campo, D. (2002). Brooklyn's vernacular waterfront. *Journal of Urban Design, 7*(2), 171–199.

Campo, D. (2013). *The accidental playground: Brooklyn waterfront narratives of the undesigned and unplanned*. New York: Fordham University Press.

Castells, M., & Portes, A. (1989). World underneath: The origins, dynamics, and effects of the informal economy. In A. Portes, M. Castells, & L. A. Benton (Eds.), *The informal economy: Studies in advanced and less developed countries* (pp. 11–37). Baltimore: Johns Hopkins University Press.

Chase, J., Crawford, M., & Kaliski, J. (Eds.). (2008). *Everyday urbanism*. New York: Monacelli Press.

Crawford, M. (1995). Contesting the public realm: Struggles over public space in Los Angeles. *Journal of Architectural Education, 49*(1), 4–9.

Crawford, M. (2008). Blurring the boundaries: Public space and private life. In J. Chase, M. Crawford, & J. Kaliski (Eds.), *Everyday urbanism* (pp. 22–35). New York: Monacelli Press.

Douglas, G. (2013). Do-it-yourself urban design: The social practice of informal 'improvement' through unauthorized alteration. *City and Community*, 1–21. https://doi.org/10.1111/cico.12029.

Ferman, P., & Ferman, L. A. (1973). The structural underpinnings of the irregular economy. *Poverty and Human Resources Abstracts, 8*(1), 3–17.

Franck, K. A., & Stevens, Q. (Eds.). (2007). *Loose space: Possibility and diversity in urban life*. New York: Routledge.

Gowan, T. (2009). New hobos or neoromantic fantasy? Urban ethnography beyond the neoliberal disconnect. *Qualitative Sociology, 32*(3), 231–257.

Hart, K. (1973). Informal income opportunities and urban employment in Ghana. *The Journal of Modern African Studies, 11*(1), 61–89.

Hondagneu-Sotelo, P. (2001). *Domestica: Immigrant workers cleaning and caring in the shadow of affluence*. Berkeley and Los Angeles: University of California Press.

Hopper, K. (2003). *Reckoning with homelessness*. Ithaca, NY: Cornell University Press.

Hou, J. (Ed.). (2010). *Insurgent public space: Guerilla urbanism and the remaking of contemporary cities*. New York: Routledge.

Hou, J. (2011). Citizen design: Participation and beyond. In T. Banerjee & A. Loukaitou-Sideris (Eds.), *Companion to urban design* (pp. 329–340). London: Routledge.

Hou, J., & Kinoshita, I. (2007). Bridging community differences through informal processes: Reexamining participatory practices in Seattle and Matsudo. *Journal of Planning Education and Research, 26*(3), 301–314.

ILO. (1972). *Employment, income and inequality: A strategy for increasing productive employment in Kenya*. Geneva: International Labour Office.

Jacobs, J. (1961). *The death and life of great American cities*. New York: Vintage Books.

Kamel, N. (2014). Learning from the margin: Placemaking tactics. In V. Mukhija & A. Loukaitou-Sideris (Eds.), *The informal American city: Beyond taco trucks and day labor*. Cambridge, MA: MIT Press.

Kinder, K. (2016). *DIY detroit: Making do in a city without services*. Minneapolis: University of Minnesota Press.

Lerner, J. (2014). *Urban acupuncture*. Washington, DC: Island Press.

Linovski, O., & Loukaitou-Sideris, A. (2013). Evolution of urban design plans in U.S. and Canada: What do the plans tell us about urban design practice? *Journal of Planning Education and Research, 33*(1), 66–82.

Loukaitou-Sideris, A. (2012). Addressing the challenges of urban landscapes: Normative goals for urban design. *Journal of Urban Design, 17*(4), 467–484.

Loukaitou-Sideris, A., & Ehrenfeucht, R. (2009). *Sidewalks: Conflict and negotiation over public space*. Cambridge, MA: MIT Press.

Loukaitou-Sideris, A., & Mukhija, V. (2016). Responding to informality through urban design studio pedagogy. *Journal of Urban Design, 21*(5), 577–595.

Morales, A. (2010). Planning and the self-organization of marketplaces. *Journal of Planning Education and Research, 30*(2), 182–197.

Moser, C. (1978). Informal sector or petty commodity production: Dualism or dependence in urban development? *World Development, 6*(9/10), 1041–1064.

Mukhija, V., & Loukaitou-Sideris, A. (Eds.). (2014). *The informal American city: Beyond taco trucks and day labor*. Cambridge, MA and London, England: The MIT Press.

Mukhija, V., & Loukaitou-Sideris, A. (2015). Reading the informal city: Why and how to deepen planers' understanding of informality. *Journal of Planning Education and Research, 35*(4), 444–454.

Mukhija, V., & Mason, D. (2015). Resident-owned, informal mobile home communities in rural California: Lessons from Rancho Don Antonio, Coachella Valley. *Housing Policy Debate, 25*(1), 179–194.

Ogbu, L. (2012). Reframing practice: Identifying a framework for social impact design. *Journal of Urban Design, 17*(4), 573–589.

Peattie, L. (1987). An idea in good currency and how it grew: The informal sector. *World Development, 15*(7), 147–158.

Portes, A., & Sassen-Koob, S. (1987). Making it underground: Comparative material on the informal sector in Western market economies. *The American Journal of Sociology, 93*(1), 30–61.

Portes, A., Castells, M., & Benton, L. (Eds.). (1989). *The informal economy: Studies in advanced and less developed countries.* Baltimore: Johns Hopkins University Press.

Rakowski, C. (1994). Convergence and divergence in the informal sector debate: A focus on Latin America, 1984–92. *World Development, 22*(4), 501–516.

Rios, M. (2014). Learning from informal practices: Implications for urban design. In V. Mukhija & A. Loukaitou-Sideris (Eds.), *The informal American city: Beyond taco trucks and day labor.* Cambridge, MA: MIT Press.

Robeyns, I. (2006). The capability approach in practice. *The Journal of Political Philosophy, 14*(3), 351–376.

Roy, A. (2005). Urban informality: Toward an epistemology of planning. *Journal of American Planning Association, 71*(2), 147–158.

Sadler, S. (1998). *The situationist city.* Cambridge, MA: MIT Press.

Sanyal, B. (1988). The urban informal sector revisited: Some notes on the relevance of the concept in the 1980s. *Third World Planning Review, 10*(1), 65–83.

Sassen, S. (1991). *The global city: New York, Tokyo, London.* Princeton: Princeton University Press.

Sassen-Koob, S. (1989). New York city's informal economy. In A. Portes, M. Castells, & L. A. Benton (Eds.), *The informal economy: Studies in advanced and less developed countries* (pp. 60–77). Baltimore: Johns Hopkins University Press.

Sen, A. (1985). *Commodities and capabilities.* Amsterdam: North-Holland.

Sennett, R. (2008). *The uses of disorder: Personal identity and city life.* New Haven: Yale University Press.

Stevens, Q. (2007). *The ludic city: Exploring the potential of public space.* London: Routledge.

Valenzuela, A. (2003). Day labor work. *Annual Review of Sociology, 29*(1), 307–333.

Valenzuela, A., Theodore, N., Melendez, E., & Gonzalez, A. L. (2006). *On the corner: Day labor in the United States.* Los Angeles: UCLA Center for the Study of Urban Poverty.

Venkatesh, S. A. (2009). *Off the books: The underground economy of the urban poor.* Cambridge, MA: Harvard University Press.

Venturi, R., Brown, D. S., & Izenour, S. (1972). *Learning from Las Vegas.* Cambridge, MA: MIT Press.

Ward, P. M. (1999). *Colonias and public policy in Texas and Mexico: Urbanization by stealth.* Austin: University of Texas Press.

Wilson, E. (1991). *The Sphinx in the city: Urban life, the control of disorder, and women.* London: Virago.

PART II

Informalizing the Formal

CHAPTER 7

DIY Neighborhoods

Emily Talen

INTRODUCTION

There are hundreds of definitions ascribed to neighborhoods. They vary by how and whether people, home, place, morphology, territory, behavior, perception, or governance is prioritized. A few are simple: "any group between the family unit and municipal government" (Cowan 2005). A 1970s-era elementary school curriculum guide offered this perfectly succinct definition: "*place, people*, and *purposes*, with the emphasis on *place*" (Providence Public Schools 1970). Or, there is the painfully uncomplicated definition of neighborhood that was offered in a 1957 editorial published in the *New York Post*: "A neighborhood is where, when you go out of it, you get beat up" (Bursik and Grasmick 2002: 5). Meanwhile, social scientists tend to equate neighborhoods with bundles of statistics.

But if the concept of neighborhood is defined normatively—that is, how the neighborhood *ought* to be—it generally involves three kinds of aspiration that they have a certain: physical quality; social make-up; and method of governance. Along the first dimension, a neighborhood ought to be identifiable and serviced, which means the neighborhood has a name, residents know where it is, what it is, and whether they belong to it, it has at least one place that serves as its center, it has a generally agreed-upon spatial extent, it has everyday facilities and services (although it is not self-contained), and it has internal and external connectivity. Along the second dimension—and bearing in mind that this is an aspirational view—neighborhoods have social diversity within them, or are at least open to its enabling (diversity could be measured in many ways, including age, family type, race, ethnicity, or income).

E. Talen (✉)
University of Chicago, Chicago, IL 60637, USA

© The Author(s) 2019
M. Arefi and C. Kickert (eds.), *The Palgrave Handbook of Bottom-Up Urbanism*, https://doi.org/10.1007/978-3-319-90131-2_7

Along the governance dimension, neighborhoods ought to have a means of representation, a means by which residents can be involved in its affairs, and an ability for the neighborhood to speak with a collective voice.

It is this last normative quality—the third dimension of the normative neighborhood—that most connects the neighborhood to the spirit and energy of DI, or bottom-up urbanism. A self-determined, self-managed neighborhood—a DIY neighborhood—is one that is in control of its own destiny, in many cases self-actualized (where local residents play a role in determining neighborhood boundaries and definition), but more importantly, that has an ability to control and govern its own affairs. On the other side, neighborhood self-management does little to address the structural basis of inequity, and if neighborhoods are put in charge of their own destinies, they can also be cast as the source of their own problems (Baily and Pill 2011; see also Durose and Lowndes 2010; Lowndes and Sullivan 2008). This chapter traces the history and the pros and cons of the DIY neighborhood, using the United States as context.

A Brief History of the DIY Neighborhood

In the USA, the attempt to improve urban life via neighborhood-based empowerment and local control—what we might today term the DIY neighborhood—has long been regarded as the antidote to the indifference and corruption of big city politics, or at least, the point of entry into wider political networks. The century-long quest for neighborhood-based political power is a venture that resonates on both sides of the political spectrum. Neighborhood associations aimed at protecting property rights and excluding unwanted uses and tenants are just as interested in bottom-up localized control as neighborhood groups aimed at social justice. Their history reflects a consistent prioritization of organizational power over political ideology. Mary Parker Follett's 1918 *The New State: Group Organization, the Solution of Popular Government* argued that civic action at neighborhood level should be the basis of civic involvement in America. Decades later, Norman Mailer ran for mayor of New York City with the slogan "Power to the Neighborhood," based on his platform for neighborhood-level governance and New York's secession from the USA as the 51st state (Trager 2004).

In the early twentieth century, leaders of the social and civic center movement had high hopes that their Neighborhood Civic Clubs would take charge of local governance. The clubs were to be headquartered in local schools or other venues, they would be "non-partisan, non-sectarian, non-exclusive," and they would be dedicated to the "presentation and discussion of public questions." One enthusiast wrote: "You who have not witnessed it cannot understand how party spirit, class spirit, and even race spirit fade out in the intense civic and community atmosphere of these neighborhood civic clubs." The civic center movement was going to be the antidote to the "problem" of democracy by helping to develop "the community spirit." It was a feeling that was "latent" in every person—it only needed "an appropriate stimulus to arouse it." Since the government had failed to develop communally-oriented institutions, the neighborhood club was needed to provide the stimulus and awaken civic spirit (University of Washington 1912: 7).

In the 1940s, neighborhood planning exhortations about size, school centrality, and other physical parameters were met with the understanding that "a properly served neighborhood necessarily involves the people who live in the neighborhood in the planning of their neighborhood." Within this physically prescribed neighborhood, one only needed "faith enough in democracy" to improve its quality (Isaacs 1945 as cited in MOMA 1944: 3). And without this collective capacity and ability to govern locally, neighborhoods could become vacuums filled by other types of control. Gangs, for example, were believed to be a form of control put in place to respond to a lack of order. In some eras, gangs might even be viewed as "an important part of a neighborhood's search for order" (Hagedorn 1991: 529).

In the wake of urban renewal failures in the 1960s, neighborhood organizing was supported by the federal government. First was the Community Action Program under the 1964 Economic Opportunity Act, followed by the 1966 Model Cities program (formally, the Demonstration Cities and Metropolitan Development Act). Both Acts were part of President Johnson's War on Poverty. Some of the features of these programs remain (such as Head Start), but the neighborhood organizing and empowerment initiatives faded out in the 1970s. President Carter's 1979 National Commission on Neighborhoods studied the issue and concluded that neighborhoods should be empowered, but the receding federal government ceased to play a lead role (Fisher 1984).

Some cities stepped in to support neighborhood governance. In the early 1980s Philadelphia convened 45 neighborhood groups to create a six-item "Neighborhood Agenda" after neighborhood leaders demanded "tangible opportunities to become involved in the planning and implementation of every program designed to meet neighborhood needs." It was a governance strategy as much as a quest for funds, as the six items—Neighborhood Jobs Bank, acquisition of vacant properties, energy conservation, crime prevention, and educational programs—involved direct decision-making by neighborhood groups. These programs continue today under the Neighborhood Transformation Initiative whose "keystone" is the issuance of $295 million in bonds by the Redevelopment Authority of the City of Philadelphia (Schwartz 1982).

Other kinds of neighborhood empowerment activities became more philanthropic. One example is the Citizens Committee of New York City (www. citizensnyc.org), founded in the 1970s to give small grants to neighborhoods for events, park clean-up projects, cookbook compilations, and nutrition classes. Funds are allocated to street block associations, community councils, garden clubs, schools, and street alliances. Transforming trash-strewn lots into community gardens is a common target of funding.

COMPLICATIONS OF AN IDEAL

While many people believe as a matter of principle that cities should empower neighborhoods to control their own destinies, there are definitional, governance, and funding issues to deal with. Some issues are practical. What are the boundaries of neighborhoods? Who counts as a resident: absentee landlords; renters; business owners? What level of authority should neighborhood councils or associations have—advisory, or more? What level of funding should be provided and who controls it? Some cities—notably Albuquerque, Minneapolis, Los Angeles, Portland—have found ways to wade through the definitional challenges and operationalize neighborhood-scale governance (Miller 2012).

A bigger issue seems to be whether and to what extent neighborhood empowerment actually occurs. In theory, neighborhoods hold significant political power. The most recent survey from the National League of Cities is that almost 50% of large (over 200,000) cities in the USA have district-based elections for mayor and council—neighborhood-scale rather than "at-large" (another 38% had a mix of at-large and district-based) (Syara 2003). In practice, neighborhood-based empowerment has struggled for over a century against citywide elections, city-manager forms of government that hold power (and resist delegating power to neighborhoods), and an emphasis on regionalism (Miller 1981; Campleman 1951).

Some of the basis of this loss of power lies in the ambiguous interest of neighborhoods to seek empowerment. On the question of who should "govern" the neighborhood, early twentieth-century neighborhood activist Mary Simkhovitch (1926) thought that the "neighborhood powers" rested in those servicing the neighborhood day by day—"the doctor, the teacher, the clergyman ... the baker, the delicatessen man, the pool parlor proprietor" (p. 12). Neighborhood governance, she wrote, is about the "interacting relationships which come both to color and to fix the habits of the population" (p. 12). Formal government, she contended, is for people to voice objections, and unless there is some specific disturbance, people are uninterested in it.

For many neighborhood residents, interest in neighborhood control especially arises when there are threats to 'the neighborhood,' such as physical change via outsiders (like developers), or social change from incoming ethnic and racial groups—unfortunately existing residents are often motivated to keep this change out of their backyards (see examples cited in Hojnacki 1979). There is an uncomfortable relationship between a neighborhood's social homogeneity, its physical definition, and its ability to self-govern. In the early twentieth century, sociologist R.D. McKenzie already thought it was of "the utmost importance" that political organization and local governance be made to "coincide as near as possible with the natural neighborhood groupings of the population." If that happened, it would be easier to establish "community safeguards against encroaching disturbing factors" (McKenzie 1922: 785, 799). At the end of the century, neighborhood-based planning initiatives were accused of legitimizing this exclusionary motivation, for instance, Seattle's

Neighborhood Planning Program began after residents complained about a proposal to change zoning to allow the development of pedestrian-oriented, higher density 'urban villages' (complete with multifamily affordable housing) proposed by the city's first black mayor, Norman Rice (Ceraso 1999).

Beyond the often-exclusionary nature of neighborhood empowerment, many people have argued that the resultant neighborhood-based democracy has been rather weak. Jesse Steiner, as president of the National Community Center Association, wrote in 1930 that the neighborhood was not the right scale for dealing with urban issues—social problems, he said, should instead be addressed at the regional scale. Hans Blumenfeld had disdain for "the neighborhood concept" because it rendered so little. He wrote: "it is hardly an adequate means to save the city, the nation, and the world from impending disaster" (Steiner 1930: 492–493).

It is true that the burgeoning metropolis had rendered the neighborhood somewhat powerless. Early twentieth-century neighborhood sympathizers blamed large cities for undermining localized political action. According to sociologist Harvey Zorbaugh, author of the 1929 neighborhood classic *The Gold Coast and the Slum*, it was the mobility of society that rendered neighborhood-based organization weak. Political organization required stability, and an expanding cosmopolitanism undermined that. Expanding social networks did not necessarily mitigate the problem. One study of neighborhood social networks in New York City used data from 1937 to show that social networks were incapable of erasing the "negative effects of adverse social and economic conditions" in New York's neighborhoods (Kadushin and Delmos 1992: 58).

Some Progressive Era reformers thought that neighborhood-based authority should be controlled to account for supra-neighborhood issues. George Hooker made the case in a 1917 edition of the *National Municipal Review*. "Real city neighborhoods," he wrote, need a "federal scheme" that conjoined "limited city neighborhood government" to "a well-considered plan" (Hooker 1917: 341–342). The challenge was that when neighborhoods were inserted into big city politics they were caught between grass-roots organizing and bureaucratic maneuvering—and they needed to be good at both. The two strategies were often in conflict. Even in cities with a strong tradition of sanctioned neighborhood organization, neighborhoods might be used for legitimacy but ultimately ignored in the name of efficiency (Jezierski 1990). Hooker's wishes were granted with the dissolution of voting blocks in condensed ethnic neighborhoods that fueled early twentieth-century politics in American cities, in lieu of broader community organizations. Even such organizations have been described as "reactive," tending to respond to issues that lie well outside of neighborhood boundaries, such as urban renewal or school busing. Neighborhood groups might aggregate, but when they do, the tie to neighborhood has a different flavor. A few, like Albert Hunter, welcomed this aggregation. As Hunter saw it, the decisions impacting neighborhoods are being "traced up the vertical hierarchy," whereby "neighborhood politics becomes national politics" (1979: 281, 285).

The community organizer Saul Alinsky believed that activism should be organized around the workplace or ethnic group, not necessarily a geographically constructed, government-defined neighborhood. In fact, he was not supportive of neighborhoods as a distinct spatial, physical, or functional place because such delineation held little political advantage. Alinsky's thinking was more in the tradition of Durkheim and Weber, who pitted the emotional "sentiments of place" against the rational thinking of "the organization of interest." In this way of thinking, neighborhoods exist not for the purpose of promoting neighborhood-based political activism, but to dissipate power and help corporations and politicians avoid action and responsibility. Broader coalitions, not small-area neighborhoods, are a more effective political force (McCann 2003).

For different reasons, progressives thought this kind of fluidity was needed if the neighborhood was to be leveraged to help the plight of poor residents. In the quest for political and collective relevance, advocates of "the neighborhood" watered down the neighborhood ideal and became instead advocates of a more spatially unconstrained form of community organizing. The Executive Director of the Industrial Areas Foundation (the organization created by Saul Alinsky in the 1940s), declared at a 1988 conference called Church and City, held in Philadelphia, that "the neighborhood as an organizing mechanism is dead" (Woods 1988: 474)! Freeing community activism from the constraints of neighborhood put the focus squarely on housing "projects" (Simkhovitch 1938: 293).

The lack of political power of neighborhoods has become painfully apparent in Europe, where governments in France, Denmark, the Netherlands, and especially the UK, have adopted the neighborhood as a cover for neoliberal public restructuring and austerity programs. Among academics, European "new localism" is critiqued as merely decentralizing responsibility, not power. It translates to short-term and incremental policies and actions at the expense of long-term and sustainable ones. In the end, neighborhood-based planning and governance seems to be an exercise in political cover, where neighborhood-based service delivery makes possible the "wholesale reform of mainstream services" (Bailey and Pill 2011). Such programs are also more easily dismantled.

Neighborhood Versus Community

A legitimate question is whether neighborhood-based political activism translates to issues that derive meaning from neighborhood scale and the experience of living in a physically defined area with a unique identity. At the very least, it would seem that the language of neighborhood implies that the issues being addressed have a limited geographic range. Neighborhood-based service delivery, economic empowerment projects, or the effort to reduce hazardous waste from neighborhoods are topics that are rooted in space and place proximities and consequences, whether articulated or not. Broader coalitions and agglomerations need not constrain themselves to these local concerns.

Untethered, many organizations that are explicitly about neighborhood find themselves drawing few distinctions between 'neighborhood' and 'community'. George Hillery's (1968) study of local societies had claimed that there was an overlap with neighborhood in many of the 94 definitions of community he uncovered—an insight that many neighborhood organizations would have found axiomatic. The National Association of Neighborhoods is the oldest (established in 1975) and largest (2500 members) organization of what it calls "the heart of the nation's community: America's neighborhoods." The definition of neighborhood it uses is entirely open-ended. Place-bound solidarity is invoked in a way that omits any consideration of spatial concepts like place, proximity, centrality, or boundedness. The mission statement of the national non-profit called Neighborhoods USA, or NUSA, also formed in 1975, has a similar quest to "build stronger communities" that have no particular requirement for neighborhood, place-based definition (Neighborhoods USA 2014).

This lack of specificity is risky, as a neighborhood is not a community, or even a 'place-based community.' There are distinctions to be made between neighborhoods as places, as social networks, as parts of the city whole, or as places of "conceptual identity" (Meegan and Mitchell 2001; see also Davies and Hebert 1993). When neighborhood and community are combined, the resultant 'neighborhood community,' describes a fairly small subset, given the definition: "a close-knit network of households [who] participate in common social activities" or "an area which contains all or most of the elements of a complete social system" ("Neighborhoods and Communities" 2013).

In fact, there is a danger in equating neighborhood with community, because it leads to the conclusion that neighborhoods without "community" are somehow "deficient, dysfunctional, and doomed" (Garrioch and Peel 2006: 665). The common theme of "community lost" can spill over to "neighborhood lost" without a clear understanding of the difference. The result is that neighborhoods are then just as likely to disintegrate under the weight of Louis Wirth's "mass society" of alienation, where the remedy is cosmopolitanism and connection to the world at large, potentially undermining the very reality of neighborhood as a physical place. From the opposite angle, there is the argument that neighborhood should never be equated with community because the exclusionary tactics of neighborhoods might infiltrate the more legitimate concerns of community-building. Neighborhoods are critiqued for being bedroom communities, for having little meaning in people's lives, for being restrictive and exclusionary. An unplaced, abstract notion of community has the ability to rise above this (Agnew 1980).

The elevation of community over neighborhood could potentially weaken neighborhood self-determination. The widespread use of technology for community engagement creates an additional strain by making the connection between neighborhood governance and neighborhood as a defined place seem somewhat irrelevant. Although technology broadens engagement more generally, it enables the ability to conceive of neighborhood in a technical but unspecified way, at odds with the hands-on governance of a place-based neighborhood.

Some activists are more inclined to embrace the indeterminacy of neighborhood and find ways to use it to empower residents. Perhaps inexactness is how the essence of neighborhood is revealed, where contestations over physical form—how boundaries nest, overlap, blur, and change, especially in relation to individualized cognitive understandings of neighborhood—determine its true meaning. For pro-action, one might avoid the loaded term of neighborhood altogether and insert 'place-frames' instead. The hope is to build identity, unite residents around a common purpose, downplay differences, and leverage *place* to inform local activism, nested in a macro political economy but operating within their own realm. In this way, the "territorial sphere" can be activated as "a legitimate and meaningful site for activism." Although potentially undermining "global activist agendas," place-frames would allow residents to make sense of daily life (Martin 2003: 747).

The suggested approach shows how far the relationship between neighborhood planning and neighborhood governance has strayed. The neighborhood association was already being dismissed as ineffective in the 1980s. Some lobbied against it, arguing that "the preparation of plans is among the least useful activities undertaken by neighborhood organizations" in part because "such plans can relate poorly to daily concerns" (Checkoway 1984: 106). Many typical 'master plans' of American cities have kept these sentiments in place. Such plans—originating in the 1950s, but still in use today—plan the city in Modernist terms of discrete functions: the 'land use element,' the 'circulation element,' the 'recreation element.' Missing is planning by neighborhood, especially in a way that bestows real power at neighborhood level.

The fact that neighborhood planning is no longer seen as a means by which neighborhoods can take control of their own destinies probably grew out of disillusionment with the authenticity and effectiveness of the neighborhood plan. Neighborhood planning, Barry Checkoway (1984: 102) warned, was really just "subarea planning in disguise." Subarea planning is the top-down decentralization of facilities and an expedient way to satisfy citizen participation requirements set by federal agencies or comprehensive planning statutes—not a resident-generated vision in which neighborhoods are assigned real decision-making authority. The attempt to manage services more efficiently by invoking neighborhood is certainly not in the spirit of bottom-up democracy. Ironically, even the supposed efficiency of neighborhood planning has been called into question since using the neighborhood as the unit of government increases the number of units to be administered, which can create more bureaucracy and cost. And, if service delivery depends on neighborhood input and consensus, this is not likely to be a very efficient process. It is challenged as backward in an age when more information-age enabled, "subtle and sophisticated tools" to manage a city might be better than governance based on neighborhood segmentation (Madanipour 2001: 180).

Even when enabled, neighborhood governance based on neighborhood planning is criticized as inauthentic. Residents are supposed to become involved in the review of proposals or plans, in the development of new plans as part of a city's overall comprehensive plan, or in neighborhood improvement activities such as cleanup campaigns or housing rehabilitation projects. But all these forms of involvement require time and energy, and few neighborhood residents might actually participate—especially if they do not feel any of their vested interests are under threat. Instead, a lack of resident involvement can quickly lead to what Sherry Arnstein (1969) labeled "tokenism." Even where there is a more concerted effort toward neighborhood-level governance, it may not be representative. An analysis of Los Angeles' 86 Neighborhood Councils found "substantial racial bias" against representation by Hispanics, putting the legitimacy of the neighborhood councils "at risk " (Jun and Juliet 2007: 54).

THE FUTURE OF DIY NEIGHBORHOODS

Regularities about neighborhood governance and engagement can be hard to pin down. Political scientists and sociologists have analyzed neighborhood governance trying to assess what variables seem to improve or diminish its political capacities—looking at variables such as degree of social networking, membership levels, or sense of community. Research results have vacillated, either finding that strong social networks determine neighborhood self-governance, or that they have no effect at all (e.g., Crenson 1978).

While there are problems with over-zealous neighborhood protectionism, neighborhoods run a risk when they lack control of their own destinies. The outcome can be a severe disjuncture between what residents want in a neighborhood and what governments impose. In Western cities, this was the legacy of urban renewal. In other parts of the world, colonialist exporters of twentieth-century neighborhood planning implemented the idea in a way that was strictly top-down, with no resident input into design or governance (Ahmed 2012).

Not all neighborhood associations are based on preserving the status quo and keeping others out. There are exemplary volunteer neighborhood organizations committed to facilitating the ability of neighbors to help other neighbors, hence maintaining an inclusive and diverse neighborhood. Many of these organizations focus on helping elderly residents stay out of retirement homes. Some of these are funded by social service organizations or governments, while some are initiated from the bottom up by residents and supported by membership dues or donations. The idea is to match those willing to help with those in need, and the latter group is usually much smaller than the former. It is termed the 'village model' if it is resident-funded and consumer-driven—a kind of service bartering system and a way for neighbors to assist each other so that they do not have to move away if their needs change.

One example is the Neighbors Assisting Neighbors, or NAN, group, an organization of 450 households comprising the Bannockburn neighborhood in Bethesda, Maryland. Early on, the group administered a survey to assess needs and willingness to volunteer. There are block coordinators assigned to fifteen households who "make sure no one falls through the cracks." Community events and programs are organized, like the Wise Elder project that connects seniors to high school students to do oral histories. There are, of course, constant funding stresses, as the village model relies on membership dues, a higher tax for low-income neighborhoods (Baker 2014: 29, 35; Scharlach 2012).

Sometimes the arts are used as the mechanism for neighborhood empowerment. In Theaster Gates' Dorchester Projects on Chicago's South Side, a cluster of renovated houses function as artist spaces, drawing in people for arts functions, selling art made of locally found objects, and reinvesting money from sales and events back into the community (in line with Boltanski and Chiapello's *The New Spirit of Capitalism* 2007; see also Reinhardt 2014). These can be interpreted as expressions of neighborhood building and engagement. However, initial review of the process-oriented ideals demonstrates that the activities group loosely, and their connection to the wider neighborhood (or its goals) are ambiguous.

Neighborhoods should focus on strengthening their institutions to improve their production capacity and make them stewards of their own well-being—producers rather than consumers. These neighborhood organizations and institutions must be trusted, which is a special challenge in high poverty neighborhoods that experience "moral and legal cynicism." McKnight and Kretzman's Asset-Based Community Development (ABCD) approach tries to integrate the organizations of a neighborhood to form an "association of associations" that together become a "unified neighborhood force" (2014: 2, 17). Similarly, sociologist Robert Sampson argued the importance of an "organizational infrastructure"—a diversity of non-profits and collective enterprises. Neighborhoods with this diversity are better able to break the cycle of decline. as over-reliance on one institution, like a church, can be a liability in the face of strong adversity (Sampson 2011). The importance of a neighborhood's "institutional base" also extends to diverse neighborhoods; institutions are seen as especially important for creating "strong cross-status ties" in mixed-income areas (McKnight 2013: 23; Rose 2000; Clampet-Lundquist 2004: 443).

Neighborhood empowerment not only hinges on institutional diversity, but also on spatial clarity. Drawing from Oscar Newman's defensible space ideas, there needs to be an explicit understanding of what neighborhood space is public and what is private and who has responsibility for each. This may require the delineation of access zones and transitional spaces, the support of environmental cues that subtly inform movement and behavior, and the explicit programming of activities and events to populate space (Newman 1972).

Besides institutional support and the strengthening of place identity, neighborhoods can draw from a wide array of legal tools that can be used to help them take control of their own spaces and destinies, even giving them the ability to implement neighborhood plans. Stephen Miller's review of "legal neighborhoods" argues that there is great promise here—legal tools provide empowerment, reduce urban alienation, maintain accountability, and, if these tools are embraced by local governments, can stimulate interest in environmentally beneficial, compact urbanism. Although the US Supreme Court ruled that the definition of neighborhood may be too vague for enforcing criminal laws, the variability of definitions for neighborhood does not undermine its legal standing for other functions. These neighborhood legal tools put non-suburban neighborhoods on a more equal footing with suburbs. Miller argues: "because the neighborhood is such a resonant institution in the minds of residents," and there is the added bonus that "neighborhood is a constituency that politicians feel comfortable serving," "failing to structure legal tools for the neighborhood is at best a missed opportunity, and perhaps even perilous."

The key to making neighborhood stronger as a legal entity is to connect and overlap the various neighborhood-scale tools at hand. Examples of neighborhood legal tools that can be overlaid include: the taxing powers of business improvement districts; code enforcement; neighborhood service centers; schools; neighborhood councils; and zoning. Miller makes the case that the combined power of these myriad, neighborhood-level tools has not been leveraged. Legal structures that operate at the small scale of neighborhood are rarely connected—instead, they empower different constituencies in the same neighborhood—business owners, renters, parents. Rather than encouraging fights between these groups, legal neighborhoods could put priority on visioning: "The more neighbors have the chance to define visions for their neighborhood, the more likely they are to care about where they live." Neighborhood legal tools can then be used to focus on "arbitration and negotiation" rather than litigation (Miller 2012: 141–142, 165).

The rise of populism in many Western countries over the past few years demonstrates that a significant part of the electorate feels a lack of control, no sense of agency, and no political voice. Social theorists argued that the growing anxiety among blue-collar workers and the middle class does not simply reflect lost income, but mostly pertains to feelings of a loss of identity and purpose. This realization has fueled calls for localized "investment in civic and social infrastructure" and more effective political engagement as a counter-reaction to a rightward shift in Western politics that diminish the hope of national urban policies (Silva 2013). How does this practically translate if not to a re-energized application of neighborhood relevance? What is needed are neighborhood-building strategies that are incremental, resident controlled, planner supported, and politically enabled.

References

A basic guide to ABCD community organizing-1.docx—A Basic guide to ABCD community organizing-1.pdf. http://www.abcdinstitute.org/docs/A%20Basic%20 Guide%20to%20ABCD%20Community%20Organizing-1.pdf. Accessed 30 June 2014.

Agnew, J. A. (1980). The danger of a neighborhood definition of community. *Community Education Journal, 7*(3), 30–31.

Ahmed, K. G. (2012, March 1). Urban social sustainability: A study of the Emirati local communities in Al Ain. *Journal of Urbanism: International Research on Place-making and Urban Sustainability, 5*(1), 41–66.

Arnstein, S. R. (1969, July). A ladder of citizen participation. *Journal of the American Institute of Planners, 35*(4), 216–224.

Bailey, N., & Pill, M. (2011). The continuing popularity of the neighbourhood and neighbourhood governance in the transition from the 'big state' to the 'big society' paradigm. *Environment and Planning C: Government and Policy, 29*(5), 927–942. https://doi.org/10.1068/c1133r.

Baker, B. (2014). *With a little help from our friends: Creating community as we grow older* (1st ed.). Nashville: Vanderbilt University Press.

Boltanski, L., & Chiapello, E. (2007). *The new spirit of capitalism* (G. Elliott, Trans.). London and New York: Verso.

Bursik, R. J., Jr., & Grasmick, H. G. (2002). *Neighborhoods and crime: The dimensions of effective community control.* Lanham, MD: Lexington Books.

Campleman, G. (1951, January). Some sociological aspects of mixed-class neighbourhood planning. *Sociological Review* (1908–1952), *43*, 191–200.

Ceraso, K. (1999, November/December). Seattle neighborhood planning: Citizen empowerment or collective daydreaming? *Shelterforce, 108.* www.nhi.org/online/issues/108/seattle.html. Accessed 3 December 2017.

Checkoway, B. (1984, January 1). Two types of planning in neighborhoods. *Journal of Planning Education and Research, 3*(2), 102–109. https://doi.org/10.1177/0739456X8400300209.

Clampet-Lundquist, S. (2004). HOPE VI relocation: Moving to new neighborhoods and building new ties. *Housing Policy Debate, 15*(2), 415–447.

Cowan, R. (2005). *The dictionary of urbanism.* Tisbury: Streetwise Press.

Crenson, M. A. (1978, August). Social networks and political processes in urban neighborhoods. *American Journal of Political Science, 22*(3), 578. https://doi.org/10.2307/2110462.

Davies, W. K. D., & Herbert, D. T. (1993). *Communities within cities: An urban social geography.* London: Belhaven Press.

Durose, C., & Lowndes, V. (2010). Neighbourhood governance: Contested rationales within a multi-level setting: A study of Manchester. *Local Government Studies, 36*, 341–359.

Fisher, R. (1984). *Let the people decide: Neighborhood organizing in America* (1st ed.). Boston, MA: Twayne Publishers.

Garrioch, D., & Peel, M. (2006, July 1). Introduction: The social history of urban neighborhoods. *Journal of Urban History, 32*(5): 663–676. https://doi.org/10.1177/0096144206287093.

Hagedorn, J. M. (1991). Gangs, neighborhoods, and public policy. *Social Problems, 38*, 529–542.

Hillery, G. A. (1968). *Communal organizations: A study of local societies.* Chicago: University of Chicago Press.

Hojnacki, W. P. (1979). What is a neighborhood? *Social Policy, 10*(2), 47–52.

Hooker, G. E. (1917, May). City planning and political areas. *National Municipal Review, VI*, 341–342.

Hunter, A. (1979, March 1). The urban neighborhood its analytical and social contexts. *Urban Affairs Review, 14*(3), 267–288. https://dx.doi//:10.1177/107808747901400301.

Jezierski, L. (1990, December 1). Neighborhoods and public-private partnerships in Pittsburgh. *Urban Affairs Review, 26*(2), 217–249. https://doi.org/10.1177/004208169002600205.

Jun, K.-N., & Musso, J. A. (2007). Explaining minority representation in place-based associations: Los Angeles neighborhood councils in context. *Journal of Civil Society, 3*(1), 39–58. https://doi.org/10.1080/17448680701390737.

Kadushin, C., & Jones, D. J. (1992). Social networks and urban neighborhoods in New York City. *City & Society, 6*(1), 58–75. https://doi.org/10.1525/city.1992.6.1.58.

Lowndes, V., & Sullivan, H. (2008). How low can you go? Rationales and challenges for neighborhood governance. *Public Administration, 86*, 53–74.

Madanipour, A. (2001, April 1). How relevant is 'planning by neighbourhoods' today? *The Town Planning Review, 72*(2), 171–191. https://doi.org/10.2307/40112446.

Martin, D. G. (2003, September). 'Place-framing' as place-making: Constituting a neighborhood for organizing and activism. *Annals of the Association of American Geographers, 93*(3), 730–750. https://doi.org/10.1111/1467-8306.9303011.

McCann, E. J. (2003, April). Framing space and time in the city: Urban policy and the politics of spatial and temporal scale. *Journal of Urban Affairs, 25*(2), 159–178. https://doi.org/10.1111/1467-9906.t01-1-00004.

McKenzie, R. D. (1922, May 1). The neighborhood: A study of local life in the city of Columbus, Ohio—Concluded. *American Journal of Sociology, 27*(6), 780–799.

McKnight, J. (2013, September 1). Neighborhood necessities: Seven functions that only effectively organized neighborhoods can provide. *National Civic Review, 102*(3), 22–24. https://doi.org/10.1002/ncr.21134.

Meegan, R., & Mitchell, A. (2001, November 1). 'It's not community round here, it's neighbourhood': Neighbourhood change and cohesion in urban regeneration policies. *Urban Studies, 38*(12), 2167–2194. https://doi.org/10.1080/00420980120087117.

Miller, Z. (1981). The role and concept of neighborhood in American cities. In R. Fisher & P. Romanofsky (Eds.), *Community organization for urban social change: A historical perspective* (pp. 3–32). Westport, CT: Greenwood Press.

Miller, S. R. (2012, February 29). *Legal neighborhoods* (SSRN Scholarly Paper). Rochester, NY: Social Science Research Network. http://papers.ssrn.com/abstract=2013565.

Neighbourhoods and communities—Social geographies—The social inclusion of certain groups in communities. https://sites.google.com/site/socialgeography1/neighbourhoods-and-communities. Accessed 16 July 2013.

Newman, O. (1972). *Defensible space: Crime prevention through urban design.* New York: Macmillan.

New York City: The 51st state. Retrieved from https://en.wikipedia.org/wiki/New_York_City:_the_51st_State.

NUSA—Neighborhoods U.S.A. http://www.nusa.org. Accessed 14 March 2014.

Providence Public Schools, RI, & Rhode Island Coll. (1970). Providence. Neighborhoods. Curriculum guide, Grades K-3. Providence social studies curriculum project. S.l.: Distributed by ERIC Clearinghouse. Cited in Bursik, R. J., Jr., & Grasmick, H. G. (2002). Neighborhoods and crime: The dimensions of effective community control. Lanham, MD: Lexington Books.

Reinhardt, K. (2014). Theaster Gates's Dorchester projects in Chicago. *Journal of Urban History.* https://doi.org/10.1177/0096144214563507.

Rose, D. R. (2000). Social disorganization and parochial control: Religious institutions and their communities. *Sociological Forum, 15,* 339–358.

Sampson, R. J. (2011). *Great American City: Chicago and the enduring neighborhood effect.* Chicago and London: The University of Chicago Press.

Scharlach, A. (2012). Creating aging-friendly communities in the United States. *Ageing International, 37,* 25–38.

Schwartz, E. A., & Institute for the Study of Civic Values. (1982). *The neighborhood agenda.* Philadelphia, PA: Institute for the Study of Civic Values. Philadelphia's Neighborhood Transformation Initiative can be found at http://www.phila.gov/ohcd/conplan31/strategy.pdf.

Silva, J. (2013). *Coming up short: Working-class adulthood in an age of uncertainty.* Oxford and New York: Oxford University Press.

Simkhovitch, M. K. (1926). *The settlement primer, Boston.* Retrieved from http://hdl.handle.net/2027/mdp.39015005273860.

Simkhovitch, M. K. (c.1938). *My story of Greenwich House* (1st ed.). New York: W. W. Norton. http://hdl.handle.net/2027/mdp.39015027424558.

Stanley Isaacs. (1945). President of the Board of Directors of United Neighborhood Houses, commenting on a MOMA exhibit on neighborhood planning. http://www.moma.org/pdfs/docs/press_archives/928/releases/MOMA_1944_0014_1944-03-28_44328-12.pdf. Accessed 25 February 2014.

Steiner, J. F. (1930, June). Is the neighborhood a safe unit for community planning? *Social Forces, 8*(4), 492–493 (University of North Carolina Press).

Svara, J. (2003). Two decades of continuity and change in American city councils. Washington, DC: National League of Cities. http://docplayer.net/145973-Two-decades-of-continuity-and-change-in-american-city-councils.html.

Trager, J. (2004). *The New York chronology: The ultimate compendium of events, people and anecdotes from the Dutch to the present.* New York: HarperCollins.

University of Washington, University Extension Division. (1912). *The social and civic center.* Seattle, WA: University of Washington. http://hdl.handle.net/2027/loc.ark:/13960/t23b6sb8v.

Woods, W. K. (1988, September 1). Neighborhood innovations. *National Civic Review, 77*(5), 473–475. https://doi.org/10.1002/ncr.4100770511.

CHAPTER 8

New Trends in Bottom-Up Urbanism and Governance—Reformulating Ways for Mutual Engagement Between Municipalities and Citizen-Led Urban Initiatives

Rosa Danenberg and Tigran Haas

INTRODUCTION

In recent years, bottom-up urban development has proliferated as an alternative to conventional top-down planning in Europe. Citizen-led urban initiatives occur in unusual places and surprising forms and shapes, particularly outside the rules and regulations of formal structures (Zardini 2008; Talen 2014). As a result, urban spaces that have been reshaped by bottom-up initiatives do not always align with the system of public planning and administration. These newly emerged dynamics among actors form both opportunities and challenges for urban governance, management, and planning (Finn 2014). This chapter addresses how bottom-up urbanism relates to urban governance in Europe. More specifically, what are the ways to develop *mutually engaging relationships* between municipalities and citizen-led urban initiatives? This chapter analyzes examples of bottom-up urbanism in two empirical cases (Stockholm in Sweden and Istanbul in Turkey) through the lens of urban governance. In these two remarkably distinctive cities, with varying governmental frameworks, the relationships between the two municipalities

R. Danenberg (✉) · T. Haas
KTH Royal Institute of Technology, Stockholm, Sweden

© The Author(s) 2019
M. Arefi and C. Kickert (eds.), *The Palgrave Handbook of Bottom-Up Urbanism*, https://doi.org/10.1007/978-3-319-90131-2_8

and eight citizen-led urban initiatives are evaluated. The chapter concludes with two reformulated ways for mutual engagements that foster a *Social Innovation* viewpoint.

MUTUAL ENGAGEMENT IN THE FORM OF GOVERNANCE ARRANGEMENTS

Since the 1990s a number of theories, approaches, and models have influenced the practice of urban planning and design. Their effects can be seen in the form of our built environments. The specific ideals dominating today's urban planning and design discourse have been examined and defined in various ways. Examples include territories of urban design (Krieger 2006), urban design force fields (Fraker 2007), integrated paradigms in urbanism (Kelbaugh 2008a, b), urbanist cultures and approaches to city-making (Talen 2005), new directions in planning theory (Fainstein 2000), and typologies of urban design (Cuthbert 2006). This chapter specifically focuses on the more recent theories and ideals of bottom-up urbanism (Finn 2014; Iveson 2013; Talen 2014; Deslandes 2013).

The bottom-up movement is the umbrella for the growth of, and interest in, small-scale urban interventions, including tactical, pop-up, or guerilla urbanism. The characteristics of bottom-up urban interventions are generally defined as locally-driven, citizen-initiated, low budget and temporary, and are aimed at improvement (Talen 2014; Deslandes 2013; Lydon and Garcia 2015; Chase et al. 2008). The diversity among the interventions is probably what unifies them most (Iveson 2013; Brenner 2015). Finn (2014) defines DIY practices as that: are initiated by citizens; increase or extend official municipal infrastructures; are produced by users who are the main beneficiaries. This definition relates well to the citizen-led urban initiatives that are central to this chapter—those that seek an engaged relationship with the municipality. Provided that these initiatives, "propose alternative lifestyles, reinvent our daily lives, and reoccupy urban spaces with new uses" (Zardini 2008: 16), the initiatives may be recognized by their confidence in the possibility of effecting change, regardless of obstacles (Iveson 2013). This possibility also refers to the challenge of changing conventional power structures, procedures, and traditional purposes (Hou 2010). Change is considered most effective when initiated from outside governmental structures, in particular from the bottom, while leaving room for professionals to participate in and actively stimulate the efforts (Brenner 2015).

Emerging dynamics between top-down and bottom-up urban actors add complexity to urban governance structures, as citizens and communities no longer want to wait for or rely on bureaucracies and, instead, take matters into their own hands (Talen 2014; Finn 2014). The consequences of bottom-up urban practices challenge formal urban actors. For instance, governmental actors experience considerable confusion and divergence regarding the meanings and implications of bottom-up urban development (Brenner 2015). In addition, taking bottom-up initiatives seriously enough and acknowledging their potential for participatory planning is a common struggle (Finn 2014; Brenner 2015). Complexity also lies in the heterogeneity of the range of bottom-up initiatives, how and by whom they are organized and initiated (Gerometta et al. 2005). Furthermore, initiators often comprise only a few informed, creative, and entrepreneurial citizens, as citizens face high barriers to entry and limited access to inclusion in the planning process (Deslandes 2013).

The discussion of the relationship between bottom-up urban actors and the government leads to a set of questions related to urban governance (Finn 2014). Urban governance is concerned with the struggle and potential for a mutually engaging relationship with citizens and their initiatives, as "governance is about the capacity to get things done in the face of complexity, conflict and social change: organisations, notably but not only urban governments, empower themselves by blending their resources, skills and purpose with others" (Kearns and Paddison 2000: 847). The involvement of more governing actors has become a crucial demand during the recent neoliberal era, since governments are deemed less capable at managing local issues and other actors are expected to step in (Taylor 2007; Blomgren Bingham et al. 2005; Gerometta et al. 2005). Therefore, new ways of governance are needed. One prospect is to utilize *governance arrangements* to promote such new ways. An urban governance arrangement "is understood as a specific setting of different actors with specific shared norms and values (institutions), who reach decisions on urban places" (Gerometta et al. 2005: 2015). The arrangement adopts innovative processes that are open to non-governmental actors, leading to increased participation (Swyngedouw 2005). The non-governmental actors are the citizen-led urban initiatives in which citizens are engaged in both inventing and utilizing tools to participate in the governance arrangements (Blomgren Bingham et al. 2005). The following case studies illustrate how bottom-up initiatives have created new governance arrangements, and the challenges they face.

CASE STUDIES

Stockholm and Istanbul are two cases with rather different contexts. Stockholm, as the capital city of Sweden with nearly one million inhabitants (SCB 2016), has been governed by a strong welfare state but has witnessed a shift toward privatization in recent decades (Engström and Cars 2013). Istanbul is the largest city in Turkey with a population of almost 15 million in 2016 that is expected to grow to nearly 17 million by 2030 (UN 2016). Istanbul is located on two continents and is governed by a neoliberal planning agenda and ruled by a centrally controlled political system (Akpınar 2014). Because the cities differ in many ways, an analysis of such remarkably distinctive cases should serve as a test bed for assessing the generalizability of the results (Yin 2009). In each city, four citizen-led urban initiatives were selected for analysis. Through in-depth interviews held in 2015 with key stakeholders from the initiatives and one municipal official in each city (plus one politician in Istanbul), the relationship between bottom-up initiators and the government was studied. What follows is a short introduction of both urban contexts and further insights regarding these relationships.

Stockholm

Sweden's urban development is characterized by a history of modern planning that underwent a number of major shifts. From the mid-twentieth century onwards, the dominance of the welfare state strongly positioned the Swedish planning process toward Modernism (Engström and Cars 2013). There was an expansion of housing provision and an abundance of legislation, policies, and public instruments that were meant to take care of its growing urban population, albeit with hardly any public input. In response to citizen demand for communicative planning, the Planning and Building Act of 1987 gave citizens the legal right to participate in planning processes (Wänström 2013). Initially, privatization combined with planning standardization continued to strengthen the municipality's planning monopoly. However, structural changes in society and further privatization have diminished the power of municipal planning and urban development. Private actors have gained more power, while municipalities increasingly face the challenge of how to best use the planning tools and instruments at their disposal (Engström and Cars 2013). Society has also become more diverse and more unequal. Although Swedish cities scored high in the *State of the World's Cities* report for 2012/2013 in terms of quality of life (UN-Habitat 2013), segregation has become a growing issue and living conditions in high- and low-income areas are diverging as a result (Legeby 2010; Andersson 2017). Many bottom-up initiatives aim to bridge these growing societal divides.

For this research, interviews were held with a strategist from Stockholm's City Planning Administration and one of the founders of each of the following four initiatives, which are ranked from stronger to weaker relationships with the municipality:

1. *Trädgård På Spåret* was founded in 2011 as a mobile, semi-temporary urban garden on abandoned rail tracks in the inner city. The initiators developed a proposal, including a self-financing model, to which the municipality responded with the offer for a land lease contract (see Fig. 8.1).
2. *Brotherhood Plaza* in Skarpnäck was proposed to the local council in 2003 by a skater as a self-design and self-build skatepark. The realization took place between 2006 and 2012 with support from local and city authorities.
3. *Cyklopen* in Högdalen was founded in 2013 as a self-built anarchist community house located in the woods on the edge of a suburb. It operates as a multifunctional cultural center without structural municipal support (see Fig. 8.2).

Fig. 8.1   *Trädgård På Spåret*, Stockholm

4. *Livstycket* in Tensta was founded in 1992 as a women's empowerment center that has minimized its interaction with the municipality and only receives funding for integration program services. The center welcomes immigrant women for language and cultural programs and is involved in the production of high quality handicrafts.

These citizen-led urban initiatives have encountered several challenges in their relationships with the municipal government. The challenges are predominantly consequences of the discrepancy between the operating organizational structures of the municipality and those of the initiatives. This is exacerbated by the strong variance between the organizational structures of the initiatives, which have mostly been dictated by the different motivations that started them. The bottom-up organizational structures that resulted include a flat democratic structure in which the users make the decisions (Cyklopen) and a structure consisting of autonomous work groups (På Spåret).[1]

---

[1] Furthermore, all of the initiatives emerged out of an explicit urge to create a place outside government structures to evade the perceived pressing neoliberal agenda in planning, which increases tension. For example, the municipal inability to provide a community skatepark prompted the Brotherhood Plaza project, and the Cyklopen project aimed to provide a cultural center that is an alternative to traditional commoditized public spaces.

Fig. 8.2 *Cyklopen*, Stockholm, Sweden

In contrast, the municipality follows bureaucratic structures, which do not always align with the initiatives. This discrepancy can generate mutual distrust in communication, transparency, accountability, and responsibilities. For example, Cyklopen perceived municipal officials as working in isolation with few incentives for taking responsibility or for being transparent toward their initiative. Furthermore, the municipality prefers to communicate with one spokesperson and struggles to handle a collective organizational structure like Cyklopen's. På Spåret's urban gardening activities are not yet taken seriously enough by the municipality, even though the initiative's founders possessed the relevant education to raise their chances of receiving municipal support. Nonetheless, the municipality seemed to prioritize solving practical problems over engaging with På Spåret's initiators in larger debates on urban transformation, which undermined a serious partnership.

At the Brotherhood Plaza, the municipality adhered to inefficient processes that slowed collaboration. The initiator was a highly motivated skater who was in need of more funding, but he was only able to receive municipal support by mobilizing the community and a local politician. Municipal funding for Livstycket is often delayed for a year before disbursement; even the city acknowledges that its inertia threatens to harm the continuity of the relationship with the initiators. The municipality initiated The Dialogue Project in 2011 to respond to complaints about the rigidness of the public participation process, according to the municipal strategist. Yet this project mainly focused on citizen engagement within the conventional structures, and not on enlarging the possibilities for bottom-up urban development. This evidence of inflexibility obstructs the support needed for entering further stages of the relationship.

Istanbul

Istanbul is a global city with growing pains. Government-led neoliberal policies aim to strengthen the service industry, technology, and financial sectors, while social values are structurally undermined (Akpınar 2014). Since the 1980s, national policies have paved the way for globalization and are preparing for closer ties with the EU's neoliberalism, resulting in more top-down and centralized urban transformation projects (Erkut and Shirazi 2014). The centralization of planning has left little to no room for public participation, it can bypass local authorities, and undermines the integrity of their holistic development of plans. Since 1985 the national government has ceded some planning responsibilities to local governments, and municipal laws introduced in 2004 and 2005 allow municipalities to implement urban renewal projects with built-in flexibility toward following the national legal system (Rodriguez and Azenha 2014). Still, centralized planning has given rise to opposition and disputes, which has prompted many of the following bottom-up initiatives.

For this research, interviews were held with a political councilor and municipal official from Kadıköy's local administration, and one of the founders of each initiative.[2] These initiatives selected in Istanbul are ranked from stronger to weaker relationships with the municipality:

1. *Design Atelier Kadıköy (TAK)* in Kadıköy was founded in 2013 by a local civil society institution (ÇEKÜL) and holds formal partnerships with the municipality (Kadıköy Municipality) and a planning consultancy (Kentsel Strateji). TAK applies strategic participatory design processes on a neighborhood level and promotes "You design, we implement!"

2. *İlk Adım* is a women's and children's center (WCC) in Nurtepe, one of the lowest-income neighborhoods in Istanbul. It opened in 2004 with the help of a nationally operating NGO and the municipality's offer for utilizing a building free of charge. It offers affordable childcare and the empowerment of women through the production of handicrafts (see Fig. 8.3).

3. *Sokak Bizim* (Streets Belong to Us) in Kuzguncuk is a public space advocacy initiated by young professionals. The NGO seeks permission to carry out campaigns and introduces and implements models for street life improvements (see Fig. 8.4).

---

[2] Both the political councilor and municipal official held seats on the TAK board.

Fig. 8.3 *İlk Adim*, a Women's and Children's Center (*WCC*), Istanbul, Turkey

4. *Music for Peace* in Edirnekapı was founded in 2005 to offer free music education to as many children as possible and to give voice to peace through music. It is financed through private sponsorship (see Fig. 8.5).

The dominance of municipal politics and the complicated governmental framework of Istanbul pose challenges to a mutually engaged relationship between the city and these citizen-led urban initiatives, which emerged as a response to a void in municipal services or priorities—such as a women's center that offers services not provided by the municipality (WCC), similar to the school with classes in classical music (Music for Peace). Furthermore, public space is generally not prioritized (Sokak Bizim) and public participation is not well incorporated in planning (TAK). Politics seem far more ingrained in all facets of the relationship of these initiatives with the municipality of Istanbul than in Stockholm. Due to the growing distrust of politicians and a fear of discontinuity in their commitments, each initiative has a well-considered level of disengagement with the municipality. For example, Sokak Bizim's initiators decided against starting a new municipal cooperation during the election season, as financial support promised by political candidates could not be trusted and engagement with one particular municipality is too politically sensitive. Music for Peace's initiators decided against engaging with the government at all, because municipal

Fig. 8.4 *Sokak Bizim* (Streets Belong to Us), Istanbul, Turkey

Fig. 8.5   Music for Peace, Istanbul, Turkey

officials have proven to be inconsistent and the initiators suspected that the municipality's support of social initiatives was only to win votes. Similarly, WCC's initiators decided after many failed attempts at municipal collaboration that it was better to remain autonomous and stay politically neutral. Conversely, TAK was partly initiated by the municipality from the outset. Furthermore, all governmental layers (local, regional, and central) have the authority to influence local planning processes, which has complicated formalizing permissions for initiatives like Sokak Bizim and for the legal structures of WCC. Applying for formal permissions is Sokak Bizim's only way to trust the municipality, whereas WCC only relies on informal agreements. This disconnection adds to the informality of the bottom-up initiatives and the growing lack of transparency in their communications with the municipality, ultimately creating an additional challenge to mutually engaged relationships.

NEW GOVERNANCE ARRANGEMENTS

Initiatives in both cases emerged from different contexts and were confronted with varying challenges. The initiatives in Stockholm, particularly, began outside Stockholm's municipal bureaucracy and its perceived neoliberal planning agenda. Nevertheless, citizens involved in the initiatives proclaim that they are too dependent on the bureaucracy; they do not fully understand how to cope with the bureaucratic hurdles and delays. This results in struggles with the municipality's responsibility, transparency, and continuity. In Istanbul, initiatives have emerged to supplement the functions and services that are either not provided or not accepted by the municipality. The initiatives face the challenges of dominant politics, a lack of transparency within municipalities, and an informality in planning practices. For the citizens involved in the initiatives, this leads to distrust and a fear of discontinuity.

Hence, this chapter concludes that these new governance arrangements do not sufficiently stimulate an environment for a mutually engaging relationship between citizens and their governments. Despite their struggles with the government, the initiatives in both cities do seek a relationship with their municipalities to better cope with the initiatives' various organizational structures. To forge a mutually engaged relationship, the initiatives must face the challenge of continually safeguarding their effectiveness and legitimacy (Talen 2014). There is no real alternative. Deciding to not engage at all with the government, framing the initiative's agendas as an alternative to governance, and promoting anti-statist and anti-planning mentalities would eventually be counterproductive in formulating a mutually engaged relationship (Brenner 2015). After all, most municipalities struggle to accept other authorities without a strong mutual understanding (Taylor 2007; Blomgren Bingham et al. 2005). This signals the pitfall in new governance arrangements. The impression exists that new governance arrangements include innovative and participatory approaches leading to rearranged ways of collaboration between municipalities and bottom-up initiatives, empowering the latter (Gerometta et al. 2005). However, governmental power persists. The redefinition and repositioning of the relationship between governmental and non-governmental actors enables the municipality to take control over whom to collaborate with and how. Thus, the ostensible innovation in governance actually enlarges the realm of governing for municipalities. Subsequently, some bottom-up initiatives are selectively empowered while others remain disempowered (Taylor 2007; Swyngedouw 2005).

NOT GOVERNANCE INNOVATION, BUT SOCIAL INNOVATION

New governance arrangements hence fall short of enabling a mutually engaged relationship between bottom-up initiatives and the government. Such a relationship cannot be applied or enforced from the top down, instead it should leave more room for innovation to stimulate citizen participation and involvement in decision-making processes. Social innovation can provide a more desirable lens for cooperation between bottom-up initiatives and cities, as it provides an, "alternative view that has inspired, but also integrated, alternative discourses and strategies for re-building and re-socializing cities and their neighborhoods" (Moulaert et al. 2007: 196). Social innovation initiatives that take place (or at least are initiated) at the local level, have an innovative character in responding to different social problems that arise, and have the capacity to generate positive impacts on various scales (Moulaert et al. 2010 in Nicolò 2012). Social innovation includes three core dimensions that promote empowerment from the bottom up: "satisfaction of human needs (content dimension); changes in social relations, especially with regard to governance (process dimension); and an increase in socio-political capability and access to resources (empowerment dimension)" (Gerometta et al. 2005: 2007). The concept emerged out of discontent with technocratic approaches to urban planning and has brought together a diversified range of counter-hegemonic movements and initiatives that challenge traditional governance relations (Moulaert et al. 2007). What sets the perspective of social innovation apart from other bottom-up theories is the strong encouragement for initiatives to become active political subjects and self-steering actors that shape and influence physical and administrative spaces outside of traditional administrative structures. Social innovation therefore argues for bottom-up initiatives to acknowledge and leverage their political power without turning their backs on established authorities. Iveson (2013) argues that governmental structures need to be disputed from a political perspective, as this perspective will "teach us how we might transform the localized possibilities that are being explored through individual practices of DIY urbanism into a wider politics of the city which challenges existing forms of authority and titles to govern" (p. 955).

The initiatives illustrate that becoming active subjects with a political approach creates the ability to increase power to participate in decision-making. In Istanbul, Sokak Bizim's initiators were able to reach an agreement over the strategy and direction of public space management with the municipality, and WCC's initiators were able to co-design local disaster plans with the city. In Stockholm, På Spåret's initiators became more involved in the area's local planning, and Brotherhood Plaza's initiator became a consultant to the project team building the skate park. In the quest for advancing a mutually engaged relationship, the initiators of Cyklopen and På Spåret argued for hiring a municipal guide to help bottom-up initiatives navigate the municipal bureaucracy. From the municipal side, the guide balances local politics with new perspectives. As a gatekeeper, the guide also addresses constraints such as transparency, responsibility, and continuity. Citizens involved in the initiatives conveniently communicate directly with the guide as their contact person in the municipality, although communication in the other direction is expected to remain difficult as many initiatives tend to have more than one spokesperson.

In Istanbul, the municipal official and the politician involved in TAK foresaw further mutual engagement through creating a local neighborhood council operated by local citizens, which could function as an umbrella for citizen-led urban initiatives. Each initiative takes a member seat on the council, which attempts to create equal empowerment among all initiatives. The neighborhood council communicates directly with the municipal guide, but it otherwise keeps its distance from the municipality. It hence formalizes the balance between autonomy and connection with the municipality. Attributing greater control and power to the local level could improve representation, affirming its socially innovative character (Gerometta et al. 2005). Also, the council stimulates a learning environment in which both citizens and the municipality can share knowledge, experiences, and best practices. This may lead to crucial cultural shifts within the municipality to adequately listen and respond to bottom-up initiatives (Taylor 2007). There is a difficulty in the ability to transition from political opposition to process-engagement in order to avoid becoming one's own enemy and not getting anything done (Taylor 2007). The perspective of social innovation does not shy away from powerful independent actors and does focus on their continuous and fruitful collaboration with politicians and public agencies to create effective structural change, while keeping a safe distance to avoid the institutionalization of independent visions.

In the aforementioned initiatives, many questions arise. Which public–private cooperation model is in the best interests of the community? How do we best balance the competing objectives of different actors and interest groups? When we look ahead, what is the collective long-term advantage of these bottom-up initiatives? Many planners and urban designers who have worked with community planning exercises and schemes (for example in the charrette process) have tackled issues of representation and, in many cases, have come to realize that public participation requirements are often too onerous to enable any significant work to be done (Onyango & Noguchi 2009; Halbur 2010; Rucker 2011). Conversely, others acknowledge that the traditional bottom-up process avoids the big mistakes of top-down planning but does so in an often inefficient and ephemeral manner (Talen 2014). Instead, many scholars merge the virtues of top-down and bottom-up planning. Future research will show whether it is possible to move beyond acknowledging the persistent weaknesses in participatory approaches and begin to understand what needs to be done to improve the process itself (Hurley 2013).

Conclusion

This chapter analyzed emerging actor dynamics in the new era of bottom-up urban development, specifically focusing on the relationship between bottom-up initiatives, initiators, and local governments. It has studied multiple ways for establishing a mutually engaged relationship between municipalities and citizen-led urban initiatives. The research, however, concludes that missing links persist in this relationship. In Stockholm, the discrepancy between bottom-up and municipal organizational structures creates mutual distrust and inflexibility. In Istanbul, the hierarchical governmental structure and the dominance of politics have exacerbated initiatives' informality and a growing lack of transparency. In both cases, the varying degrees of engagement between the initiatives and the municipalities illuminate the rather mixed results of new governance arrangements, in which stakeholders clarify their role in a participatory process. It can be concluded that new governance arrangements alone are not adequate for creating opportunities for citizens to partake in participatory methods or to be involved in decision-making processes. The apolitical nature of these arrangements negates the need for structural engagement. Hence, adopting the more political view of social innovation theory furthers the way for mutual engagement. This view departs from the local level and recognizes the ability for structural change, it also focuses on the necessity of transferring power and authority to the local level. The political perspective of social innovation reformulates mutual engagement between initiatives and government by introducing political liaisons, such as a municipal guide or a neighborhood council. By acknowledging the inherent political nature of bottom-up initiatives and framing them as social innovations, initiatives turn into active subjects in the municipal process and a learning environment is encouraged. However, questions persist about public engagement and the efficiency of the planning process in these bottom-up processes. Who participates and how? What are the motivations for participation? How do participants interact? How do facilitators perform? What is the influence of local systems and institutions? Further research is needed to help answer these questions.

List of Interviewees

Stockholm:

1. One of the founders of Cyklopen on June 9, 2015, outside the house in Högdalen, Stockholm
2. One of the founders of Trädgård På Spåret on June 9, 2015, at the community garden in Södermalm, Stockholm
3. The initiator of Skatepark Brotherhood Plaza on June 8, 2015, on a terrace in Östermalm, Stockholm
4. The founder of Livstycket on June 4, 2015, in the center in Tensta, Stockholm
5. Municipal strategist from City Planning Administration, City of Stockholm on June 22, 2015, at Stadsbyggnadskontoret, Stockholm.

Istanbul:

6. One of the founders of WCC İlk Adim Cooperative on May 11, 2015, at the center in Nurtepe, Istanbul
7. One of the founders of Sokak Bizim on May 20, 2015, at their office in Kuzguncuk, Istanbul

8. One of the founders of Music for Peace on May 8, 2015, at the music school in Fatih, Istanbul
9. The initiative's representative of TAK on May 6, 2015, at the atelier in Kadıköy, Istanbul
10. The political councilor and municipal official, part of the TAK board, on May 12, 2015, Kadıköy, Istanbul.

Interviewer on all occasions: Rosa Danenberg
The interview with WCC İlk Adim Cooperative was held by Irem Anik and translated by Tuba Kolat.

References

Akpınar, I. (2014). Legal and institutional context of urban planning and urban renewal in Turkey: Thinking about Istanbul. In G. Erkut & R. Shirazi (Eds.), *Dimensions of urban re-development—The case of Beyoğlu, Istanbul* (Ch. 2.2). Berlin: Technische Universität Berlin.

Andersson, O. (2017). Segregationen är inbyggd och avsiktlig. *DN* [Online]. Available at: http://www.dn.se/kultur-noje/ola-andersson-segregationen-ar-inbyggd-och-avsiktlig/?forceScript=1&variantType=large. Accessed June 27, 2017.

Blomgren Bingham, L., Nabatchi, T., & O'Leary, R. (2005). The new governance: Practices and processes for stakeholder and citizen participation in the work of government. *Public Administration Review, 65*(5), 547–558.

Brenner, N. (2015, April 1). Is "tactical urbanism" an alternative to neoliberal urbanism? *MoMa* [Online]. Available at: http://post.at.moma.org/content_items/587-is-tactical-urbanism-an-alternative-to-neoliberal-urbanism. Accessed May 13, 2015.

Chase, J., Crawford, M., & Kaliski, J. (2008). *Everyday urbanism*. New York: The Monacelli Press.

Cuthbert, A. (2006). *The form of cities: Political economy and urban design*. Oxford: Blackwell.

Deslandes, A. (2013). Exemplary amateurism—Thoughts on DIY urbanism. *Cultural Studies Review, 19*(1), 216–227.

Engström, C. J., & Cars, G. (2013). Planning in a new reality—New conditions, demands and discourses. In M. J. Lundström, C. Frederiksson, & J. Witzell (Eds.), *Planning and sustainable development in Sweden* (Ch. 1). Stockholm: Föreningen för Samhällsplanering.

Erkut, G., & Shirazi, R. (2014). Introduction. In G. Erkut & R. Shirazi (Eds.), *Dimensions of urban re-development—The case of Beyoğlu, Istanbul* (Ch. 1). Berlin: Technische Universität Berlin.

Fainstein, S. S. (2000). New directions in planning theory. *Urban Affairs Review, 35*(4), 451–478.

Finn, D. (2014). DIY urbanism: Implications for cities. *Journal of Urbanism: International Research on Placemaking and Urban Sustainability, 7*(4), 381–398.

Fraker, H. (2007). Where is the urban design discourse? [To rally discussion]. *Places, 19*(3).

Gerometta, J., Haüssermann, H., & Longo, G. (2005). Social innovation and civil society in urban governance—Strategies for an inclusive city. *Urban Studies, 42*(11), 2007–2021.

Halbur, T. (2010, April 26). Andres Duany wants to reform the public process. *Planetizen* [Online]. Available at: https://www.planetizen.com/node/43935. Accessed February 28, 2017.

Hou, J. (2010). (Not) your everyday public space. In *Insurgent public space: Guerilla urbanism and the remaking of contemporary cities* (pp. 1–16). New York: Routledge.

Hurley, J. (2013). The public process and new urbanism. In E. Talen & Center for the New Urbanism (Eds.), *Charter of the new urbanism* (2nd ed.). New York: McGraw-Hill Education.

Iveson, K. (2013). Cities within the city: Do-it-yourself urbanism and the right to the city. *International Journal of Urban and Regional Research, 37*(3), 941–956.

Kearns, A., & Paddison, R. (2000). New challenges for urban governance. *Urban Studies, 37*(5–6), 845–850.

Kelbaugh, D. (2008a). *Three urbanisms: New, everyday, and post*. In: T. Haas (Ed.), *New urbanism and beyond: Designing cities for the future*. New York: Rizzoli.

Kelbaugh, D. (2008b). Introduction. Further thoughts on the three urbanisms. In D. Kelbaugh & K. K. McCullough (Eds.), *Writing urbanism*. New York: Routledge.

Krieger, A. (2006). Territories of urban design. In J. Rowland & M. Malcolm (Eds.), *Urban design futures*. London: Routledge.

Legeby, A. (2010). From housing segregation to integration in public space: A space syntax approach applied on the city of Södertälje. *The Journal of Space Syntax, 1*(1), 92–207.

Lydon, M., & Garcia, A. (2015). *Short-term action for long-term change*. Washington, DC: Island Press.

Moulaert, F., Martinelli, F., Gonzlezález, S., & Swyngedouw, E. (2007). Introduction: Social innovation and governance in European cities—Urban development between path dependency and radical innovation. *European Urban and Regional Studies, 14*(3), 195–209.

Nicolò, F. (2012). Can neighbourhoods save the city? Community development and social innovation, by Frank Moulaert. *Urban Research & Practice, 5*(2), 293–295 [Online]. Available at: http://dx.doi.org/10.1080/17535069.2012.691630 Accessed June 14, 2015.

Onyango, J., & Noguchi, M. (2009). Changing attitudes of community through the design charrette process. *The International Journal of Neighborhood Renewal, 1*(3), 19–30.

Rodriguez, M. A., & Azenha, A. L. (2014). Understanding the urban context. In G. Erkut & R. Shirazi (Eds.), *Dimensions of urban re-development—The case of Beyoğlu, Istanbul* (Ch.3.). Berlin: Technische Universität Berlin.

Rucker, D. (2011, December 2). Why Duany is wrong about the importance of public participation. *New Geography* [Online]. Available at: http://www.newgeography.com/content/002046-why-duany-wrong-about-importance-public-participation. Accessed February 28, 2017.

SCB. (2016, December). *Key figures for Sweden* [Online]. Available at: http://www.scb.se/en/finding-statistics/statistics-by-subject-area/population/population-composition/population-statistics/#_Keyfigures. Accessed February 28, 2017.

Swyngedouw, E. (2005). Governance innovation and the citizen: The Janus face of governance-beyond-the-state. *Urban Studies, 42*(11), 1991–2006.

Talen, E. (2005). *New urbanism & American planning: The conflict of cultures.* New York: Routledge.

Talen, E. (2014, September, 2). Do-it-yourself urbanism: A history. *Journal of Planning History,* 1–14 [Online]. Available at: http://jph.sagepub.com/content/earl y/2014/09/01/1538513214549325.abstract#corresp-1. Accessed March 24, 2015.

Taylor, M. (2007). Community participation in the real world: Opportunities and pitfalls in new governance spaces. *Urban Studies, 44*(2), 297–317.

UN-Habitat. (2013). *State of the world's cities 2012/2013* [Internet]. New York: Routledge. Available at: https://sustainabledevelopment.un.org/content/documents/745habitat.pdf. Accessed December 2, 2015.

United Nations. (2016). *The world cities in 2016—Data booklet* [Internet]. Available at: http://www.un.org/en/development/desa/population/publications/pdf/urbanization/the_worlds_cities_in_2016_data_booklet.pdf. Accessed April 14, 2017.

Wänström, J. (2013). Communicative planning processes—Involving the citizens. In M. J. Lundström, C. Frederiksson, & J. Witzell (Eds.), *Planning and sustainable development in Sweden* (Ch.13). Stockholm: Föreningen för Samhällsplanering.

Yin, R. K. (2009). *Case study research.* London: Sage.

Zardini, M. (2008). A new urban takeover. In G. Borasi & M. Zardini (Eds.), *Actions: What you can do with the city.* Montreal: Canadian Centre for Architecture (co-published by SUN).

CHAPTER 9

The Self-Made City—Urban Living and Alternative Development Models

Kristien Ring

Worldwide, more and more people are moving into urban centers. Unfortunately, many new urban developments suffer from a lack of: affordability; choice; diversity; care for the quality of the built environment; or care for social and environmental sustainability (Bloomfield 2017; Davis 2016; Demographia 2017; Nations 2014; Robertson 2017). The success of our cities in the future will depend on finding new solutions for growth that offer an urban alternative to suburban sprawl, but with the qualities that people desire. We need new strategies for densifying our cities that actually improve and bring benefits to the surrounding neighborhoods with each new development. An increasing number of people take development into their own hands in the Self-Made City, realizing customized, community-oriented and high-quality, affordable housing that they dwell in themselves.[1] This chapter describes how an increasing amount of Germany's new housing stock is developed from the bottom up.

---

[1] Kristien Ring, AA PROJECTS (ed.) in collaboration with the Senate Department for Urban Development and the Environment of Berlin, *SELF MADE CITY. Berlin: Self-initiated Urban Living and Architectural Interventions*, Jovis Publishers, 2013. The publication *Self Made City*, is a quantitative and qualitative analysis of over 125 projects from Berlin, Germany that serves as a basis of this text, in addition to further research including over fifty lectures, workshops, and conferences, attended by the author in cities worldwide between 2013 and 2017.

K. Ring (✉)
School of Architecture and Community Design, University of South Florida, Tampa, FL, USA

© The Author(s) 2019
M. Arefi and C. Kickert (eds.), *The Palgrave Handbook of Bottom-Up Urbanism*, https://doi.org/10.1007/978-3-319-90131-2_9

Resident-led development of housing fits into a wider trend of bottom-up urban interventions that activate unused spaces with temporary interventions and provide much-needed urban amenities. Many new bottom-up initiatives create places that have a story and motivation behind their creation, making people identify with them (Berlin 2007; d'Alençon et al. 2017; Ziehl et al. 2012). Through private initiatives, solutions are created that otherwise would not have been available, or possible. While many bottom-up projects are ephemeral, a trend toward permanent improvement is taking place. New models of collaborative housing development now offer both long-term solutions and are more affordable. This is in contrast to current housing markets in most major cities, which are shaped largely by profit-driven developments (Dürr and Kuhn 2017; Förster et al. 2016; Krämer and Kuhn 2009; Lafond and Honeck 2012).

Many of these collaborative developments are initiated by families who want to stay in urban centers. Today, having a family often means moving to a detached single-family home, outside of the city, and having to drive a car for every aspect of daily life (Montgomery 2013). Chance meetings are rare, as are collaborations or sharing responsibilities with neighbors. Families that stay in urban centers often struggle to balance their tighter private living quarters with a high-quality urban lifestyle. Self-organized development enables families to share amenities that would otherwise not be affordable, such as rooftop terraces, playrooms, guest rooms, and even swimming pools, gyms, libraries, and workshops. Projects in a self-made city not only create living spaces that are based on the real needs of the owners, but also bring new neighbors into existing communities that have a vested interest, and contribute positively to the micro economies and sustainable, resilient urban developments. In a self-made city people combine their resources and create their own shared niches and oases. Self-made in this context means that the people who have initiated the projects actually live there, long-term; essentially, a group of people become their own investors and developers.

The self-determined design of space, building, living, and working has produced an exemplary architectural diversity and quality, that has taken shape as the *Baugruppe*, and as co-housing, co-operatives, co-working spaces, or other project forms. The German Baugruppe, or building group, has a long tradition of self-initiated, shared responsibility building. In Berlin alone between 2004 and 2014 well over 400 projects were completed as a Baugruppe. Although the origin of community-oriented housing dates back to the early nineteenth century, it is the co-housing settlements of the 1960s in Scandinavia, the Netherlands, and later in the United States, that we consider the key background of today's developments. Most co-housing settlements were built as private initiatives that were financed and run by their residents, centered around a specific way of life. In this way, they were often seen as closed communities. The Baugruppe has developed as a more open and pragmatic solution. In the 1990s, German cities such as Tübingen and Freiburg developed new neighborhoods through this evolution of private building collectives (McCamant and Durrett 2011: 39–50, 247–270). There is no 'typical' Baugruppe model; every project differs in its financing, social make-up, the wishes and desires of the group, and the project's resulting architectural and urban qualities. Whereas the Baugruppe is owner-based, co-op associations are a membership and rental-based alternative. A self-made city is not necessarily about self-build, but rather about self-determined ways of life and initiating project forms that combine ownership with urban community.

Architects are playing a leading role in initiating and managing these participative projects, as well as in facilitating alternative solutions for the development process with government and city planning officials. The most significant and innovative built examples of Baugruppe developments have been initiated by architects. On the surface, many of these examples are pragmatic solutions for family living in urban locations, which stack and combine single-family homes. But, on closer inspection, it is clear that the close collaboration between the architects and the clients has resulted in projects packed with special features and spaces that foster social interaction. Key are the common spaces and public places in which people not only come together, but also have the feeling that they belong and can have a hand in what happens there—that they can do something. The spirit of bottom-up interventions is reflected in the participatory process and the chance to configure, determine, and design a place of one's own.

Commonalities of the projects analyzed and discussed in this chapter show what level of quality can be achieved architecturally. Furthermore, when a high number of collaborative projects are built in close proximity, we can start to measure the benefits for the surrounding neighborhoods and the city. The self-made city is more than a formal architectural style, it is a new way of developing our cities for a greater benefit—one that involves collaboration between planning departments, municipal decision-makers, architects, investors, future residents, and neighbors. The processes and projects described in this chapter illustrate urban densification that improves urbanity with suitable, affordable, living and working spaces, and planning that meets our growing ecological challenges. They demonstrate resilience, not only in the built environment, but also socially, in terms of the people living there and how society interacts and evolves, upon which the future success of our cities will ultimately hinge.

Drivers and Qualities of the Self-Made City

The owner-occupied and self-initiated projects of a self-made city demonstrate three key drivers behind their success that enable them to balance density, urbanity, social qualities and individuality (Ring 2015; Wohnen 2012).

1. **Urban densification**: The walkable, polycentric city of close-knit neighborhoods is a model that stands for high-quality urban life. In our, mostly built out, city centers, densification that includes a vital mix of uses is an important strategy for fostering urbanity (Sonne 2017). Intelligent densification reduces land use, generates additional living space, improves the quality of existing green and (underused) open spaces, and utilizes existing infrastructure. Successful densification projects can also activate the street zone and contribute to the walkability of the city by adding to its diversity of spaces and uses.

2. **Enhancing neighborhood qualities**: New living spaces, additional street level uses, and improved green and open spaces can benefit both existing and new residents of a neighborhood. The concepts and strategies presented in this chapter create a specific mix of uses, foster a social mix, support interaction between new and established neighbors, and strengthen the social network of the community. Successful projects generate a vital mix and create synergies and acceptance of densification within the existing urban context.

3. **New and adaptable forms of living**: We need suitable forms of living in the city that meet the diverse needs of the people living there now, but that can also be adapted to new conditions over time. The examples in this chapter introduce new living typologies, the ways in which they are adaptable, (new) types of circulation, and various kinds of shared spaces.

These success drivers of self-made city projects result in five key qualities that all projects demonstrate. The examples in this chapter explain all of these qualities.

1. **Open spaces**: Green, open, and community spaces are vital to good neighborhoods. Spaces in which people can come together, or meet by chance, help to foster a sense of community identity. Identifying with a community or neighborhood encourages people to take responsibility for the place where they live. In Berlin, for example, every Baugruppe project has a garden that is not only shared among all house members, but also is often open to the public, so the entire neighborhood profits from the green and surrounding urban spaces. The possibility to, for example, garden or feel at home in an outdoor space is essential, and this can also happen among neighbors with the advantage of sharing responsibilities.

2. **Shared amenities**: Common shared spaces—such as rooftop terraces, playrooms, guest rooms, common kitchens, party spaces, or even saunas—are amenities that are made feasible by sharing costs. They also help to bring residents and neighbors together. The advantages of living closer to others and the higher quality of life that is made possible

becomes tangible through such amenities. More and more often, young families see the advantages of sharing childcare responsibilities within a house community and in keeping social ties that they would give up if they moved outside of the city.[2] The same is true for older people, who can escape the isolation of former family dwellings in a suitable form of dwelling and living in close proximity to the social infrastructure that self-made projects can create.

3. **Mixed uses**: The vitality of urban spaces is determined by the mix of uses in each building and within a neighborhood. A good mix of uses fuels interaction and ultimately decreases the need to commute, which contributes to micro economies in a neighborhood and increases the quality of life. Interaction is encouraged when spatial connections are made between a building and the surrounding environment; particularly when the ground floor zone creates interaction spaces with certain uses, and when house residents engage in the community through festivities or common activities. Self-initiated projects most often create a great mix of spaces with uses that, particularly on the ground floors, are public and add to the urban quality of the neighborhood.

4. **Customization**: Self-made projects are personalized solutions providing spaces that can be adapted to suit changing needs over time, and that allow people with specific needs to find a place in the city. For example, these spaces can allow multi-generational living, barrier-free standards, or an environmentally aware way of life. Therefore, these projects can accommodate living situations in urban areas that do not exist on the normal real estate market. Clever solutions for flexibility show how buildings can be made adaptable over time to accommodate different uses and economic situations, allowing spaces to be divided off and rented-out or even used in completely different ways. Projects also facilitate a growing tendency toward co-working spaces and offices in the lower floors, or with a mix of small-scale commercial spaces that contribute to the diversity of the urban landscape. They also allow for novel floor plans. In Berlin, nearly every Baugruppe apartment chose to have an open kitchen, when the conventional real estate market offered few to no apartments of this configuration. The self-initiated process often raises the bar on ecological systems and building standards. Architects, tenants, and owners willingly explore new green technology, carefully balancing its advantages and disadvantages. Many resulting projects are leading the way in environmental sustainability by employing, for example, high-rise timber construction and passive house design. Several different multistory wooden construction solutions are now certified in Germany as a result of Baugruppe experimentation.

---

[2] In Berlin, on average, 48% of the people living in Baugruppe projects (constructed between 2004 and 2013) are families with children, as compared to only 6.4% of families living in apartments in the city as a whole.

5. **Lower cost**: Long-term affordability helps to create stable neighborhoods. In self-initiated projects, the future users (or owners) decide what to invest in and where money can best be saved, redefining the quality-to-price ratio. On average, Baugruppe projects in Berlin are 20 per cent cheaper than conventional projects of similar building standards. The savings are made mainly by eliminating the developer's profit, but this does mean that the group needs to assume the developer's role. The significant effort involved in creating these projects means that people want to reap the benefits by staying there for the longer term. This has led to a very low turnover rate, adding to social stability and, to an extent, decreasing the spiraling cost of urban living. Finally, alternative models for financing and ownership have offered a new level of long-term affordability within a non-profit ideology.

SELF-MADE AS A MODEL

With hundreds of recently completed examples, the self-made city has matured into a fully-fledged alternative development model. However, its solutions have gradually become more complex on every level, as have the goals and diversity of the inhabitants. By its nature, each self-initiated project represents a very specific, tailor-made solution. Nevertheless, there are common self-made city strategies and elements. The following section outlines how these strategies and elements can be transferred to projects of every size in cities across the globe. First, we discuss the legal and property definition of self-made projects, which allow initiators to secure land and realize the project. Subsequently, we demonstrate how the goals of self-initiators will determine all other aspects of the project—what motivation the project initiators have, who they are, and what their specific needs are. This includes, for example, developing a concept for a mix of uses and diversity of users, as well as architectural solutions for different ways of life and the idea of community. Meeting these goals ensures that the five qualities of self-initiated projects drive sustainable urbanism.

Legal Forms of Bottom-Up Building Initiatives

Creating more suitable living situations with long-term affordability is usually the main motivation behind self-initiated projects. The most trusted and well-known legal or organizational forms for alternative development are the co-op association, co-housing, and the Baugruppe. Co-op associations can accommodate a broad range of people and viewpoints, and the nature of the projects may change as members of the group change. The co-housing and Baugruppe models have grown over the past years by focusing on commonalities; they tend to be made up of people with similar backgrounds and mindsets and they often have a special social focus. These groups are especially novel due to their focus on sharing spaces and amenities.

The tendency to build shared spaces has grown significantly since the Great Recession of 2008. There was a change in the way many citizens viewed personal and shared spaces. With an increase in the kinds and amounts of shared spaces, the average size of apartments has shrunk somewhat so owners want to reduce the number of things they own and the space they personally need to take care of.[3] Co-housing and Baugruppe projects respond to this trend. Definitions may vary from country to country, but co-housing and the Baugruppe are mainly understood as owner-occupied developments in which each member of the group owns their own individual dwelling, as well as a percentage of the shared spaces in common ownership.

Co-housing projects tend to focus on sharing a common house and kitchen, and occupants share the preparation of daily meals. Tenants tend to form an organization prior to the commencement of development in which interested parties invest, which in turn pays for the construction costs. This investment entitles members to the ownership of their own dwelling after the construction is completed. The Baugruppe model also includes shared spaces but tends to be more pragmatic about their nature. Baugruppe projects are generally complete, separate dwelling units that are simply stacked as a multi-story building in an urban context. The Baugruppe also organize before construction, but as a simple legal association of the parties. Both co-housing and Baugruppe models allow the owners to sell their property for normal market value, provided that the group has not made any other agreements. Conversely, in a co-op association, the association is the legal entity that owns the real estate and members of the association buy shares, usually in direct relation to the size of the apartment they will occupy. Additionally, a monthly rent is paid to the association, which is usually low compared to market rates, and the members participate in operating and managing the building. When selling, it is the association shares that are sold and, normally, the member is not entitled to any financial gain, since associations are commonly non-profit and have the ideology of maintaining low prices. In many countries developers have worked with co-housing and Baugruppe organizations to complete projects. This can be effective in terms of simplifying organizational and legal work, particularly at the beginning of the project when securing the site. However, since the developer earns a percentage for their investment, the cost savings of the project diminish (Kleilein et al. 2011).[4]

---

[3]Accompanying the Urban Living exhibition at the DAZ German Architecture Center in Berlin in June 2015, a series of discussions looked at these themes. For more information, see http://aa-projects.eu/de/urban-living-y-table-talks-at-the-dazurban-living-y-table-talks-at-the-daz/.

[4]Developers are responding to the self-made trend. The Nightingale model from Australia is a fixed-profit, ecological, developer: http://nightingalehousing.org. In Switzerland, the Stiftung Edith Maryon develops group housing: http://www.maryon.ch.

Land

The most difficult point in a self-initiated project is the purchase of buildable land. Groups are often not quick enough to be able to compete with professional investors in desirable urban markets, particularly if each member needs to have a separate bank loan in order to buy the site. Especially in the negotiation phase, investors can make far quicker decisions than groups. However, if the group pays cash for the land, already having an agreement with a bank to use it as collateral for the further financing of the project, the site procurement can be reasonably quick.[5]

Common (and currently becoming more popular) in many European countries, is the long-term land lease. Especially in cities where real estate is very expensive, land lease can be an effective tool in facilitating affordable housing, socially oriented projects, and more suitable mid-market housing. The city, or private owners, can retain ownership of the land and lease the right to build on the property over the period of, for example, 100 years. This has the advantage of reducing the group's initial cost of investment by spreading land payments over a long period. The land owner retains ownership, future control of the site, and collects continual revenue. A city, in this way, can connect the distribution of land use to achieving goals that it would not be able to with normal investors and by selling the land. For example, projects that offer benefits to the surrounding neighborhood, such as multiuse spaces, open park spaces, or even facilities that are accessible to a wider public. Also, in retaining ownership, the city not only has more flexibility in the future use of the site, but it also retains the capital for long-term gain as opposed to short-term profit. This is a good way for a city to provide financial support to key projects. In cities such a Hamburg, Vienna, and Munich, when city-owned sites are developed, at least 20% of the land is designated for Baugruppen and further percentages for other kinds of socially oriented projects. Groups often compete in an investor competition that is not just based on money, but also on the quality of the architectural and ecological design and on their contribution to the surrounding community.

THREE SELF-MADE EXAMPLES FROM BERLIN

The following three built examples show the potential for self-initiated, urban, residential construction to improve the quality of urban ways of life. They demonstrate how self-made city projects achieve the five qualities of sustainable urban development. They illustrate the three drivers for self-initiated housing: offering new typologies and forms of urban living; contributions to the community and improved neighborhood qualities; improving the urban environment through densification. Furthermore, the examples show legal

---

[5] This necessitates that all members use the same bank. In urban environments, the cost of land usually makes up around 20–30% of the cost of the project.

and conceptual processes for self-made city development and a comparative analysis of the resulting mixed-use and shared spaces. The range of projects shows how the model has been scaled up to enable large developments.

Oderberger Mix

In which hybrid, flexible architecture supports a vital mix, activating the urban environment (Figs. 9.1 and 9.2).

- Initiator, development concept, architects: Antje Buchholz, Jack Burnett-Stuart, Michael von Matuschka und Jürgen Patzak-Poor—BAR Architekten
- Legal form: Modified Baugruppe. 50% rental, 50% owner occupied
- Project timeline: site purchase (auction) 2007; start 2007; completion 2010

Goals and Model

The Oderberger Mix project shows how economic uncertainty can lead to a solution that offers adaptability and financial security. In 2007, this site was up for auction, next to a decommissioned and unused former city swimming pool, in an area where few ground floor spaces were rented, most stood empty or were underused. The BAR architects' idea was maximum adaptability; to be able to use the building in different ways, depending on what would be feasible. Correspondingly, the goal was to foster a mix of different functions and people in the building that would add to the diversity of the neighborhood. The architects bought the site and then decided to use a 50/50 model of ownership and rental: individual living spaces are privately owned; the mixed-use bottom floors remain in shared ownership and are rented to third parties.

Densification

Within the form of a typical perimeter-block building, a sophisticated puzzle of varying volumes allows for flexibility of space and use, both now and in the future—a mix of use, volume, area, and financing models.

Neighborhood Qualities

A ground floor restaurant, shop, and an art space activate the street frontage and, as popular neighborhood venues, strengthen social networks. The art space is non-commercial and open to the public. The mix of uses within the building is analogous to the surrounding city, contributing to its urbanity.

Forms of Living

Four of the five apartments can be divided up, with a 40 m² module that can be divided off, to make potentially nine separate apartments. A modular

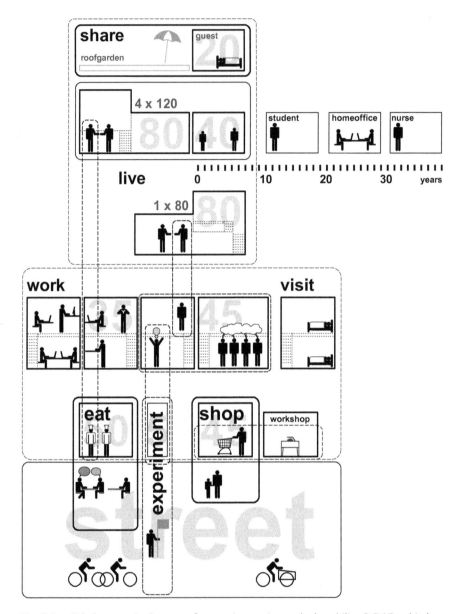

Fig. 9.1 Oderberger mix diagram of spaces, interaction and adaptability © BARarchitekten

system of various volumes and uses allows for future flexible spatial arrangements, including live-work spaces. On the first floor, double height studios can be used as offices or short-stay apartments. A guest apartment is located on the shared roof terrace.

The project utilizes ecological building systems and high-quality, energy-saving construction, such as highly insulated timber panel elements, a wood

Fig. 9.2 Oderberger mix interior of live-work space © Andrea Kroth

pellet central heating system, a decentralized ventilation system with heat recovery, and a soakaway for roof drainage.

Individual and Shared Spaces
Individual spaces: 5–9 units + 5 studios + 1 bar or restaurant + 1 commercial space + 1 music studio
  Shared spaces: 1 public art space + garden + roof-top terrace with guest apartment
  Total common space: 155 m² + 100 m² shared courtyard garden

Big Yard

A clever planning solution for a difficult infill site provides alternative urban housing typologies and a convivial atmosphere within the shared courtyard garden.

- Initiator, development concept, architects: zanderroth architekten gmbh, Berlin
- Legal form: Baugruppe (Civil Law Association)
- Project timeline: start 2006; site purchase 2007; completion 2010

Fig. 9.3 Big Yard development by Zanderroth Architects, a family-oriented living community © Simon Menges

Goals and Model

The Big Yard project offers a model for families who otherwise would not be able to find adequate living space in the city. The architects took the initiative to create the design for this site before a Baugruppe association had been formed. Advertising the design on web platforms for real estate and those for Baugruppen, they were able to gather together enough interested parties to commonly finance the purchase of the site. Finally, each unit was customized with the owners, but the architects remained in control of the overall design. This design has won several prestigious architectural awards, such as the BDA Preis Berlin 2013 and the Deutschen Architekturpreis Award 2011. After completion, all titles were separated and the group formed an owners' association. This project has given 72 parties affordable but high-quality living within a tight-knit community (Fig. 9.3).

Densification

The project provides urban infill in an innovative urban form, providing high amenity, family-sized apartments within the inner city. Despite it being a north-facing site, the architects managed to design optimal day-lit apartments by raising the height of the buildings to the north. The courtyard is an enclosed space that is particularly attractive as a safe place for children to play.

Neighborhood Qualities

At street level there are individual entrances to the four-story townhouses. Living spaces also have direct access to the large, shared, central garden, which creates the heart of the community. Alcove areas allow for individual privacy.

Forms of Living

Three different architectural typologies, all designed around open-plan living spaces, each with a private terrace space, offer high-quality living. There are four-story townhouses to the street, and a rear tower that contains both split-level garden houses and penthouses. The townhouses contain three bedrooms, a workspace at street level, a terrace to the shared garden, a balcony, and a private roof terrace. The garden houses have two bedrooms, a workspace and a living space with direct access to the garden. The penthouses have four bedrooms, a terrace off the living space, and direct access to a shared roof garden.

The project utilizes low energy-use building construction, eco-friendly materials and district heating.

Individual and Shared Spaces

Individual spaces: 45 units with live–work spaces

Shared spaces: Common amenities include a roof terrace, shared garden, summer kitchen, sauna, and a guest apartment.

R50—Co-housing

This project explores the spatial translation of private and common needs and has a high potential for fostering urban interaction and spatial variety.

- Initiator, development concept, architects: ifau und Jesko Fezer | HEIDE & VON BECKERATH
- Legal form: Baugruppe (Civil Law Association)
- Project timeline: site purchase (investor competition) 2010; start 2011; completion 2013

Goals and Model

An architect group initiated this project to apply for a building site being offered by the Berlin Senate in an investor competition. The fixed price site was awarded based on the quality of the design and the overall concept. The group proposed to open the garden surrounding the building to the public, and to establish a large shared space that could be used (at a nominal rent)

Fig. 9.4   R50—Co-housing development, an incubator for improving urban interaction and qualities in the neighborhood © AndreaKroth

by the neighborhood for events. In this way, the Senate could be assured that the neighborhood would accept the new building, which was to be a densification of a post-war housing estate. The initiators documented their complex decision-making process, illustrating the importance of this process in meeting their goal of building high-quality apartments that remained affordable. During the process, the group asked themselves what kind of spaces and activities they might be able to share, if they might make their individual apartments smaller by sharing more, or if they could save money by, for example, choosing all the same standard finishes and bathroom fixtures. The group shares many flexible spaces in the building, as well as the wraparound balconies and a roof terrace. If all of these spaces are counted as shared spaces, they account for 22% of the usable area of the building. After completion of the building, the titles were divided and an owners' association was formed (Fig. 9.4).

Densification
The solitary block creates a new, denser relationship within a post-war, modernist, housing estate context. The building re-establishes a physical urban

street front and programmatically addresses the street with commercial and shared spaces on the ground floor.

Neighborhood Qualities

The community-shared space on the ground floor of the building can serve local initiatives as a meeting and event space. Existing green spaces and the new garden merge, allowing neighbors to travel through while providing new activity spaces.

Forms of Living

The layout of each apartment was highly individualized to meet each family's preferences and needs, while their lightweight interior walls also make them adaptable. An integrated design approach allows for cost-efficient building elements, such as a modular facade. A catalogue of simple, but flexible, finishing standards was developed together with the group. A variety of shared spaces are provided, including a community room, garden, roof terrace and walkways, which wrap around each building level and also serve as balconies.

Individual and Shared Spaces

Individual spaces: 19 units (15 families, total of 64 inhabitants)

Shared spaces: a community multi-use space, guest space, workshop and laundry, roof terrace, continuous exterior walkway around each level, a mixed-use-studio, office space

Total common space: 600 m^2 + shared garden 1483 m^2

IMPACT OF SELF-MADE PROJECTS

Over just a decade, self-initiated projects in Germany matured from small urban infill buildings with just eight or ten apartments to projects with more than 400 living units that encompass entire new urban quarters. Self-initiated projects have become more than a niche market; in 2010 in Berlin these made up 12% of all construction activity (Ring 2013: 206–207). Several cities, such as Hamburg, Munich, and Freiburg, support Baugruppe, co-housing, and co-op associations by offering building sites at a fixed price in investor competitions. These and many other cities have recognized the qualities that can be achieved by self-initiated and owner-occupied developments and use the Baugruppe projects strategically to help in the development of new quarters. Currently, citizens and cities are recognizing that land in the urban core must be used wisely, with long-term social and financial goals in mind, and that the people who live there must have a vested interest in the place—as opposed to too much luxury development, which so often leads to apartments that stand empty most of the year.

Even outside of Germany, an increasing number of cities are recognizing the value of self-initiated projects and are re-establishing policies that facilitate collective building. In England, for example, the national government requires each relevant public authority to keep a register of individuals and associations in their area who are seeking to acquire serviced plots of land for self-build and custom projects (Government 2016). In Helsinki, Finland, most of buildable land is still owned by the city itself, which is now piloting a self-build project. In Perth, Australia, a new, self-initiated, development process is modeled after—and even called—the Baugruppe, and two new pilot projects are currently being planned—with a landholder—that follow high sustainability goals.[6] In the Netherlands, the self-initiated model is called CPC (collective private commissioning, or CPO in Dutch) and legally enables a group to act as a developer (Housing 2013; Vergunst and Peborde 2016). Recent successful projects have been completed in Amsterdam, Rotterdam, Almere, and Enschede, to name just a few. Austria and Switzerland also have a long and ongoing tradition of high-quality, co-op association, housing projects. Ultimately, self-made projects have the potential to increase the quality, affordability, and viability of urban living, while contributing to the sustainability of the city. Citizens can go from consumer to pioneer by driving new designs for living and ways of life.

References

Berlin, S. f. S. (2007). *Urban pioneers: Berlin: Stadtentwicklung durch Zwischennutzung; temporary use and urban development.* Berlin: Jovis.

Bloomfield, R. (2017). *London's affordability crisis.* Retrieved from https://www.homesandproperty.co.uk/property-news/londons-affordability-crisis-after-years-of-overwhelming-house-price-hikes-buyers-will-welcome-news-a110641.html.

d'Alençon, P. A., Bauerfeind, B., & Konrad, D. (2017). *Handbuch und Planungshilfe—Ephemere Stadtentwicklung.* Berlin: DOM Publishers.

Davis, S. (2016). The housing affordability crisis: Can it be solved? *Berkely Blog.* Retrieved from http://blogs.berkeley.edu/2016/06/27/the-housing-affordability-crisis-can-it-be-solved/.

Demographia. (2017). *13th Annual demographia international housing affordability survey: 2017.* Belleville, IL: Demographia.

Dürr, S., & Kuhn, G. (2017). *Wohnvielfalt. Gemeinschaftlich wohnen—im Quarter vernetzt und sozial orientiert.* Ludwigsburg: Wüstenrot Stiftung in Kraemerverlag.

Förster, W., Menking, W., Bitter, S., & Aedes am, P. (2016). *Das Wiener Modell: Wohnbau für die Stadt des 21. Jahrhunderts= The Vienna model: Housing for the twenty-first-century city.* Berlin: Jovis.

Government, D. f. C. a. L. (2016). *Self-build and Custom Housebuilding Act 2015 (as amended by the Housing and Planning Act 2016).* London: United Kingdom Government Legislation.

---

[6]For more information on Australian Baugruppen, see https://baugruppen.com.au/index.html and https://www.landcorp.com.au/Our-Work/Innovation-Through-Demonstration/Baugruppen/.

Housing, D. A. S. o. (2013). *Samen Bouwen/Building together*. Rotterdam: nai010 uitgevers.

Kleilein, D., Ballhausen, N., & Geipel, K. (2011). Wir Reden Über Grund und Boden, Weil Wir Darum Gekämpft Haben. *Bauwelt, 102*, 8–15.

Krämer, S., & Kuhn, G. (2009). *Städte und Baugemeinschaften*. Stuttgart and Zurich: Krämer.

Lafond, M., & Honeck, T. (2012). *CoHousing cultures. Handbook for self-organized, community-oriented and sustainable housing: Id22*. Berlin: Jovis-Verlag.

McCamant, K., & Durrett, C. (2011). *Creating cohousing: Building sustainable communities*. Gabriola: New Society Publishers.

Montgomery, C. (2013). *Happy city: Transforming our lives through urban design*. New York: Macmillan.

Nations, U. (2014). World urbanization prospects: The 2014 revision, highlights. Department of Economic and Social Affairs. *Population Division, United Nations*.

Ring, K. (2013). *Self-made city—Berlin: Self-initiated urban living and architectural interventions*. Berlin: Jovis.

Ring, K. (2015). *Urban living: Strategies for the future, Berlin*. Berlin: Jovis.

Robertson, J. (2017, May 2). 'Eye-watering prices': Australia's housing affordability crisis laid bare. *The Guardian*.

Sonne, W. (2017). *Urbanity and density in 20th century urban design*. Berlin: DOM Publishers.

Vergunst, M., & Peborde, I. (2016). Co-housing in Europe #2: Vrijburcht, Amsterdam. *New Europe—Cities in transition*. Retrieved from https://citiesintransition. eu/interview/co-housing-in-europe-2-vrijburcht-amsterdam.

Wohnen, S. f. S. u. (2012). *IBA 2020—Dokumentation der Vorarbeiten*. Berlin: Senatsverwaltung of Berlin.

Ziehl, M., Osswald, S., Hasemann, O., & Schnier, D. (2012). *Second hand spaces: Recycling sites undergoing urban transformation*. Berlin: Jovis.

CHAPTER 10

Informal Place-Making: Mosques, Muslims, and Urban Innovation in Germany

Petra Kuppinger

December 29, 2016. On this mild winter afternoon I headed to the Salam Mosque complex.[1] For more than ten years, I have regularly visited this vibrant spiritual, social, and commercial center and never cease to be surprised at its continuously changing Turkish–German commercial landscape. It was a busy Thursday and a convenient time for shopping or family outings, as many people do not work between Christmas and New Year. The license plates of parked cars indicated visitors from a radius of about 100 kilometers, including cars from neighboring France. I went to the Sultan Restaurant, a restaurant and shisha lounge that had recently opened. Two floors of their building accommodate a sizable restaurant and another (with a separate entrance) is a shisha lounge. The interior was neatly renovated and combined shiny, state of the art, light fixtures with austere dining tables, colorful velvet armchairs at coffee tables by the windows, and Ottoman-style wall paintings depicting hospitality scenes. The image of a life-size man with a turban pouring coffee was facing me. There were shelves with decorative coffee pots. The windows included shelves with potted plants. Much thought, energy, and money had been invested in the decoration of this formerly bland store. About 70 patrons were seated in the first floor, and the basement (the same size as the first floor) accommodated another 40 visitors (I did not venture into the shisha lounge). Guests included young and old, families, couples, and groups

---

[1] All places and personal names, unless otherwise indicated, are pseudonyms.

P. Kuppinger (✉)
Monmouth College, Monmouth, IL, USA

© The Author(s) 2019
M. Arefi and C. Kickert (eds.), *The Palgrave Handbook of Bottom-Up Urbanism*, https://doi.org/10.1007/978-3-319-90131-2_10

of friends. There were women with headscarves and women without. Almost all guests were Turkish-Germans. In front of me sat a young family with a toddler enjoying dinner. Close by was a fashionably dressed forty-something couple. Further down, two tables were pushed together to accommodate a large, three-generation, family gathering.

The Sultan Restaurant, the Salam Mosque complex, and the neighboring businesses are located in an industrial quarter in Stuttgart. Since the early 1990s the Salam Mosque and a steadily growing assortment of businesses within its compound and in adjacent buildings have dramatically altered the social, cultural, and aesthetic characteristics of this previously nondescript manufacturing street. Located in a former industrial facility, the mosque—the undisputed center of this vibrant landscape—is not recognizable as such. One enters the mosque and community complex through a large metal gate that bears witness to the premises' industrial past. Beyond this gate (marked by three flag poles flying the Turkish, mosque association, and German flags) grouped around a yard are the men's and women's prayer rooms and the upstairs community facilities. To the rear of the yard is a restaurant. Outside and to the left and right of the front gate are a travel agent, book store, insurance agent, undertaker, jewelry store, a housewares store from which colorful merchandise spills onto the sidewalk, and a modest Islamic fashion shop. A large bakery and restaurant protrude into the sidewalk, and a supermarket is set back slightly from the street (the jagged front bespeaks the piecemeal expansion of this former industrial facility). To the rear of the mosque complex (it occupies the width of a block) are another restaurant, a barber shop, and a supermarket. All these businesses rent their premises from the mosque association. Not part of the mosque complex, but immediately south of it, is another store of housewares, jewelry, and carpets, sporting a tent-like structure by its side entrance where bargain dishes, suitcases, blankets, and clothes are exhibited. Finally, there is a row of stores (perpendicular to the street) including another furniture and housewares store, two modest Islamic women's fashion stores, and a bridal outfitter. The Sultan Restaurant occupies the end of this row. Other businesses are north of the mosque.

On evenings, weekends, and holidays, the two city blocks around the mosque complex attract thousands of visitors and transform into a lively social, cultural, economic, and spiritual landscape. Mostly Turkish–Germans, but also ethnic Germans, other Muslims, and non-Muslims of diverse backgrounds rub shoulders. Old and young visitors, visibly pious or not, enjoy different aspects of the mosque complex and its environs. On weekends, individuals and families come to pray, socialize, eat, attend communal meetings or public events, do their grocery shopping, buy Islamic fashion, have a family meal, buy items to furnish the apartments of newly-weds, or to watch soccer matches of the Turkish league. Weddings are celebrated in the communal hall. On weekdays, some employees of the neighboring industries eat their lunches at the various eateries. Women come to learn Qur'an recitation, socialize, or occasionally prepare food together in the mosque's women's quarters. In the mosque's yard male retirees socialize.

Looking at this spatiality, a number of questions arise: How did this pulsating space emerge? Was this a planned transformation? What ideas guided the development of this vibrant and successful social, cultural, spiritual, and economic spatiality? Which elements and processes mediated the transformation of this space from a socially inhospitable to a popular space? What role does religion play in this spatiality? This chapter examines the Salam Mosque complex and its creative place-making. It explores how changes in the material and cultural environment of this complex reflect creative interactions between urban spatial features and regulations, religious rules and sensitivities, ethnic preferences, and commercial considerations. It illustrates how spatial, religious, cultural, and commercial dynamics interact in the complex. These processes have produced a unique, faith-inspired spatiality. Without blueprints, urban planners, or business consultants, the mosque community, its board of directors, entrepreneurs, visitors, customers, and worshippers have created a largely unplanned spatiality that attracts many visitors.

Based on ongoing ethnographic fieldwork in Stuttgart since 2006, this chapter analyzes cultural innovations and creative place-making at the Salam Mosque complex and debates urban vernacular creativity. It explores the neglected role of minority religions in spatial and cultural processes in secular cities; situates religion in the realm of urban innovation; and analyzes informal Muslim cultural negotiation as a dynamic that (re)shapes spaces, civic life, and culture. I argue that to understand urban transformations in global cities, it is necessary to look beyond glitzy downtowns and to explore invisible vernacular spatialities and place-making activities. I demonstrate the relevance of faith-inspired place-making in contemporary cities.

INFORMAL PLACE-MAKING

The use of cultural anchors has been central in plans of urban renewal. Museums, historical streets or markets, and convention centers are said to attract businesses and further residential development. If successful, culture-centered projects are projected to increase surrounding real estate values. For less prominent cities, such projects promise visibility in the cut-throat competition of global cities, and the resulting recognition is profitable for cities and entrepreneurs. Poster children of such planning feats include New York's South Street Seaport and the Guggenheim Museum in Bilbao (e.g., Sorkin 1992; Harvey 1989). However, some of these projects have failed, as Michael Moore (1989) illustrated for "Autoworld," the desperate attempt of Flint, Michigan, to revive its downtown with a car museum in the wake of industrial decline. To be profitable cultural anchors need to sustain the flow of wealthy residents, visitors, and investors. Related processes of gentrification have been widely debated and frequently criticized (Zukin 1989; Abu-Lughod 1994;

Lloyd 2006). Less has been said about the creative role of vernacular cultures and the informal contributions of ordinary residents in recent urban transformations. Equally sparse attention has been paid to the role of religion in vernacular urban place-making and the resulting larger transformations. The role of houses of worship as potential anchors for urban change has rarely been discussed in secular urban contexts.

Starting in the 1990s, scholars explored vernacular cultural and spatial creativities (Zukin 2010; Gibson 2012; Lippard 1997; Brandt 1994; Hallam and Ingold 2007). They examined the transformative potential of ordinary people's creative activities and challenged dominant discourses and definitions of art and creativity that are often contemptuous of popular aesthetics and creativity (Chappell 2012; Mattausch and Yildiz 2009; Gibson and Connell 2011). They questioned the central focus on profitability, and, instead, emphasized quality of life or livability concerns for all residents, including lower class, marginalized, and immigrant communities.

Informal place-making includes individuals' efforts to remake or embellish homes for their own or their neighbors' aesthetic pleasure (Edensor and Millington 2009, 2010). Roberto Bedoya examines the "Chicano practice of Rasquachification," in which homeowners decorate their houses in bright colors and (re)use mundane artifacts (2014: 1). Reflecting on the working-class barrio in San Francisco where he grew up, Bedoya argues that the colorful brightness in Rasquache represents "an aesthetic of intensity that confronted our invisibility, our treatment of *less than*" (ibid.: 2, emphasis in original). Becoming visible is part of a quest for social and spatial justice as marginalized groups claim urban spaces. Frequently perceived as a threat to mainstream society, these marginalized groups offer their aesthetic forms in part to "provide a counterframe to gentrification" and dominant spatial imaginaries (ibid.: 3). Bedoya explains that Rasquache practices are "rooted in resourcefulness and adaptability" and have the "capacity to hold life together with bits of string, old coffee cans, and broken mirrors in a dazzling gesture of aesthetic bravado" (ibid.). It is an aesthetic of "making something of nothing, of the discarded, irreverent and spontaneous" (ibid.: 5), which can produce "ultravisibility" that challenges dominant aesthetics and processes of exclusion. Bedoya highlights the "creative resilience found in communities of color" (ibid.: 1) and emphasizes their "art of *poiesis*, or making in the sense of transforming the world" (ibid.: 4, emphasis in original). This creative place-making constitutes organic processes that are not guided by profitability, but rather represent people's quests to embellish and claim spaces in the city.

Examining interactions between the "mundane and creativity" in community gardening in the UK, David Crouch argues that creativity "emerges in cultural improvisation" and reflects "the experience or practice of doing" (2010: 132). When faced with diverse places, people, circumstances, and possibilities, individuals and communities seize unique moments and creatively respond to them. Thus, "creativity can happen unexpectedly in the everyday, the ordinary and mundane, because each of these is open to accident, variety, disruption and change" (ibid.: 133). Treading a complex territory between the old/conventional and the new, innovative and exciting—between established ways and risky experiments—people improvise in unexpected ways. Their improvisations often remain invisible and ephemeral. Yet some take hold, draw more participants and resources into their orbit, and visibly transform spaces. The results of such creativity are frequently overlooked, because they do not fit official urban designs. They might showcase un-hip and unprofitable aesthetics and challenge "the commonly presumed distinction between 'art-creativity' and the emergent creativities performed in everyday life" (ibid.: 140).

Vernacular creativity can radiate beyond its immediate context. Paul Milbourne examines community gardening in marginalized quarters and demonstrates that "vernacular forms of creativity [...] are contributing to the reinvention and, in some cases the regeneration of these places" (2010: 142). They remake spaces and "create new forms of sociality and conviviality" (ibid.). They might not produce hip aesthetic formations, upscale social and cultural venues, or profitable spaces, but they appeal to their stakeholders' sense of beauty and to social needs. Milbourne analyzes a back-alley gating scheme in metropolitan Manchester, where after the first alley was gated off, one woman embellished her alley space with potted plants. Others followed and added more plants. Soon, nearby alleys copied these efforts, which helped in the regeneration of the quarter as residents transformed bleak alleys into welcoming green spaces where neighbors socialized. These examples illustrate that informal activities and creative projects transcend small spaces and connect with similar processes elsewhere in the city.

FAITH-INSPIRED PLACE-MAKING

The continued relevance of religion in contemporary cities has been widely debated (Orsi 1985, 1999; Livezey 2000). Scholars highlight the role of immigrant and minority religions (Stepick et al. 2009; Warner and Wittner 1998). When recognizing the urban engagement of faith-based communities, it is important to understand whether such groups only negotiate spaces for themselves or if their activities affect their urban environs; how faith-inspired immigrant spaces interact with spatialities beyond their houses of worship; and how faith-based spaces position themselves vis-à-vis dominant spatial discourses, politics, and aesthetics.

Much has been written about recent struggles to construct mosques in European cities (Cesari 2005; Tamimi Arab 2013; Kuppinger 2015: 34). Less has been said about place-making and the creation of successful Muslim spatialities once a mosque exists (Idriz 2010; Kuppinger 2017). How do Muslims and other minority religious groups create spaces in cities? Chantal Saint-Blancat and Adriano Cancellieri (2014) examine the impact of a religious procession of Filipino immigrants on both the local Catholic community and the city of Padua. They illustrate how accessibility, temporary appropriation, and visibility are relevant elements in a groups' quest for belonging. Claiming space and visibility constitute important elements of empowerment for "socially invisible actors" (ibid.: 647). Thomas Tweed (1997) explores the link between religion, identity, and place for diasporic communities. Examining the construction of a Cuban–American shrine in Miami, he notes that immigrants often "map the landscape and history of the homeland onto the new urban environment" (ibid.: 10). Tweed observes that many exiles turn to religion in search of stability in their new cities. Through their houses of worship, they spatialize their belonging to their old and new homes. Houses of worship anchor new arrivals in cities. Dwyer et al. (2016) analyze the impact of the highly visible presence of immigrant faith communities on what became known as the "highway to heaven" in suburban Vancouver. A stretch of suburban highway, designated for use by religious assemblies, ended up as the home for more than a dozen diverse faith communities, which dramatically changed the architecture of the environs and allowed for new forms of cooperation between religious groups.

Whether by claiming spaces momentarily and becoming visible in the city, or by building sizable houses of worship, faith-inspired activities and edifices illustrate transformative processes triggered by religious constituencies. Often starting in backyards and hidden spaces, immigrant faith communities, through their increasing engagement, visible events, and more noticeable spaces, insert themselves into cityscapes. Their faith-inspired activities are increasingly seen (Saint-Blancat and Cancellieri 2014) and heard (Garbin 2012; David 2012). Individuals become visible as parts of a larger, faith-inspired constituency (Tarlo 2010) and slowly settle into their own spatialities (Shah et al. 2012; Dwyer et al. 2012). As faith-based communities negotiate urban participation and material spaces, there is little to guide their steps. Especially for groups who are the first of their faith community, these negotiations are journeys of trial and error. Informal processes of localization are framed by multilayered interactions within communities and with larger political, historical, legal, cultural, and social dynamics. A community's financial circumstances also play an important role. In long-term processes, without help from mainstream society and its institutions, many immigrant faith communities become engaged urban constituencies. They rent or buy spaces, invest in those spaces, make them comfortable for insiders and hospitable to outsiders. They become creative actors who (re)make spatialities.

THE SALAM MOSQUE COMPLEX

Traveling north from downtown Stuttgart, the busy urban highway eventually makes a slow descent. Buildings that line this long slope have dramatically changed in recent decades. Once home to small factories and workshops, this stretch is now lined by glitzy car showrooms, office buildings, a hotel, a car wash, an electronics market, a McDonald's, and a discotheque/restaurant. Few buildings are more than 20-years-old, and the construction frenzy continues to swallow up older structures. Turning left at the bottom of the showroom mile, one enters an industrial district that is similarly in flux. In 1992, a mosque association bought a defunct industrial complex in this quarter. After renovations, part of this compound became the Salam Mosque and community center, which opened in 1993.[2] The center includes spacious upstairs community facilities with offices, a large multipurpose hall, meeting rooms, and a sizable women's section with its own kitchen, office, meeting rooms, and large prayer/meeting room (in addition to the downstairs women's prayer room). The prayer rooms and community facilities use about 40% of the complex. The rest is rented to commercial tenants. Looking at the dynamic commercial landscape that surrounds the mosque today, it is hard to imagine that the mosque association struggled in the 1990s to convince entrepreneurs to rent premises in the site.[3] Located in a gray industrial district that was in the midst of industrial restructuring in the 1990s, with no residential quarters nearby, it seemed unlikely that this complex would become an attractive commercial location. Eventually, the mosque association persuaded two entrepreneurs to open a supermarket and a restaurant. The mosque, argued the association, would attract the pious who would frequent these businesses. This model of a mosque surrounded by small businesses was familiar to Turkish visitors and hence seemed feasible, even in a German city.

The mosque association and the new businesses saw the mosque as the anchor that would make businesses viable. They hoped that men who came to pray would buy groceries or eat in the restaurant. They expected that Turkish workers and employees in the surrounding factories might frequent the complex and also recruit non-Turkish and non-Muslim colleagues. In the mid-1990s, this was wishful thinking supported by hope. No feasibility studies were conducted. No media blitz or advertising campaign accompanied the opening of the mosque and the two businesses. It was a trial and error project that vaguely drew on similar experiences in other cities (e.g., Mannheim; Kessner 2004). Initial hopes proved right. The restaurant and supermarket weathered the challenge of building a customer base (both businesses still exist).

Most of the initial customers were men. Many came to pray. Others came to eat and shop close to their jobs. Some elderly men came to socialize around the mosque. A men's clothing shop that offered sweaters, dress pants, and sports jackets, catering to the needs of middle-aged and older Turkish men, opened in the mosque's backyard. Many of these men never felt quite at home in the city, and a store next to the mosque, run by a co-patriot and that

---

[2] I have written elsewhere in more detail about the Salam Mosque as part of the larger mosque-scape in Stuttgart (2010, 2014) and the details of the mosque's complex early transformation processes (2011).

[3] Details on the history of the mosque complex are based on an interview with a mosque board member, June 2007.

catered to their tastes was a perfect solution for their sartorial needs and social sensitivities. A rent-a-garage (where one can fix one's own car, renting the shop's tools) opened close to the mosque's main gate. For years its oil spills and tools decorated the sidewalk. The restaurant, which is divided into two sections, a coffee shop and a restaurant, aimed at an all-male clientele in the coffee shop, where men could watch Turkish soccer league matches. These initiatives were successful in attracting more men to the complex. Businesses prospered and a bakery, undertaker, travel agent, import/household store, and a jewelry store opened in the complex. Soon women started coming in larger numbers, and the first women's clothes store opened in a tiny room in the mosque's backyard.

As the Salam Mosque complex transformed, the local Turkish–German community underwent its own changes. By the late 1990s the first generation retired and many returned to Turkey. Their children and grandchildren developed different social habits and practices, and material needs and tastes (often ignored by mainstream businesses). Simultaneously, the Turkish–German grocery economy changed from small neighborhood stores to larger supermarkets. The local market for *halal* and Turkish ethnic foods consolidated.[4] Such changes were reflected in the constantly changing landscape of the mosque complex. Younger men no longer bought their clothes in the sheltered environment of the mosque. In 2007, the men's store closed and was replaced by the women's fashion shop that had been tucked away in the backyard. A few months later the rent-a-garage closed. Both premises were taken over by stores for women's modest Islamic fashion. In the early 2000s shopping had turned into a women's or family activity that was increasingly tied to leisure or entertainment. The complex's commercial scenery changed from being all-male to one catering for women and families. None of this had been planned. Transformations occurred as the Turkish–German community's lifeworlds evolved.

This changing commercial landscape is best-illustrated with the example of the first housewares store (in the mosque complex). This business had started as an import-export store offering dishes, pots, pans, sheets, towels, and curtains at bargain prices that attracted Turkish–German customers and other bargain hunters. Initially, this shop was housed within bare concrete walls with merchandise displayed on simple shelves. In 2010, the store eliminated bargain offerings and started to focus on more upscale wares, specializing in carpets, curtains, elaborate and pricy china sets, and glittery wall decorations with religious calligraphy. In 2013, the store expanded, took over the (male) billiard café, and considerably embellished its interior and its storefront. Where once plastic plates were sold for one euro, now, fancy, 84-piece china sets are sold for 599 euros. These changes and the proliferation of similar businesses signify established lifeworlds where Turkish-Germans marry and live (in style) in Stuttgart.

----

[4] *Halal* is food permitted by Islamic law.

With more businesses, the Salam Mosque complex attracted an increasingly regional clientele, especially on weekends, when one could see families doing their weekly shopping and eating in restaurants. Some smaller stores closed or changed hands, and more businesses opened adjacent to the complex. By 2016, customers could purchase dishes, carpets, furniture, or curtains in six stores. In addition to changing shopping patterns and habits, social and cultural preferences further transformed the complex. Visitors no longer come only to pray and shop, but to linger and enjoy. This transformation has facilitated the opening and success of more restaurants, bakery/coffee shops, and fast food outlets (*döner kebab*).

The Bakery, a vast facility that sells baked goods and doubles as a coffee-shop and restaurant, is an established business that illustrates changes in the mosque complex and its constituency. The Bakery's space reflects its industrial past—huge and with high ceilings. These premises never seem crowded, even when dozens of people are there shopping and eating. In 2006, the Bakery was a bare industrial hall with little decoration and few comforts. Only one small wall included a painting of an idealized Ottoman peasant scene. It had a long counter to the left, where Turkish fast food and baked goods were on display. Behind the counter, women rolled, formed, and filled dough. Beyond this workspace, there was an extension to the left. The right side of this extension accommodated a station with Turkish breads and rolls. On the left side was a long glass counter with endless trays of sweets (this extension remains unchanged). Overall, the premises felt bare, betraying their unappealing industrial architecture. This feeling was particularly pressing in the winter when, despite the ovens, the hall remained cold and patrons kept their coats on while eating. In 2008, the Bakery started to renovate and reorganize its premises. A second ceiling was constructed to turn the front part (toward the street) into a two-storey, cozier space. Comfortable upholstered chairs and wooden dining tables replaced older furniture. The tables were more evenly distributed and positioned closer together, potted plants were placed between tables and by the windows, and walls were painted and decorated. Most of the (savory) food production was moved to the side opposite the entrance. Large images of food items now decorate the wall above this food counter. The Bakery attracts diverse visitors at different times of the day and week. In the morning, women who have been shopping, alone or in small groups, stop for snacks. On work days between noon and 2 p.m., men of different ethnicities come from nearby factories for lunch. In the afternoon and early evenings, families visit the Bakery. Some customers walk straight to the bread counter at the back, buy bread, and leave. On Saturdays, from the late morning, the Bakery is packed with families, some pushing in their carts from the adjacent supermarket to buy bread or eat a snack. Recently the Bakery has gained a reputation for its *lahmacun* and similar foods.[5]

---

[5] *Lahmacun* is a flat bread topped with a ground beef mix, rolled up with tomatoes and other ingredients.

Often there are long lines of patrons waiting for these foods (the bakery is self-service).

More examples could be told to illustrate the remarkable, creative place-making and cultural production that has unfolded at the Salam Mosque complex since the 1990s. Building regulations limit what the mosque, its commercial tenants, and neighboring businesses can do to their premises. Despite these limitations, the mosque complex and its visitors turned these environs into a unique and unprecedented spatiality. While building exteriors remained largely unchanged, some interiors have changed dramatically (mosque, Sultan Restaurant, Bakery). They became hospitable spaces in which patrons no longer shop quickly, but linger and socialize in comfort. The mosque community, local entrepreneurs, and ordinary people turned this industrial quarter into an extraordinary spatiality that attracts people from across the region.

Informality, Vernacular Creativity, and Successful Spatialities

Unlike its neighboring industrial and commercial facilities, the Salam Mosque complex is not governed by neatly designed plans and fine-tuned guidelines formulated in distant corporate headquarters. Instead the complex is a vibrant vernacular space that evolved using organic creative devices. Without formal planning and lavish resources to cover renovations, the mosque complex and its environs developed into a lively and profitable environment. The mosque's board of directors and courageous entrepreneurs, vaguely guided by similar complexes in Turkey and emerging ones in Germany, took one step at a time and hoped they could turn these premises into a spiritual and commercial center. Their own experiences, needs, tastes, and financial means framed the project. Not surprisingly, these men initially produced a male-dominated landscape that reflected their lifeworlds. They miscalculated what their long-term clientele and its needs would be. Women soon started to dominate as customers. Stores targeting men were replaced by women's fashion stores that benefited from the fact that the sartorial mainstream neglected pious Muslimas. As more women came, housewares and related stores proliferated. Customers traveled from further away, combined shopping with family outings, thus creating a demand for more eateries.

The first generation of Turkish migrants had conducted their lives envisioning their return to Turkey and saved their money toward this goal. Their children and grandchildren departed from the associated frugal lifestyles and spent their money differently as they married and settled locally. Many furnished their homes with locally bought, Turkish-made furniture, carpets, and decorations. The proliferation of furniture stores bespeaks these changes. Since the 1990s the mosque complex and its businesses have continuously adjusted to the changing needs of their Turkish–German constituency.

Stores came and went. Premises were enlarged, renovated, and improved. Stores specialized, and some became pricier. Food stalls, coffee shops, and restaurants turned the complex into a family leisure destination.

The Salam Mosque complex is the result of ongoing social, cultural, spiritual, and economic negotiations. Its aesthetic expressions, cultural artic- ulations, and commercial transformations reflect the ideas and needs of its tenants and visitors. They are mediated by Stuttgart's changing cityscape and its cultural, social, economic, and spiritual transformations. The complex is framed by municipal regulations and the economic calculations of its busi- ness owners. In the absence of corporate feasibility studies, one-size fits all aesthetic designs, or guidelines for profit maximization, the complex's spa- tial, cultural, and commercial aspects are mediated by creative interactions between spaces, people, and religious sensitivities. Success or failure are the result of trial and error and of sensible predictions about developments in and around the complex. They reflect the unfolding project of Muslim Stuttgart.

The complex's success is anchored in the mosque, which frames this spa- tiality, even for those who do not come to pray. The presence of the mosque and the fact that the mosque association is the landlord of some businesses create a vaguely Muslim spatiality. Barbara Metcalf (1996a, b) explains that Muslim space is characterized by shared practices that result from the "ways [...] Muslims interact with one another and with the larger community" (1996b: 2). She identifies "sacred words and normatively enjoined practices as the core of cultural elaboration, transformation and reproduction" of Mus- lim spaces (ibid.: 5). She notes that physical spaces are not the central fea- ture here, rather practices are decisive. Mundane features like "sounds, smells, and practices" or merchandise like *halal* food products or prayer clocks help define Muslim spaces (ibid.: 9). The mosque and its association set param- eters for the commercial spaces. No pork or alcohol are sold. Stores and eateries beyond the complex also situate themselves in this sphere of Mus- lim sensitivities and regulations. In Stuttgart, where many women who wear headscarves face difficulties finding employment, businesses at the complex employ many women with headscarves (and without). The jewelry store sells religious pendants, and the housewares stores offer decorations with religious calligraphy. Depending on one's location (e.g., the central housewares store or the family restaurant), one can hear the call to prayer. One supermarket has a cold room full of *halal* sausage and meat products and a *halal* meat counter. On Ramadan evenings, patrons sit in front of meals, awaiting the call to prayer from the mosque or their cell phones. In short, whether one is pious or not, one cannot escape the religious artifacts, sounds, and symbols at the complex, as mundane activities unfold under the faith-inspired umbrella of Muslim sensitivities and practices, while not forcing visitors to participate in faith activities. The Salam Mosque complex is a faith-inspired space that is a cherished home space to many Turkish-Germans (pious or not). As Met- calf notes: "In the very act of naming and orienting space through religious practice, we see a kind of empowering of Muslims and a clear form of resistance to the dominant categories of the larger culture" (1996b: 12). A sense of pride and confidence can be seen, especially in the presence of employees wearing headscarves.

The Salam Mosque complex bespeaks the successful localization of Turkish-Germans (and other Muslims) in Stuttgart. This process did not unfold in isolation and the mosque and surrounding businesses are eager to maintain links with the city. The mosque community is involved in interfaith activities, accommodates visits from schools, and cooperates with the city (e.g., open nights). Entrepreneurs, in particular the owners of restaurants, are eager to attract diverse patrons and especially the lunch crowd from surrounding companies. One restaurant has a large window decoration announcing its *Mittagstisch* (lunch menu). Another restaurant has posters on a neighboring fence advertising its breakfast and lunch specials.

CONCLUDING REMARKS

More than two decades after it opened, the Salam Mosque complex continues to be a spatiality in the making. This does not result from faulty planning or mismanagement, but illustrates demographic, social, and cultural changes in the local Turkish–German and Muslim communities, the city, and beyond. While businesses are changing, the complex's larger spatiality is solidly established as a faith-inspired one that is open to all. The Salam Mosque is the complex's heart, which sets the tone for surrounding spaces without overwhelming them or excluding the not-so-pious. The mosque is the anchor that made the complex possible and successful. In the quarter century of its existence, and by way of largely informal strategies and activities, the mosque complex has evolved into a lively and prosperous space that attracts thousands of visitors. While the complex's architectural substance remains largely unchanged, tenants and owners in and around the complex have changed and embellished their premises. One restaurant built a wooden deck for additional outdoor seating. Others added murals to advertise their wares or replaced industrial windows with attractive shop windows. These small changes give the area a more colorful and welcoming appearance. More than material changes, it is the people and their activities that make this space. The hustle and bustle on Saturday illustrates the successful localization of an ethnic and faith-based community. This unique spatiality emerged as the result of informal social, cultural, and commercial negotiations. Its success illustrates the continued relevance of religion in global cities and is an example of how houses of worship can be anchors for urban transformation. It highlights the powerful role of ordinary culture (Williams 2011 [1958]) and vernacular creativity, as immigrant and faith-based groups claim spaces and visibility in the face of considerable resistance and resentment in a German city.

References

Abu-Lughod, J. (1994). *From urban village to east village*. Cambridge: Blackwell.

Bedoya, R. (2014, 15 September). Spatial justice, rasquachification, race, and the city. *Creative Times Report*.

Brandt, S. (1994). *How buildings learn*. New York: Penguin.

Cesari, J. (2005). Mosque conflicts in European cities: Introduction. *Journal of Ethnic and Migration Studies, 31*(6), 1015–1024.

Chappell, B. (2012). *Lowrider space*. Austin: University of Texas Press.

Crouch, D. (2010). Creativity, space and performance: Community gardening. In T. Edensor et al. (Eds.), *Spaces of vernacular creativity* (pp. 129–140). London: Routledge.

David, A. (2012). Sacralising the city: Sound, space and performance in Hindu ritual practice in London. *Culture and Religion, 13*(4), 449–467.

Dwyer, C., David, G., & Bindi, S. (2012). Faith and suburbia: Secularisation, modernity and the changing geographies of religion in London's suburbs. *Transactions of the Institute of British Geographers, 38*(3), 403–419.

Dwyer, C., Tse, J., & Ley, D. (2016). "Highway to heaven": The creation of a multicultural, religious landscape in suburban Richmond, British Columbia. *Social and Cultural Geography, 17*(5), 667–693.

Edensor, T., & Millington, S. (2009). Illuminations, class identities and the contested landscapes of Christmas. *Sociology, 43*(1), 103–121.

Edensor, T., & Millington, S. (2010). Christmas light displays and the creative production of spaces of generosity. In T. Edensor et al. (Eds.), *Spaces of vernacular creativity* (pp. 170–182). London: Routledge.

Garbin, D. (2012). Marching for God in the global city: Public space, religion and diasporic identities in a transnational African church. *Culture and Religion, 13*(4), 425–447.

Gibson, C. (Ed.). (2012). *Creativity in peripheral places*. London: Routledge.

Gibson, C., & Connell, J. (2011). Elvis in the country: Transforming place in rural Australia. In C. Gibson & J. Connell (Eds.), *Festival places: Revitalising rural Australia* (pp. 175–193). Bristol: Channel View Publications.

Hallam, E., & Ingold, T. (Eds.). (2007). *Creativity and cultural improvisation*. Oxford: Berg.

Harvey, D. (1989). *The condition of postmodernity*. Malden: Blackwell.

Idriz, B. (2010). *Grüss Gott Herr Imam!* München: Diederichs.

Kessner, I. (2004). *Christen und Muslime – Nachbarn in Deutschland*. Gütersloh: Gütersloher Verlagshaus.

Kuppinger, P. (2015). *Faithfully urban: Pious Muslims in a German city*. New York: Berghahn Books.

Kuppinger, P. (2017). At home in the multi-cultural city: Islam and religious place-making in Stuttgart, Germany. In D. Garbin & A. Strhan (Eds.), *Religion and the global city* (pp. 205–221). London: Bloomsbury.

Lippard, L. (1997). *The lure of the local*. New York: The New Press.

Livezey, L. (Ed.). (2000). *Public religion and urban transformation*. New York: New York University Press.

Lloyd, R. (2006). *Neo-Bohemians: Art and commerce in the postindustrial city*. New York: Routledge.

Mattausch, B., & Yildiz, E. (Eds.). (2009). *Urban recycling: Migration als Grossstadt Ressource*. Basel: Birkhäuser.

Metcalf, B. (Ed.). (1996a). *Making Muslim space in North America and Europe*. Berkeley: University of California Press.

Metcalf, B. (Ed.). (1996b). Introduction: Sacred words, sanctioned practices, new communities. In B. Metcalf (Ed.), *Making Muslim space in North America and Europe* (pp. 1–30). Berkeley: University of California Press.

Milbourne, P. (2010). Growing places: Community gardening, ordinary creativities and place-based regeneration in a Northern English city. In T. Edensor et al. (Eds.), *Spaces of vernacular creativity* (pp. 141–154). London: Routledge.

Moore, M. (1989). *Roger and me*. Dog Eat Dog Production.

Orsi, R. (1985). *The Madonna of 115th street*. New Haven: Yale University Press.

Orsi, R. (Ed.). (1999). *Gods in the city*. Bloomington: Indiana University Press.

Saint-Blancat, C., & Cancellieri, A. (2014). From invisibility to visibility? The appropriation of public space through religious ritual: The Filipino procession of Santacruzan in Padua, Italy. *Social and Cultural Geography, 15*(6), 645–663.

Shah, B., Dwyer, C., & Gilbert, D. (2012). Landscapes of diasporic religious belonging in the edge-city: The jain temple at Potters Bar, outer London. *South Asian Diaspora, 4*(1), 77–94.

Sorkin, M. (1992). *Variations on a theme park*. New York: Macmillan.

Stepick, A., Rey, T., & Mahler, S. (Eds.). (2009). *Churches and charity in the immigrant city*. New Brunswick: Rutgers University Press.

Tamimi Arab, P. (2013). Mosques in the Netherlands: Transforming the meaning of marginal spaces. *Journal of Muslim Minority Affairs, 33*(4), 477–494.

Tarlo, E. (2010). *Visibly Muslim*. Oxford: Berg.

Tweed, T. (1997). *Our lady of the exile*. New York: Oxford University Press.

Warner, R. S., & Wittner, J. (Eds.). (1998). *Gatherings in diaspora*. Philadelphia: Temple University Press.

Williams, R. (2011 [1958]). Culture is ordinary. In I. Szeman & T. Kaplan (Eds.), *Cultural theory* (pp. 53–59). Malden: Wiley Blackman.

Zukin, S. (1989). *Loft living*. Brunswick: Rutgers University Press.

Zukin, S. (2010). *Naked city*. Oxford: Oxford University Press.

CHAPTER 11

Typologies of Bottom-Up Planning in Southern Europe: The Case of Greek Urbanism During the Economic Crisis

Konstantinos Serraos and Evangelos Asprogerakas

INTRODUCTION

Spatial planning and management in Greece has transformed to embrace two trends that involve essential elements of bottom-up procedures. The first concerns policies generated by official authorities, either aimed at expanding citizens' participation opportunities, or to offer wider opportunities for a more active involvement of the private sector. The second primarily involves non-formal urban planning initiatives produced by the citizens themselves.

This chapter explores the types of bottom-up interventions in the spatial planning process in Greece and attempts to categorize, understand, and evaluate them, while drawing attention to relevant European experience and practice. Initially, the basic characteristics of the Greek planning system are presented, focusing on available governance schemes. It presents a selection of case studies with the aim of creating a typology of bottom-up initiatives in Greece primarily based on the actors involved and the formal or informal character of the initiatives. This chapter also describes their level of integration in the mainstream planning process, their on-the-ground characteristics,

K. Serraos (✉)
School of Architecture, National Technical University of Athens (NTUA), Athens, Greece

E. Asprogerakas
Department of Spatial Planning and Regional Development, University of Thessaly, Volos, Greece

© The Author(s) 2019
M. Arefi and C. Kickert (eds.), *The Palgrave Handbook of Bottom-Up Urbanism*, https://doi.org/10.1007/978-3-319-90131-2_11

and their main short- and long-term results. The basic research for this chapter uncovered the interaction between civic society initiatives and responses by official authorities, especially during times of economic crisis. This chapter also connects these initiatives to trends and dynamics that are being shaped by the European and international scene concerning issues of spatial management and new governance structures.

OFFICIAL TOP-DOWN PLANNING IN GREECE: BACKGROUND AND PROBLEMS

In the post-war era urban planning has become a crucial tool in the hands of national and local authorities in Europe. Authorities tried to manage conflicts among varying interests concerning the use of space, while simultaneously attempting to promote the development of endogenous dynamics in a way that was beneficial to society. In Greece, after the dictatorship fell in 1974, an official top-down spatial planning system was organized and implemented for many decades on a regional and an urban scale. In the late 1970s/early 1980s a comprehensive two-level spatial planning system was launched.[1] The system was complemented a decade later with new tools covering all spatial scales[2]; ranging from the regional and national planning level up to urban design plans (Serraos 1998; Angelidis 2000).

By the early twenty-first century, Greece was equipped with a complete top-down urban planning system in terms of tools and interconnections, which was expected to effectively fulfill its role in the spatial planning process. Unfortunately, this expectation could only partially be fulfilled, since several fundamental weaknesses emerged. Specifically, the inefficiency of the spatial planning process as a substantial reconciliation tool for conflicting interests regarding the use and management of space became apparent. Important reasons for this phenomenon were the complex and lengthy procedures, delays in launching the National Cadastre (including forest maps, archaeological cadastre, agricultural land maps, and others), and a lack of funding. An indicative result of these problems is the fact that, after 30 years, clear and binding land use controls (within and outside settlements) are still not completed, while the building control system also shows serious inefficiencies (Serraos 2015; Koudouni 2014).

It can easily be concluded that the existing Greek spatial planning system needs to transform by establishing a framework within a reasonable time based on two objectives. On the one hand, spatial planning should focus on sustainable urban development, and on the other hand on the protection of valuable natural and cultural environments. More specifically, this planning revision should focus on: (a) reducing planning preparation time and plan-

[1] Laws 947/1979 and 1337/1983.
[2] Laws 2508/1997 and 2742/1999.

ning cost; (b) improving its effectiveness in the spatial management process; (c) clarifying its binding effect toward citizens, lower planning levels and specific actions associated with land management (protection/exploitation); and (d) involving citizens and other actors more effectively through an enriched dialogue and participation process.

STAKEHOLDER INVOLVEMENT WITHIN FORMAL PLANNING

There has been a policy shift from central to local authorities to create opportunities for citizen participation in the urban planning procedure. Authorities have developed contemporary formal spatial governance schemes that introduce relations among a wide range of actors while defining and implementing policies for spatial planning and urban development, such as the Reactivate Athens planning event, organized by the City of Athens (Mpouzali 2014). Furthermore, reforms of the official spatial planning system aim to provide opportunities for more 'flexible' private investment activity (e.g., location of economic, productive, and general investment activities). This is done through the introduction of, sometimes controversial, procedures that allow and promote peculiar bottom-up spatial development based on market dynamics. For example, Special Spatial Plans for the development of strategic investments were introduced in 2014.

The first participation schemes were introduced at local planning level as long ago as 1923. In a more comprehensive way consultation and participation processes were incorporated into the Greek planning system in 1983 as mandatory procedures for the approval of urban plans at all levels.[3] A pivotal tool has been the Urban Neighborhood Committee as a body that formulates proposals and opinions for the municipality, focusing on urban and functional problems at the neighborhood level and the implementation of urban plans. However, this institution has never functioned effectively as it has not been sufficiently supported by local authorities, politically or financially.

The spatial planning framework, as amended in the 1990s, continued to urge the active participation of citizens and social actors in the planning process.[4] Nevertheless, in practice the discussion and exchange of views with stakeholders only takes place at the public presentation of urban plans prior to finalization. Thus, as the consultation process is integrated at an already advanced stage of planning, the possibility for active citizen participation remains quite limited.

More attention to participation and consent among stakeholders is paid in urban regeneration processes that fall under Projects of Integrated Urban Interventions. The enabling legislation for these projects encourages broad consultation among residents, agencies, and local NGOs and should include

[3]Law 1337/83.
[4]Law 2508/97.

the systematic recording, encoding, and evaluation of views. Nevertheless, this process relies on local authorities, without providing specifics on formal implementation mechanisms (Asprogerakas 2016). Consultation with the public is provided during the process of a Strategic Environmental Assessment for spatial plans and programs in accordance with Directive 2001/42/EC of the European Union. For this purpose, several methods may be used as convenient, such as digital surveys, public hearings, conferences, and open discussions.

Introduced in 2010, the Municipal Consultation Committee also functions as an advisory body on the local level.[5] It consists of 25–50 members who are representatives of local stakeholders and citizens and are selected by ballot. Furthermore, an online consultation procedure is generally provided for each act of legislation through an official governmental website.[6,7]

Cities across Europe are exploring and developing new formats for public involvement in spatial planning and management procedures. On the one hand, these initiatives achieve a broader acceptance of the planning procedure by citizens, leading to an improved practicability of spatial plans. On the other hand, they aim to boost public support, a lack of which usually leads to delays or even cancellations of urban interventions. In the context of recent interuniversity cooperation, selected German case studies of planning reform and innovation have been analyzed. Worth mentioning are: Hanover's Development Plan 2030 (*Stadtentwicklungskonzept* 2030), which was based on a newly developed 'dialogue' system between citizens, politicians, and public administrators (LHH 2014)[8,9]; the innovative consultation for traffic calming in Augsburg/Hochzoll (Urbanes Wohnen eG 2009)[10]; and the engagement of local enterprises in Business Improvement Districts (BIDs) in Hamburg (Kreutz 2009; Asprogerakas 2014).[11]

---

[5] A quite new institution according to the law 3852/2010.

[6] Law 4048/2012.

[7] This website is www.opengov.gr.

[8] Research was conducted in the framework of the joint Greek–German academic programme IKYDA 2015, "New forms of governance and democratic legitimacy of urban planning decisions. Urban planning cultures in the frame of civil society initiatives and protests," supported by the Greek State Scholarship Foundation (IKY) and the German Academic Exchange Service (DAAD), and carried out by the Urban Planning Research Laboratory/National Technical University of Athens (Prof. Konstantinos Serraos) and the Institute of Environmental Planning/Leibniz University Hannover (Prof. Frank Othengrafen).

[9] The dialogue was based on rules already formulated by the administration, including various actions of a direct and representative citizen participation, organized by thematic field according to a flow chart and a three-year timeframe.

[10] Allocation of the entire study area into sub-study areas and public consultation for six months according to five predefined steps.

[11] For instance, *Neuer Wall*: encouragement of the involvement of local private enterprises in shaping self-funded management schemes.

To fit the framework of the bailout agreed between Greece and the 'troika' of the EU, ECB, and IMF, the Greek national government has introduced a radical overhaul of the legislative spatial planning framework.[12] The new framework emphasizes 'flexibility' and also, unfortunately, a fragmented accommodation of economic, productive, and general investment activities (Serraos 2015). This overhaul also introduces bottom-up processes for spatial development, based on locally emerging dynamics initiated by specific investment requests. To this aim, new Special Territorial Plans can supersede current spatial plans, especially in terms of permitted land use and building conditions and restrictions. Significant opposition from the academic, scientific, and professional communities have led to limited improvements to this framework, but they could not reverse its general philosophy.[13]

SPATIAL BOTTOM-UP ACTIONS

Public patience for formal planning has been decreasing for years, as citizens face deteriorating conditions in densely built Greek cities, plagued by economic, social, environmental, and functional issues. Simultaneously, Greece's degree of sensitivity to these issues has been significantly improving in recent years, especially as Greeks increasingly exchange views and experiences with other countries. Moreover, Greece's recent unprecedented and prolonged fiscal crisis has intensified its urban problems, urging citizens to be more actively organized. In this context, new unofficial actors have arisen and are interacting with each other, either cooperating or clashing. This trend has followed that in many other European countries and has mostly become visible in the capital, Athens, Greece's major economic, social, and population hub. In projects such as the Ellinikon Metropolitan Park, the pedestrianization of Panepistimiou Avenue, the Navarino Park, and the Citizen's Network for Athens' Inner City, the priority has been to secure outdoor, public, green recreation spaces, a goal in which citizen action has played a key role. Additionally, the actions of existing NGOs (e.g., WWF Hellas and the Hellenic Society for the Environment and Cultural Heritage) are of great significance as they closely observe Athens' spatial planning developments and will not hesitate to officially intervene using all available legal ways.

First Level of Action: Informal Collectives

A characteristic case of action coming from individual citizens' groups is Navarino Park, which is a large open area of about 1500 m² in the center of Athens. The plot, a property of the Technical Chamber of Greece (TCG), has been functioning for years as an outdoor car parking lot. At the end of

[12] Law 4269/2014 Spatial and Urban Reform/Sustainable Development.
[13] Law 4447/16.

the 2000s TCG planned to erect a new block of offices to serve as their head-quarters. Threatened with losing this space the Initiative Committee of the Citizens of Eksarcheia Neighborhood was formed to activate the citizens living and working in this area, toward the goal of making the plot an area for common use. At the same time, more groups arose and were activated that finally decided to squat in the area demanding, "the parking to become a park". In 2009, they finally removed the hard surface, provided soil, and planted trees (Fig. 11.1a, b).

The squatters still occupy the area and self-organize to manage and use the space. They discuss ideas and suggestions, while subsidiary work groups engage in planning, planting, and organizing a playground, accompanied by the appropriate infrastructure. According to the squatters, they aspire to keep the plot as a neighborhood garden which, through hosting part of the social life of the citizens beyond any profit and property framework, functions as an area for playing, strolling, getting in touch, communicating, working out, and creating, regardless of age, origin, educational, social, or financial back-ground. Today, the park has accumulated both positive and negative comments; it has intrigued politicians and citizens thanks to its anti-authoritarian political nature; but it has also established itself in the consciousness of citizens and passersby as a free green area for recreation within the highly dense, built web of the city (Ismailidou 2011).[14]

Second Level of Action: Formal Citizens' Associations

Apart from bottom-up interventions initiated by individual informal groups of citizens tackling specific spaces, there are some interesting examples of bottom-up initiatives generated by more formal associations of citizens. In contrast to informal collectives, these associations function within the legal framework of the state and force the relevant administrative bodies to fulfill their commitments in a well-structured and legal way.

Such an example is the creation of a new public green space known as KAPAPS Park in the Ano Ambelokipoi neighborhood in central Athens. This park is part of a broader area that was expropriated in 1937 for a contemporary hospital, which never was built. Instead, in the 1950s, a Center of Rehabilitation of Handicapped Citizens (KAPAPS) (EINAP 2005) opened on the site, while at the beginning of 2000 the construction of a square, a playground, and a center for road behavior with an underground car park were scheduled but never implemented. Insistence upon underground parking was the primary reason the above project proposal failed. Residents formed the Ano Ambelokipoi Citizens Association and joined forces with the local authorities to reactivate the park plan through formal procedures, including

---

[14]For more information, visit http://parkingparko.espivblogs.net and http://www.attiko-prasino.gr.

Fig. 11.1 **a** Navarino Park before (*Source* K. Serraos, March 1987). **b** Navarino Park after (*Source* K. Serraos, November 2017)

Fig. 11.2 General view of KAPAPS Park as redesigned by the City of Athens (*Source* E. Asprogerakas, August 2017)

open letters, documentation, and an actual physical presence before the relevant bodies (http://www.anoampelokipi.gr/parko-kapaps). Together, they achieved an agreement between the mayor of Athens and the local council, giving the green light for the completion of an extensive, free, and public green space (Fig. 11.2) (Angeli 2014).

Third Level of Action: Networks of Citizens' Associations and the Role of Civil Society

The targeted and locally focused bottom-up actions of informal collectives and formal associations of citizens are successful at times but often lack coordination and a common strategy concerning their functions and actions. The organization of broader civil society networks has been maturing, with a function and structure similar to their peers in other European countries. Their influential role is proving to be much more significant than the one demonstrated by discrete individual collectives and associations. As a result, these networks can be more influential on political decisions involving their various fields of interest.

An example of a successful bottom-up network organization is the recently inaugurated Network of Organizations and Citizens for the Historical Center of Athens. This network unifies ten different associations, collectives, and citizens' movements that focus on certain neighborhoods in the oldest and most central part of Athens. The network was founded in 2014 and functions under the supervision of the Greek environmental NGO, the Hellenic Society of the Environment and Cultural Heritage. It reinforces demands by citizens and organizations for a sustainable historical center in Athens.[15] The vision of the network focuses on: respecting the special character of the historical center of Athens, its local residents, public spaces and public and private properties; providing free access to the center; ensuring healthy residence and entrepreneurship; providing a contemporary infrastructure and high-quality public spaces; and preventing crime; among others.

The network serves as a repository and generator of knowledge on current issues and structural problems, at the same time functioning as a legal framework for relevant authorities, structures, and procedures. It also provides the most up to date and reliable information to citizens on issues that concern them; it works on the elaboration of well-structured suggestions, and cooperates with the Athens Municipality, other administrative bodies, and involved stakeholders, to promote their views, recommendations, and proposals while supervising formal actions. Simultaneously, the network informs and sensitizes citizens about the historical center of Athens and organizes events in its various neighborhoods (Konstandatou 2015).

Fourth Level of Action: Engaging Official Bodies Through Online Platforms

Online platforms have proven to be an excellent method for interaction between citizens and governments. Syn-Athina is an innovative example of governmental organizations aiming to activate Athens' citizens. It is a program initiated by the Municipality of Athens intended to facilitate and benefit from the citizens' groups involved in the improvement of the quality of city life. It was created in July 2013 and functions based on a web platform.[16] The site gives citizens' groups the opportunity to promote their proposals and get in touch with other groups, sponsors, and public or private bodies. The Municipality itself has the duty to spot and leverage the most influential actions. This has resulted in the successful execution of the proposals of 280 citizens' groups. The platform won the 2014 Mayor's Challenge of the Bloomberg Philanthropies Organization and a one-million euro award.

The Municipalilty of Korydallos has initiated a similar platform to engage citizens in creating an integrated cultural policy at local level, called DemoCU.[17] This initiative focuses on specific target groups, such as young people, people with special needs, Roma, immigrants, elderly people, and NGO members. Through this platform interested users have access to all information regarding culture and athletics in Korydallos. They can also exchange views on the official municipal Cultural Structures and Infrastructures plan.

[15] For more information, see http://istorikokentro.gr and http://www.ellet.gr/node/662.

[16] This platform can be accessed via http://www.synathina.gr.

[17] This platform can be accessed via http://democu.gr.

Fig. 11.3 General view of the Stavros Niarchos Cultural Center (*Source* K. Serraos, June 2016)

Fifth Level of Action: Engaging Benefactors

Two significant private sector urban initiatives have fueled the discussion on the renaissance of Athens, both funded by large philanthropic foundations. The first initiative was announced in 2006 and implemented by the Stavros Niarchos Foundation Cultural Center (SNFCC), based on an idea dating back to the 1990s to connect Athens to the sea. The project includes the construction and complete outfitting of new facilities for the National Library of Greece and the Greek National Opera and was developed on the site of the Athens old horse racing track, owned by the Greek State and located 4.5 km south of the center of Athens, on the edge of Faliro Bay. A non-profit, limited liability company, launched by the Stavros Niarchos Foundation, undertook the responsibility to plan, implement, and cover all the costs of the SNFCC (about 500 million euros). After completion, the SNFCC was donated in February 2017 to the Greek State, which now undertakes its full control and operation (Fig. 11.3).

The second case is the Re-think Athens project, a multifaceted intervention aimed primarily at contributing to the revitalization of the center of Athens. The core of the project was an idea dating back to the 1980s related to the pedestrianization of a major avenue in the center of the city (Panepistimiou Avenue). The Public Benefit Foundation Alexander S. Onassis, in collaboration with the Ministry of the Environment, facilitated an international architectural competition to generate the necessary studies for the project. The general proposal includes new transportation facilities (including the extension of a tram line) and the transformation of the Avenue to an open, public, "linear square" for pedestrians, cyclists, and public transport (Furuto 2013). The intervention constitutes a new vision for the city center. Even though its funding (almost 100 million euros) is not secured, the project is the subject of current debate among urban planners and citizens.

CONCLUSION

The economic crisis and its aftermath of governance reform has prompted Greece to transform its spatial planning structure into a more multifaceted, and possibly contradictory, bottom-up process. On the one hand, protests and demands by citizens (claiming higher quality public and green open spaces and demanding protection for natural resources and natural landscapes) have increased. On the other hand, the new spatial planning system aims for a more neoliberal structure of economic and urban development. In this context, the recent changes to the institutional framework, which either directly relate to the planning system, or have a significant impact on it, tend toward the weakening of rules.[18] This is clearly aimed at increasing 'flexibility' toward future private investment initiatives, which unfortunately are often not properly coordinated by Greece's current planning system. The ultimate purpose is clearly to increase possible economic benefits and contribute to overcoming the economic crisis.

Citizens, on their side, have become far more active stakeholders in the formal planning process, overturning decisions to which they do not consent. By joining forces in informal collectives, formal associations, and powerful networks, citizens' movements have become more successful in making their voices heard by exchanging information and knowledge. Along the same lines, interesting initiatives by official local actors with an experimental character, target this civil activation, constituting a clear trend toward the formation of a future integrated framework that would help to form stronger relationships of trust between governments and citizens.

Meanwhile, municipal authorities increasingly realize that urban planning requires a broad acceptance, which can be provided either by developing and integrating new substantive and multilevel participative structures into the formal planning process, or by shifting toward additional and increased bottom-up civil engagement structures, simultaneously with more and more intensive dialogue.

Recent experience in Greece teaches that the above issues are important tools in the hands of politicians to prevent possible negative reactions from civil society, to help planned interventions become more widely accepted and therefore more feasible—a shared benefit for politicians, administrators, citizens, and the city.

REFERENCES

Angeli, D. (2014, October 26). Cement to become and KAPAPS park. *Journal of Journalists.* Available at: http://www.efsyn.gr/arthro/tsimento-na-ginei-kai-parko-kapaps (in Greek).

Angelidis, M. (2000). *Spatial planning and sustainable development.* Athens: Symmetria Publications (in Greek).

[18] Laws 4269/14 and 4447/16.

Asprogerakas, E. (2014, June 27–28). Business improvement districts: Prospects and questions in relation with the open trade centers. In *"Urban and Regional Development: Modern challenges", 12th Conference of ERSA—GR, Conference Proceedings*. Athens: Panteion University (in Greek).

Asprogerakas, E. (2016). Approaches of integrated urban interventions in Greece: Tools and governance elements. *Aeihoros* (under publication).

Furuto, A. (2013). *Re-think Athens winning proposal/OKRA*. Weblog entry on Arch Daily. Available at: https://www.archdaily.com/338001/re-think-athens-winning-proposal-okra.

Ismailidou, E. (2011, April 5). Navarino: The park—Parking completed two years of life. In *To Bima*. Available at: http://www.tovima.gr/society/article/?aid=393785 (in Greek).

Konstandatou, E. (2015, May 19). Organizations network joining forces for the historic center of Athens. In *To Bima*. Available in: http://www.tovima.gr/society/article/?aid=705618 (in Greek).

Koudouni, A. (2014). *Management and spatial planning powers: Institutional issues: Examples*. Athens. Available at: http://www.arch.ntua.gr/course_instance/1180 (in Greek).

Kreutz, S. (2009). Urban improvement districts in Germany: New legal instruments for joint proprietor activities in area development. *Journal of Urban Regeneration and Renewal, 2*(4), 304–317.

Landeshauptstadt Hannover (LHH), Geschäftsbereich des Oberbürgermeisters. (2014). *Mein Hannover 2030, Stadtentwicklungskonzept 2013, Spielregeln für die Beteiligung, Zeitrahmen 2014 bis 2016* (in German).

Mpouzali, L. (2014). Reactivate Athens: 101 ideas. *Journal AZ/62/2014*. Available at: http://www.onassis.org/onassis-magazine/issue-62/reactivate-athens (in Greek).

Serraos, K. (1998). The urban plans in accordance with the latest regulations in Greece. In A. Aravantinos, *Urban planning for a sustainable development of urban space*. Athens: Symmetria Publications (in Greek).

Serraos, K. (2015). Recent regulations for the reform of the spatial planning system and the land-use planning. In D. Melissas (Ed.), *Land uses: Setting the space for the development*. Athens: Department of Urban and Regional Planning, School of Architecture NTUA (in Greek).

Union of Hospital Doctors of Athens—Piraeus (EINAP). (2005). *The public hospitals in Athens and Piraeus: Course in time*. Athens (in Greek).

Urbanes Wohnen eG. (2009). *Ein Verkehrskonzept für Hochzoll, Ergebnisse der Planungswerkstatt*. Augsburg: Stadt Augsburg (in German).

Internet Sources

City of Athens/Vice Mayoral Office for Civil Society and Innovation. SynAthina project: http://www.synathina.gr.

Cultural Association Ano Ampelokipon: http://www.anoampelokipi.gr/parko-kapaps.

Green in Attiki: http://www.attiko-prasino.gr/Default.aspx?tabid=1296&language=el-GR.

Landeshauptstadt Hannover (LHH): www.meinhannover2030.de.

National Technical University Athens (NTUA)/Department of Geography and Regional Planning/School of Rural and Surveying Engineering, Municipality of Korydallos, Network for Employment and Social Care. "DemoCU project": http://democu.gr.

Organizations and Citizens Network for a sustainable historic center: Open discussion: http://www.ellet.gr/node/662.

Organizations and Citizens Network for the historic center of Athens: http://istorikokentro.gr.

Re-think Athens winning proposal/OKRA: https://www.archdaily.com/338001/re-think-athens-winning-proposal-okra.

The Greek Open Government Initiative: www.opengov.gr.

Their parking, our park: http://parkingparko.espivblogs.net.

PART III

Formalizing the Informal

CHAPTER 12

Lean Urbanism in Central Africa

Stephen Coyle

When the final night falls on us as it fell upon our parents, we shall retire to our modest home earth-sure, secure that we have done our duty by our people; we met the challenge of history and were not afraid.
Ghanaian poet Kofi Awoonor, who was killed during the Westgate mall attack in 2015

Building and improving the resilience of villages, towns, and cities in developing regions of the world and in Africa specifically, demands a lean approach: small-scale, incremental developments that require fewer resources to incubate and mature. The pathway to Lean Urbanism in Central Africa, viewed in this chapter through the lens of its physical scale, development patterns, regulatory standards, and zoning, requires time-tested strategies and innovations in design and construction. Constraints on building resource-efficient, durable, and healthy communities differ from the financial, bureaucratic, and regulatory obstacles in the United States and the European Union that precipitated the Lean Urbanism movement. For example, the Gabonese authorities created a roadmap for economic emergence and diversification that includes the adoption of progressive "SmartCode" land use and development standards.[1]

---

[1] The Report: Gabon 2012, (P. 129). Google Books Result. https://books.google.com/books?isbn=1907065695. Under the SmartCode system, public services are designed to be integrated into building plans to ensure orderly urban development.

S. Coyle (✉)
Woodland, CA, USA

© The Author(s) 2019
M. Arefi and C. Kickert (eds.), *The Palgrave Handbook of Bottom-Up Urbanism*, https://doi.org/10.1007/978-3-319-90131-2_12

An absence of context-sensitive land use regulations leaves watersheds or mangrove estuaries unprotected and degrades potential areas for sewage treatment and stormwater recharge. Western-trained planners often configure auto-dependent developments that rely on massive earthmovers to flatten contours and fill drainage basins to accommodate superblocks with arterial roads that plough through a traditional, walkable neighborhood. Despite a decline in institutionalized corruption, an unscrupulous speculator could still construct a housing tract in a flood plain through an illegal transaction; the local farmer, lacking access to higher ground, might build along a seasonal creek that overflows its banks.

This chapter first identifies the daily activities of local populations and the hurdles and pitfalls they encounter within their physical environments. In response to these challenges, a series of lean strategies, tools, and techniques are recommended to assist policymakers, urban planners, architects, engineers, and developers in working through, over, and around the barriers and impediments the benefit these people. Through lean applications over time, experts and non-government organizations, governments, and businesses can improve the livability of Central African neighborhoods and communities.

Lean and Local

Building most often begins at the local scale of the family, the home, and along pathways to and from the village center. A seven-year-old African girl fetches water in plastic containers with other youngsters, drawing from a common water spigot where villagers congregate for potable water, bathing, and socializing. The young girl returns along a worn dirt path to her family home made of wood, clay brick, or concrete block, built by her father and friends. Daily life at the local, near-tribal scale for most Africans, remains simple, repetitive, and prone to infrequent but formidable floods and disease, punctuated by periodic funerals, marriages, and political campaigns.

Only a few relatively affluent residents own cars, so most residents walk everywhere or catch a ride. Outhouses or latrines line small creeks that tend to overflow during sustained rains, an extremely lean living arrangement that unfortunately breeds disease and dysentery, partly compensated by a lifestyle filled with physical activity that keeps villagers relatively fit (Huber 2015). At the local scale, a bird's eye view encompasses a mosaic of rudimentary structures in an apparently random setting of unpaved paths and streets, appearing as an organic and slowly evolving habitat (Fig. 12.1).

For example, a teenage Gabonese boy takes a minibus from his *quartier*— the city neighborhood composed of multiple adjoining villages—to attend his secondary school near the market center of Libreville, the regional province of Estuaire and the capital city. The washboard-surface roads, clogged with packed buses, taxis, private cars, and motorbikes, are lined with Africans of all ages walking precariously close to the passing vehicles. An occasional bicyclist hugs the shoulder of a road obstructed with utility poles, signs, and

Fig. 12.1 View southwest over the Akanda Commune, a growing middle class settlement on the Gabon Estuary, with the Pongara National Park across the bay in the far distance

vendors; the thoroughfares often provide no space for a bike, and scant room for pedestrians (Godard 2013).

The boy lives in a three-story, walk-up apartment surrounded by relatives in concrete block and wood homes, but only his building contains a bathroom in each unit. Despite the simple living and traveling accommodations, a smartphone connects him with friends, music, and a culture beyond the edge of this slice of West-Central Africa. The youngster's neighborhood, about 800 meters across, spans a tree-canopied creek, thick with undergrowth, a busy road and more aggregations of houses and apartments. A football pitch, small shops, and a primary school anchor the center.

As the boy's minibus leaves the neighborhood for the city and approaches his school, the two-and three-story concrete block apartments and shops that line the road give way to concrete slab and column, high-rise buildings that provide housing and offices for ex-pats and wealthy Gabonese. Most contain plumbing—sinks, showers, and toilets—although none connect to a long-promised municipal sewer system. Instead, they rely on holding tanks pumped out periodically or, on larger sites, tanks are connected to drain fields that occasionally overflow the effluent into ravines or creeks.

The newer buildings stay cool through individual air conditioning systems, although the inadequate window ventilation creates a steam bath interior during the occasional power failures. Many of the newer, owner-built homes and shops rely on mechanical cooling, lacking the large windowless ventilation openings of the older traditional buildings, though the more upscale versions employ windows with screens to obstruct the malarial mosquitos. The boy's school, however, smartly relies on the prevailing breezes that flow through continuous louvers and screened windows sun-and-rain protected by extended roof eaves. Traditionally designed schools offer one of the few examples of a disappearing building vernacular responsive to the climate and an unreliable electrical grid (Ebohon 2016).

Fat at the Top, Lean at the Bottom

At the city scale, the undulating hills determined the layout of the 19th and early 20th century, French colonial-period streets. Linear ridge roads attract the most expensive real estate lining each side with housing compounds and high-end apartments. The property values decline with the slope from ridge-line to valley as they do around the world. The creeks and drainways that snake along the bottom often overflow, flooding the unfortunates living at the physical and economic base of the city (Fig. 12.2).

The teen's science teacher commutes to school squeezed in a taxi with four other passengers. The aged Toyota Corolla belches gray smoke with every push of the gas pedal, navigating an intensifying urban traffic after bumping along the muddy road by her remote family hamlet along the banks of a small estuary north of the city. In this northern river-laced region, where the lowlands flood

Fig. 12.2 A "catch all" drainage ditch flanked by self-built homes, snakes through a semi-rural village in an area within the Libreville Commune subject to periodic flooding

once or twice a year, her parents and relatives survive by selling fish from the mangrove-lined creeks along the estuary, angling from their handmade wooden skiffs. From these same logs, they carve tribal masks and jungle animals to sell to the French garrison soldiers and the few tourist shops in the city. Their gardens provide greens and root vegetables, supplemented by banana trees and coconut palms, providing a lifestyle that is both simple and relatively self-sufficient.

While most individual homes are built from sawn logs with corrugated roofs, the villagers aspire to live in concrete block dwellings. Although these structures signify durability and permanence, the blocks perform poorly in this hot, humid climate, neither breathing nor cooling. Handmade on the ground by pouring a weak concrete mix into plastic molds, the monolithic base of each that supports two fragile webs prevents grouting of the block walls, and the pouring or pumping of concrete slurry into the block cores around vertical 'rebar' to make a solid, reinforced wall. Despite the solid base, blocks often crumble during construction. Although weak walls make for easy penetration for pipes and cable, mishaps result in large holes and damaged wall sections. The use of weak, cast-in-place concrete mixes and inadequately reinforced concrete columns and floors cause occasional building failures (Windapo and Rotimi 2012).

On her endless commutes to and from her village, the teacher observes, left to right, the consequences of building cheaply—broken, rusting railings and cratered concrete—that contrast with the two, three, and four stories of, apparently, more durable structures. The frailties and incompetence evident in construction and infrastructure, she understands, symbolize the kinds of unscrupulous condition that could undermine her students' prospects. Like her peers, she worries about her monthly compensation, at risk during cycles of governmental financial uncertainty (Fig. 12.3).

The African girl longs for indoor plumbing, paved sidewalks and streets. The teenager wants to move up the economic ladder and secure the funds to attend university, a necessity for decent employment, for buying a motorbike, and, later, a house for his future family. The science teacher longs for the intellectual stimulation of city life where restaurants, clubs, and a university provide urbanity, professional challenges, and level of culture unavailable in the village. The teacher, the teenager, and girl, each within their own context, endure daily difficulties. They learned early to enjoy the pleasures of a modest, family-oriented, African life that can unfold modestly at local, community, and regional scales (Simmons 2015).

FROM BUSH TO BOULEVARD

The teen and teacher observe the growing urbanization during their daily transition from neighborhood and hamlet to the city, a sequence of environments from dirt path to paved boulevard. They find an abundance of culture and sustenance in both contexts, one simple, the other complex. The jungle's natural bounty remains a relatively dependable source of nourishment for a rural population that lacks financial security, by living lean and consuming small. Healthcare, education, and employment cluster in a city that exchanges

Fig. 12.3 Children and adults circulate on elevated walkways that circumvent trash and water in flood-prone basins that house families unable to afford the rents of apartments along the ridges

polluted air and water for an influx of people and goods from the country-side. The trade-off for sharp-edged urbanity, the realm of big buildings, companies, commerce, and SUVs, is a dependency on a continuous diet of non-renewable resources.

For many Central Africans, each successive environment, from jungle to central city, holds a promise of progress. The immersive, rural environment provides the basics of food and water; the metropolis prioritizes higher incomes, superior education, and better work opportunities. The sequence of immersive environments along Central Africa's rural-to-urban transect begin with tribal settlement patterns of clustered wood dwellings surrounding a common social space or a central water spigot. A covered porch doubles as a beer and soda café, a third place for congregating. Simply sitting together under a huge okoumé tree canopy on plastic chairs reinforces the bonds of community.

In the quartier, or neighborhood, busy streets lined with restaurants two tables deep and open-front shops selling hardware, snacks, and sodas, compete with sidewalk vendors merchandizing on wooden crates under umbrellas (Price Waterhouse Cooper 2016). Apartments and offices accessed by stairways to common balconies rise two to five stories above shops. Roughly formed concrete sidewalks, often too narrow for side-by-side foot traffic, force people to circumvent cars that claim the sidewalk as an extended parking space and block the way for pedestrians. Ironically, both pedestrians and the randomly parked vehicles offer a lean means of traffic calming.

The city boasts multitudes of people, traffic, and bigger, taller, and often grander structures that increase the chaos where, bigger, is not always better. Plumbed buildings often rely on timely pump-outs of sewage to collection tanks; others simply dump wastewater in an adjoining creek. While improvements continue, years are likely to pass before sanitary pipes run under streets from buildings with plumbing to new wastewater treatment plants.

Five-meter wide shop fronts give way to 60-meter long supermarkets, hotels, and elevator-served, high-rise apartments and offices. SUVs roll through signal lights and stop signs when heavy rains force the police to seek shelter. Despite an onslaught of decrepit taxis, buses of all sizes, shiny sedans swerve around multiple accidents on unpredictable motorways, African motorists remain generally respectful of pedestrians—except for the drivers distracted by smartphones. The city maintains a working relationship with rural cousins who supply much of their food and water, but for how long and at what cost as resources dwindle?

THE SCALE OF LEAN URBANIZATION

On a continental scale, UNICEF's Generation 2030 Africa reports that the current one billion-plus population will double within 35 years and the 18-year-old and younger population will increase to almost one billion. During the same time period, over 50% of Africans will be living in cities, taxing the supporting infrastructure and natural resources, from water to energy. From 1950 to 2016, Africans moving to cities increased from 14 to 40%, with 50% projected mid-2030s.

One of the biggest drivers of urbanization, and a continuing trend according to the *African Economic Outlook 2016*, is the growth of towns and intermediate cities. The areas of these expanding cities include incrementally built shanty towns that receive few resources for survival, as they often squat, cheek by jowl, with metropolitan towers. Rarely designed with any discernable urban form, the impacts from these impromptu developments suffer a lack of sanitation and clean water. Such missing infrastructure stresses the occupants, the natural resources, and the environment that support the upscale population centers, while the impromptu settlements provide much of their labor. However, employing smartly planned, small-scale, and incremental development can reduce some of these adverse consequences and increase community assets from accessible schools to local employment clusters. Through lean planning served by essential infrastructure, organically built urban neighborhoods can incubate and mature into relatively self-sufficient communities (Fig. 12.4).

Smaller-scale projects, with their contextually responsive intensities and densities, adapt to life's uncertainties and environmental, economic, and

Fig. 12.4 Professional Gabonese planners, engineers and geographers astutely apply the new Gabon SmartCode's "transect-based neighborhoods" within the PK5-12 Redevelopment Plan

social forces, with resilience and a unique African beauty. Living graciously with dignity while facing both extreme uncertainty and poverty (a condition imposed on proportionally few Americans), demands a lean lifestyle. Africans find comfort and even joy in the simplest daily practices of sitting together or just walking and talking. Central Africa appears to manifest a larger generation gap than might be expected in a tribal culture, despite a persistent respect for elders. Although many baby-boomer Americans sought lifestyles quite different to that of their parents, that 'generation gap' decreased with the millennials. Similar to their US counterparts, young Africans represent a source of great positive energy and openness, and a propensity for risking comfort for the sake of economic progress.

By simply speaking less and listening, the outside development interventionist will find common ground with both African youth and elders who, in turn, will help build, occupy and manage and the settlements. Contributing to a region-wide demand for the development and repair of worthy places, those with a sufficiency of income will continue to select living arrangements based on their desires, fears, aspirations, and finances. The African building industry can deliver a range of housing, from Western-style subdivisions to low-, medium-, and high-rise apartments and exclusive compounds. With fewer choices, those less affluent and poor still care about their living arrangements and require housing choices for families large and small.

These represent our 'target market' who need the benefits of lean actions and decent outcomes.

Since, literally and figuratively, higher ground attracts greater wealth, poorer Africans will occupy the low lands, basins with creeks that occasionally flood, or along the slopes on each side, constructing or improving storm drainage remains a priority, followed by improving human and vehicle connectivity and mobility via streets and pathways. The topography, the rolling hills, and the clusters of existing structures often prevent construction access to trucks, backhoes, front-end loaders, and bulldozers. Taking the leanest approach and designing buildings, drainage, and thoroughfares that can be built by hand with minimal equipment, creates jobs for the semi-skilled and provides opportunities for training in construction.

Lean Urbanism on a Regional Scale

Proper African urbanization demands the construction of many thousands of dwelling units and essential non-residential spaces. One worthy outcome is simply determining when and where not to build. A lean sequence of actions for planning at the regional scale in sparsely populated areas subject to urbanization would be:

1. First, determine the location and contours of watersheds and any significant conditions from fault lines to flood plains.
2. Identify all existing road, rail, river, and port transportation corridors, established human and wildlife pathways, and propose new connections or crossings where feasible.
3. Outline the network of riparian corridors and water bodies, forests and plains, and other resource lands and agricultural areas based on soils and topography.
4. Locate and circumscribe the center and boundaries of each new population center by estimating its 'climax condition' or maximum build-out; consider likely growth patterns from crossroad to hamlet, village, town, and city, leveraged by the armature of mobility corridors and existing settlements.
5. Delineate density gradients, highest at the center, descending to the edges, defined by adjacent resource lands, 'greenbelts' or adjoining development.

In populated regions, this five-step process should target existing development when planning for new growth, while protecting or enhancing the surrounding natural resource lands. The population thresholds for each community type will vary from physical conditions to social and economic contexts.

The open source SmartCode Sector Plan (Coyle 2015), an elegant, regional tool, allocates land uses and habitats by sector or zone, providing a lean means for creating regional plans. If the mapping process is hindered by inadequate digital data, such as GIS mapping or aerial imagery, drones provide a lower cost means of photographing inaccessible areas. Where lidar topographical imagery is unavailable, field surveying with a transit, theodolite, or wye level, periodically recording distances and ground elevations, can estimate contours and site features. A 'lean team' can verify aerial images by walking the ground, noting features, and recording coordinates by GIS or analog devices.

Lean Urbanism on a Neighborhood Scale

For those without the means to buy a home or hire a builder, new housing in Central Africa is often self-built (GA Collaborative Design 2014). Home financing, a relatively recent option, is rarely available for the lower income market. Another challenge is finding, affording, and securing a legal building lot in the city or town, where jobs are as essential as building materials. In countries with national ownership over much of the land, the government can lease or extend ownership to land developers, builders, and self-builders. Alternatively, the government can let local or community-scale institutions or private entities control and deliver land for moderate and low-income family housing. Not infrequently, multiple entities might claim ownership of a single parcel. Thus, self-builders and contractors require legally titled lots, even those on hillsides and in flood plains.

If the government owns the land in perpetuity, ground leases or granting long-term land titles can leverage private development. The risk is not the length of the lease but the possibility that the government might take back the land. Along with the protection of public spaces, such as streets and other rights of way, the security of land tenancy is a primary concern that must be addressed to prevent intentional or unintentional illegal building (Allen 2010). The government can further subsidize a development by reducing or eliminating the transfer cost of the land title. Non-government organizations, in cooperation with the government, can create land trusts that maintain ownership of property to enable the construction of new housing without the expense of acquiring individual parcels by each builder.

A lean program for both middle- and lower-income families and individuals, administered by the village chief or neighborhood *mairie* trusted by the people, begins with a simple, dimensioned street, plot and block plan. After a layout of blocks and lots, often within larger public land demarcations, the government can provide housing credits or other direct subsidies to families who can rent, self-build, or purchase. As an alternative, the government can directly distribute free or discounted land titles. Families or individuals can then to choose to:

- self-build on a lot with utility services in the street
- build on a lot that includes a foundation (ideal on hillsides or floodplain) or a shell and core (e.g., toilet, sink, and range top)
- rent or purchase a building, a single house, or a multifamily unit, that requires construction by a private contractor or government housing developer
- rent or buy in the private market.

The large majority of forced-resettlement families are renters or squatters (Sud 2017). Occupancy of a three-bedroom house or apartment, despite its location in a lower density neighborhood, can accommodate a dozen people or more, a high density per square meter. Unanticipated multiple occupancies require adequately sized sanitary sewer treatment systems. However, the disincentive for intentionally building at lower densities includes the rapid consumption of valuable land and the necessary extension of road and utility infrastructure to serve spread-out dwellings. Although urbanization in Africa continues to reshape its settlements, low-density sprawl consumes resource lands and destroys biological resources and fragile habitats. This expansion of human activities into the natural environment has reduced, fragmented, damaged, and isolated fresh water sources, and interrupted and degraded watersheds. Any development density less than 20 units per gross hectare, an area calculation that includes adjacent streets, should be strongly disincentivized by policy and financing constraints (Fig. 12.5).

While a long-functioning neighborhood may require minimal regulations for its expansion, a traditional, walkable module of growth consists of roughly 800 meters from edge of one neighborhood to another, even along market streets. Walking 400 meters for a bus or longer to shop for daily needs represents no hardship for Africans who lack private vehicles. Blocks, streets, and neighborhoods within this physical catchment area are the bricks and mortar of community building, independent of motor vehicles. When built out and linked to adjoining neighborhoods, the pattern forms a healthy and resilient city (Oyeyemi et al. 2016).

Lean Urbanism on a Local Scale

In Africa, adequate infrastructure, particularly water supply and waste management, supports life and deters death. One lean approach to collecting and treating wastewater is the micro-wastewater treatment system. Small clusters of plumbed residences, or a common toilet facility for homes lacking indoor plumbing, connect through simple gravity-flow clay or concrete pipes to a common septic tank and effluent drainage field. With no moving parts or power, the system can be built with local labor and hand tools. Following the eventual construction of a municipal treatment facility, the septic tank, now a large manhole, can connect to new wastewater pipe mains.

Fig. 12.5 The Akanda and Ntoum Commune SmartCode Sector Plans, by simply identifying where and where not to build, help protect fragile, mangroves-lined estuaries and streams from unrestricted, damaging development

In reality, even when large treatment plants are built, villages may wait decades for the arrival of underground wastewater mains or lift stations with sufficient energy to pump wastewater to the treatment plant.

At the building scale, in subtropical climates, pier-supported structures are superior to the ubiquitous slab-on-grade foundation that can absorb moisture or flood water during heavy rains. Between the two, narrow crawl spaces under minimally elevated foundations provide a home for insects, snakes, and rodents. Meanwhile, pier foundations that are raised to a higher head level protect living spaces from ground water and increase seismic stability. Providing a practical way to build on slopes, the foundations can be excavated and constructed by hand although the required stairs may limit the occupancy of those with ambulatory constraints.

The prevailing breezes found in most of Central Africa temper the hot and humid climate, a timeless adaptation in building design that reduces energy consumption and increases comfort. Rarely employed to cool modern buildings in an era of cheap energy, few new developments orient blocks to capture trade winds. Air conditioning is considered essential for those who can afford it, despite an often unreliable electricity grid. Employing building and block-scale cooling, evaporative cooling and shading designs offers both climate adaptation and carbon mitigation, lean interventions that do not rely on expensive technology, high-tech materials, or complex designs. Simple techniques, utilized on a small enough scale and deployed quickly, allow locals to see and feel the benefits provided in short feedback loops. Lean strategies, tools, and techniques such as passive cooling, limit or reduce the unintended, often adverse consequences of modern buildings, such as energy disruptions and costly electrical bills, all while improving physical comfort in the short and long term (Fig. 12.6).

Time-tested technology can alleviate the hardships of power deficiencies. A simple, solar, thermosyphon, hot water system uses the sun to expand heated water to generate heated water flow from roof panel to interior tank, without electricity. Many Central Africa lowlands contain lakes, ponds, and streams. A small uphill pond or reservoir can become a power source. During peak insolation, photovoltaic panels power a low voltage pump that fills the pond or reservoir. When the released water flows by gravity downhill, low-head hydroelectric generators capture the potential energy, or pressurize building supply plumbing. The elevated water storage, a battery that utilizes stored energy, bridges temporal electrical peaks and troughs, and minimizes dependency on generators or external power supplies. From building energy and water systems to constructing streets and footpaths, working on a local, lean scale offers money-saving opportunities and leverages natural energy sources.

Fig. 12.6 French colonial-period homes and the traditional African wooden dwellings were designed to capture the prevailing breezes off the Gabon Estuary, unlike the modern structures dependent on mechanical cooling

Building durably with greater permanence represents a means of adaptability and versatility. For instance, constructing exterior walls from locally sourced and formed clay brick, then parging the walls with protective, sacrificial lime or clay stucco, reapplied from time to time, offers an inexpensive, hand-applied way to build. Despite prioritizing permanent materials and construction methods, site realities often warrant techniques that deliver short-term functionality in anticipation of more permanent, longer-term remedies when conditions improve. These interim, handmade interventions can be beautiful in their successional vernacular.

The Shift to Lean Urbanism

From natural resource lands to the metropolis, an urbanizing Central Africa requires smarter, leaner measures to accommodate growing populations, and to protect watersheds, agricultural areas, and indigenous habitats from drought, flooding, and cultural erosion. The least desirable city lands and environmentally vulnerable areas remain the settlement destinations for poor migrants. Rather than a liability, new population centers can become healthier and more prosperous with proper design and leaner development.

With smaller-scale interventions, the construction of blocks, streets, and neighborhoods can be phased incrementally, consistent with available financing. By "going slow to go fast" (Davis and Atkinson 2010), local laborers can join the process whereby suppliers can contribute a flow of nearby crushed stone and sand. Lean urbanism in Africa will lower the barriers for participation at the tribal or village scale and increase a sense of ownership in both the process and the outcomes. Customizable, step-by-step development, built upon shared goals and strategies, will deliver an accessible, resilient, and confidence-building model for a growing African population.

For a rapidly urbanizing Africa, three critical questions deserve consideration before addressing the demand for better city making. First, determine where to act, at which scale, and in what sequence. Second, in order to plan and develop efficiently and effectively, identify how to best accommodate available resources, however constrained. Each community-building project must be contextually calibrated, with proposed interventions customized for the place and people.

The third critical question is how best to rapidly train local labor and leverage the supply and quality of construction materials. Africa's educational institutions strive to inculcate the young in the value of, and techniques for, protecting the continent's human and physical resources. The students need opportunities to deploy their intellects and utilize their physical stamina, locally and regionally. Africa's young citizens can, over time, build healthy communities by leveraging the tools of lean urbanism (Satterthwaite 2015). However, the challenges of public and private corruption subvert the goals of building good places. For example, peeling back layers of deeply embedded, corrosive behavior during periods of large construction expenditure, for example, require smart transactional techniques, such as paying for labor and materials directly and publicly through the respected village chief.

LEAN DEVELOPMENT OBJECTIVES

With a customizable, community-building template, lean urbanism can be deployed throughout sub-Saharan Africa at regional, community, and neighborhood scales. Teams composed of native professionals with a few experienced advisors, can plan and code or regulate development intensities, standards, and uses across entire provinces, townships, or urban areas while minimizing external financial assistance and expertise. Small scale, development or redevelopment interventions leverage the opportunity to train, employ and involve local trades and industries (Sefa 2000).

Beyond the protection of habitats and the building of durable housing, shops, and streets, a lean approach:

- Enables policymakers, local chiefs, conservationists, economists, hydrologists, and transportation and development planners to collaborate with geography, planning, and urban design students to forge a shared vision, a physical, financial, and regulatory plan for their nation's growth, despite small budgets, minimum technology, and reliance on local talent.
- Targets smaller, substandard, informal neighborhoods or settlements as areas of intervention in an open-ended vision that focuses on incremental, ongoing improvements.
- Employs small-scale neighborhood development that requires fewer resources to initiate, incubate, and mature, allowing do-it-yourself building within economical, low-tech means (Davidson et al. 2003).
- Establishes and manages an open-access, open-source process, allowing more people to meaningfully participate in development (Cotula et al. 2009).
- Creates, deploys, and helps manage locally accessible tools and techniques for all to use, so design and development take less time, reduce resource dependency, and provide pathways to work around financial, bureaucratic, and regulatory obstructions.
- Leverages entrepreneurial and employment opportunities through lean training programs that increase the capabilities and availability of local labor and services.
- Increases the quantity and quality of building and landscape product sources, outlets, and fabrication sites that provide locally accessed materials, including wood, bricks, concrete and shade trees.
- Builds resilient civic amenities and infrastructure capacity, such as playing fields that double as temporary storm water detention and micro-renewable power systems.

Following goal setting and before strategic and action planning, the team must investigate which interventions are underway or completed, and their outcomes. These existing measures are the starting points for identifying how, where, and when to launch, expand, or improve on past successes.

LEAN STRATEGIES AND TACTICS

Often, the leanest ideas precipitate the most durable and resilient outcomes, within a contextual threshold. For example, a relatively high-density neighborhood of 50 dwellings per hectare reduces land costs per dwelling, but only to a point. Consider the infrastructure and soil bearing capacity, and the cost and complexity of structures that local labor can build with available materials. Vernacular public spaces, from informal plazas to football pitches that double as stormwater detention, provide cost effective trade-offs for higher intensity neighborhoods and amenities for the larger community. What is most important is finding a density 'sweet spot,' finding and optimizing a nexus of economic, environmental, and cultural forces, adjusted upwards for future demands (Alter 2014).

By calibrating the speed and sequence of development from initial construction to occupancy in order to match available resources, smaller increments of infrastructure can align activities to optimize the flow of development. The conventional economies of scale approach requires the construction of large or multiple sites in single phases, demanding a corresponding scale of mobilization and front-loaded costs in people, financing and equipment.

By planning streets, utilities, and blocks with a range of potential housing types, the flexibility allows the development to effectively respond to changing markets demands. Rather than planning for similar, cookie-cutter housing, a variation of lots multiplies size and cost choices for families and individuals. The lean, compact, and walkable neighborhood with expanded lot size diversity offers a range of housing types, such as apartments interspersed with shops and employment spaces (Truter 2016).

The residual value of land and buildings are enhanced following the establishment of a lean, walkable neighborhood. Residents appreciate dwelling units in neighborhoods that look, feel and function as a healthy ecology far more than units in conventional, auto-oriented, production housing developments, particularly in later phases. Even absent motor vehicle dependency, escalating rents and sale prices represent both a challenge for continued affordability and a sign of success for investors, similar to a commercial sector in continuous churn as shops and storefronts evolve upwards.

Communities benefit from mixed-use development through a wider, deeper sales and property tax base, whenever they are levied. Public maintenance and infrastructure operating costs can be reduced in higher-density development, not through economies of scale, but through shorter utility pipe runs and reduced equipment demands of lower technological infrastructure that serves a compact physical area.

TEN STEPS TO LEAN DEVELOPMENT

While local populations will respond to credible, sustained attention paid to their concerns and needs, 'out-of-town' planners and developers must recognize a neighborhood's skepticism accreted from past promises that often went unfulfilled. By clearly communicating the project or program goals, the measures of success and the milestones, and by distributing legible maps, drawings and explanatory handouts, the consultant builds a foundation of trust for all subsequent engagements and activities.

The lean approach to design and building, simply explained and differentiated from previous miss-steps and mistakes, should offer realistic descriptions of the potential benefits and risks. Outlining the opportunities for continued involvement and monitoring by stakeholders builds confidence in the impacted populations by providing a sense of control over the outcomes. This simple technique empowers local civic and business leaders to engage their constituents in a joint effort within a development process where the risks of failure are lower.

A 'lean team' could include city and regional planners, geographers, land and building developers, policymakers, engineers, architects, and hydrologists. The leadership, however, does not necessarily require technical expertise but rather one's ability to forge and implement a comprehensive vision, informed by the people and the place. The following ten steps describe the phases and tasks necessary to initiate, research, plan, and execute construction or redevelopment of an underserved neighborhood or settlement, in manageable increments:

1. Confer with lenders, investors, government officials, and other key stakeholders to select or consider opportunities for regional, community, or neighborhood development interventions. Explain the lean alternative, progressing in small sequences that offer larger, short-term benefits and reduce risks.
2. Within each selected jurisdiction, assemble a development team consisting of officials and local civic and business representatives that share the values necessary for long-term success. Establish a development commission tasked with organizing and administering the development activities, or at the very least, an oversight of the process. Include the poorest, most at-risk stakeholders through collaborative and value-driven planning that constrains a tendency to separate neighborhoods by income and social status.
3. Organize and delineate, task by task, a design, development, financing, and implementation plan, and a schedule that identifies who leads and assists, and the resources required. Plans should identify the financing and investment necessary for funding mobilization and on-going development, from roads and flood protection to building construction (Hass 2009; Sy 2016). A robust financing and investment program that specifies funding and accounting procedures for

each element of the project should include the procurement of goods, worker training, supplies, and outsourced services. Always develop a Plan B that accounts for unanticipated events and conditions.

4. Create the conceptual, schematic, design, and construction drawings, specifications, and regulatory development standards sufficient for their proper cost estimating, bidding, contracting and execution, in coordination with the key stakeholders. Cross-functional planning sessions with key disciplines and perspectives at the table, allow a diversity of viewpoints to inform the plan while adhering to the core lean principles and strategies.

5. Draft zoning maps or regulating plans at the parcel or lot scale that describe *where to build* and *at what intensities*, while development standards provide the parameters for *how to build*. Produce phasing plans that forecast *when to build*, prioritizing an optimal sequence of construction with pilot projects that test the efficacy of the plan. Design the housing, shops, schools, and other public and private buildings in incremental phases that add to the community's sense of self-realization and self-sufficiency. Civic amenity maps help determine where best to site facilities such as schools and parks. The guiding principles show *why we build* this way.

6. Design a network of new and/or improved roads, streets, and walkways, with bridges or box culverts at creek crossings. Engineer the utility services—water, sanitary sewer, electricity, and storm drainage— that serve re/development areas—for construction by local labor with materials where feasible. In struggling urbanized areas, the provision of potable water and wastewater treatment, with storm water and flood protection infrastructure, represents a locally-serving and life-saving intervention. Consider that, in areas already serviced, existing connections and equipment may be unsafe or inadequate (Chirisa 2008).

7. Increase the quantity and improve the quality of local construction jobs by establishing training and supervision programs that instruct, in the classroom and in the field, the construction methods and materials for specific trades thereby preparing candidates for upcoming employment opportunities and deliver self-building skills. Supply-chain programs train fledgling entrepreneurs and established suppliers on the processes and standards necessary to provide project materials and equipment, and how to fabricate items, such as trusses.

8. Maintain ongoing educational venues for those impacted by, or who might impact, the development, while candidly sharing the potential benefits, drawbacks, and feedback options. Demonstrate the value of and techniques for protecting agricultural and natural environments; illustrate the value of leveraging low-tech, renewable energy sources. Establish or improve solid waste collection and recycling through a management system that targets construction refuse, an ongoing business opportunity. Demonstration projects provide education by example—installing a locally built, solar, hot water system—and help generate and maintain community support by illuminating the daily specific, inspirational actions by local men and women.

9. Finalize and deploy an Action Plan, with detailed instructions, workarounds to frequent obstructions, and keep the Plan B ready. When funding constraints require revisions to the construction schedule, adjust schedules and scopes of work, such as street and utility installations. Each construction task requires supervision, management, and organizational oversight. Although staff or local managers can fulfill many of these roles, some tasks may require special technical assistance, so those assets should remain 'on call' (Coyle 2011).

10. Establish a comprehensive monitoring, assessment, and reporting system (Ofori 2007) to identify specific hurdles and obstacles while facilitating the generation of solutions. Launch a monitoring program that outlines the methods for observing and reporting on the progress and performance of actions. Timely identification of both problems and corrective actions leads to better control of the desired outcomes. Individual construction projects monitored and reported on regularly can avoid deviations from the action plan and permit course corrections and modifications in the field that save time and money, such as adjusting utility alignments.

LEANER INPUTS, HEALTHIER OUTPUTS

When lots are fully platted, recorded, and ownership protected adds to the local economy (Clos 2012). When development plans are grounded in core values and time-tested practices that reflect local, building traditions, informed through continuous dialogue with the community; when streets and utilities are properly installed, when lots are fully platted, recorded and ownerships protected, Africans will respond immediately with new homes and shops. This approach will trigger a virtuous circle of neighborhood self-improvement that incentivizes the ubiquitous entrepreneurial spirit that permeates local cultures and stimulates regional economies (Clos 2012).

Successful, lean solutions are those employed at minimal risk with relative efficiency, simplicity, and with economic feasibility, by the average community resident. An initial investment is necessary, as well as the strategies and techniques described above. What has worked best in the long run? While we cannot accurately predict the future, we can look back in time to multiple generations of human settlement to identify those lean strategies, tools, and techniques that allowed certain places to survive, endure, and even prosper over the centuries, while avoiding actions that caused settlements to decline or fail outright.

The leanest remedies offer broad applicability over a wide range of circumstances and diversity of environments, for both developers and the eventual end-users. Africa needs a lean, incremental approach for urbanizing cities and villages. Creating many small opportunities will help build resilient and healthier African communities, places built by and worthy of its people. Living with a leaner, resource-efficient footprint provides the only means to achieve long-term sustainability.

Conclusion

Urban density, along with other determinants of urban form, strongly shapes local environmental conditions such as air quality, walkability, and access to green space, all of which have a bearing on the well-being of urban residents. Moreover, developing effective strategies to adapt to and mitigate climate change in urban areas requires looking beyond aggregate statistics on population, physical extent, and resource use. In our study, the large range of potential future patterns of urban development in most of the developing world indicates that these regions can gain a lot in energy savings by encouraging higher urban densities. With growing urban extents and urban populations, how urban areas are configured spatially will matter for the reduction of energy use and associated GHG emissions, with significant implications for the global sustainability.

References

Allen, K. (2010, August 16). The land rush doesn't have to end in a poor deal for Africans. *The Guardian*.

Alter, L. (2014, April 16). Cities need Goldilocks housing density—Not too high or low, but just right. *The Guardian*.

Chirisa, I. (2008). Population growth and rapid urbanization in Africa: Implications for sustainability. *Journal of Sustainable Development in Africa, 10*(2), 361–394.

Clos, J. (2012). *For sustainable cities, Africa needs planning*. Africa Renewal Online. http://www.un.org/africarenewal/magazine/april-2012/sustainable-cities-africa-needs-planning. Accessed November 29, 2017.

Cotula, L., Vermeulen, S., Leonard, R., & Keeley, J. (2009). *Land grab or development opportunity? Agricultural investment and international land deals in Africa*. London: International Institute for Environment and Development.

Coyle, S. J. (2011). *Sustainable and resilient communities: A comprehensive action plan for towns, cities, and regions*. Hoboken: Wiley.

Coyle, S. J. (2015). *Plan de Secteurs, Ntoum, Gabon*. Town-green.com/project/ntoum-sector-plan-plan-de-secteurs. Accessed November 29, 2017.

Davidson, O., Halsnæs, K., Huq, S., Kok, M., Metz, B., & Sokona, Y. (2003). The development and climate nexus: The case of sub-Saharan Africa. *Climate Policy, 3*(1), s97–s113.

Davis, R., & Atkinson, T. (2010, May). Need speed? Slow down. *Harvard Business Review*.

Ebohon, O. J. (2016). *Sustainable construction in sub-Saharan Africa: Relevance, rhetoric, and the reality* (Agenda 21 for sustainable construction in developing countries Africa Position Paper). www.irbnet.de/daten/iconda/CIB660.pdf. Accessed December 27, 2016.

GA Collaborative Design. (2014). *The Masoro project: Self-built houses in Rwanda*. www.indiegogo.com/esi/en/projects/the-masoro-project-self-built-houses-in-rwanda#. Accessed November 29, 2017.

Godard, X. (2013). *Sustainable urban mobility in 'Francophone' sub-Saharan Africa*. United Nations Global Report on Human Settlements.

Hass, K. B. (2009). *Managing complex projects that are too large, too long and too costly.* Complex Project Management. https://www.projecttimes.com/articles/managing-complex-projects-that-are-too-large-too-long-and-too-costly.html. Accessed November 29, 2017.

Huber, B. R. (2015, February 25). *Sub-Saharan Africans rate their health and health care among the lowest in the world.* Princeton: Princeton University's Woodrow Wilson School of Public and International Affairs.

Ofori, G. (2007). Construction in developing countries. *Construction Management and Economics, 25*(1), 1–6.

Oyeyemi, A. L., Kasoma, S. S., Onywera, V. O., Assah, F., Adedoyin, R. A., et al. (2016). Adaptation and reliability of a built environment questionnaire for physical activity in seven African countries. *International Journal of Behavioral Nutrition and Physical Activity, 13*(33).

Price Waterhouse Cooper. (2016). *Retail and consumer outlook: Sub-Saharan Africa.* https://www.pwc.co.za/en/publications/prospects-in-the-retail-and-consumer-goods-sector.html. Accessed November 29, 2017.

Satterthwaite, D. (2015). *Urbanization in sub-Saharan Africa: Trends and implications for development and urban risk.* University of Oxford Urban Transformation weblog. http://www.urbantransformations.ox.ac.uk/blog/2015/urbanization-in-sub-saharan-africa-trends-and-implications-for-development-and-urban-risk. Accessed November 29, 2017.

Sefa Dei, G. J. (2000). African development: The relevance and implications of 'indigenousness'. In B. L. Hall, G. J. Sefa Dei, & D. G. Rosenberg (Eds.), *Indigenous knowledges in global contexts: Multiple readings of our world.* Toronto: University of Toronto Press.

Simmons, K. (2015). *Sub-Saharan Africa makes progress against poverty but has long way to go.* Pew Research Center Factank. http://www.pewresearch.org/fact-tank/2015/09/24/sub-saharan-africa-makes-progress-against-poverty-but-has-long-way-to-go. Accessed November 29, 2017.

Sud, N. (2017). *Rapid urbanization is pushing up demand for housing in sub-Saharan Africa.* http://www.ifc.org/wps/wcm/connect/news_ext_content/ifc_external_corporate_site/news+and+events/news/trp_featurestory_africahousing. Accessed November 29, 2017.

Sy, A. (2016). Impediment to growth. *Finance & Development, 53*(2), 26–27.

Truter, J. (2016). *3 mega trends in Africa's commercial property development.* WSP—Parsons Brinckerhoff weblog. http://www.wsp-pb.com/en/WSP-Africa/Who-we-are/In-the-media/News/2015/3-mega-trends-in-Africas-commercial-property-development2. Accessed November 29, 2017.

Windapo, A. O., & Rotimi, J. O. (2012). Contemporary issues in building collapse and its implications for sustainable development. *Buildings, 2*(3), 283–299.

CHAPTER 13

Planning and Unplanning Amman: Between Formal Planning and Non-traditional Agency

Luna Khirfan

Although Jordan's capital, Amman, is steeped in history, it was intermittently inhabited to the point that by the late 1800s it was but an archaeological ruin where a few Bedouin and Circassian tribes had settled (Munif 1996).[1] Since its appointment as the capital of Transjordan in 1921 (Rogan 1996), Amman's urban growth has been in a continuous state of flux due to regional geopolitical factors. Yet one factor has remained constant: a disparity between Amman's east and west parts that ensues from their different socio-economic and demographic compositions, urban morphologies, physical and social infrastructures, and varying levels of urban services.

In this chapter a discussion of the research methodology is followed by a review of historical and contemporary urban planning in Amman, set against the backdrop of national and regional political dynamics. With the support of census data, the myth of Amman's supposed Palestinian East versus Jordanian West division is dispelled, which has long been perceived as impacting formal planning. It then delves into the interplay of urban governance between the state, the market, and the non-traditional agency of civil society actors and reveals how, by generating new forms of urban governance that favor market forces and that exclude and alienate civil society, neoliberal tendencies have, since the 1990s, instigated unplanning processes in Amman. Unplanning is

---

[1] Circassians were displaced from the northern Caucasus with the Russian conquest and immigrated to the Levant. The first group arrived in Amman around 1879.

L. Khirfan (✉)
University of Waterloo, Waterloo, Canada

© The Author(s) 2019   201
M. Arefi and C. Kickert (eds.), *The Palgrave Handbook of Bottom-Up Urbanism*, https://doi.org/10.1007/978-3-319-90131-2_13

defined as: a regression in the formal planning processes through one or more failed outcomes and/or processes; unjustified deviation from and/or reversal of successful formal (and even informal) planning processes and/or outcomes; unlearning from or altogether rejecting the cumulative knowledge of formal (and often informal) past and current planning experiences; and wavering planning policies, initiatives, and/or processes that are discontinued, paused, halted, or even rescinded.

The last part of the chapter delineates how, in response to these (formal) unplanning processes, civil society actors react, often in non-traditional ways, and rise as influential agents in the formal–informal power networks of Amman's urban governance dynamics, especially after the Arab Spring uprisings. It distinguishes four types of such reactions: coerced apathy, revolt, subversion, and innovative negotiation.

A Grounded Theory Methodology: A Non-linear, Idiosyncratic, and Longitudinal Process

The ideas, data, and analyses presented in this chapter emerged from nearly a decade working on three empirical research projects in Amman between 2007 and 2017 that investigated knowledge exchange between Canadian planners from Toronto and local planners from the Greater Amman Municipality (GAM) (2007–2014), Amman's new urban landscape regarding the trickling of Gulf petrodollars into real estate projects (2011–2017), and Amman's contested public realm vis-à-vis development, inequality, and mobility (2013–2015).[2] Because I was embedded in Amman's context for months at a time over the span of a decade, and because of the bricolage of data I draw upon, and, more importantly, because the notions and patterns discussed herewith emerged fortuitously rather than through a priori hypothesizing, the research methodology typifies a grounded theory approach in reference to when "the researcher seeks to enter a setting without preset opinions or notions, lets the goings-on of the setting determine the data, and then lets a theory emerge from that data" (Groat and Wang 2002: 180). Surely, as I investigated other lines of inquiry, I was devoid of "preset opinions or notions" (ibid.: 180) about the interactions between the formal planning agencies and civil society in Amman. Instead my theorization about "unplanning" Amman and about civil society's "non-traditional agency" emerged through an iterative process of data collection, coding (i.e., analysis), and memoing (sorting and resorting the data, leading to theory building) (see Groat and Wang 2002). Following Anselm Strauss' (1987: 10–11) recommendation, this research also benefitted from my "experiential data" that extended beyond "technical knowledge and experience derived from research" to include my "personal experiences" in Amman.

---

[2] I received funding for the first two research projects from the Social Sciences and Humanities Research Council of Canada (SSHRC) and from Columbia Global Centers, Amman and for the third research project from the Arab Council for Social Science (ACSS).

The primary data include nearly two dozen in-depth interviews with senior planners at GAM and at the Amman Institute, Canadian planning consultants contracted by GAM, city councilors, senior Jordanian policy-makers (e.g., members of parliament), and citizen activists working individually or with non-governmental or community-based organizations. At two points in 2010–2011 and in 2014–2015, I facilitated nearly two-dozen focus group sessions with citizen groups in Amman who represented a wide range of socio-economic and demographic backgrounds. Additionally, I conducted two different on-line survey questionnaires: one in 2011 that focused on Amman's Petrodollar funded urban developments; and another in 2014 that focused on mobility and inequality. I also drew on secondary data sources, including census data, historic and contemporary planning documents, and print and on-line media and social media outlets.

As these data accumulated and as my observations of Amman's urban landscape continued, my personal and professional experience with Amman prompted me to conduct exploratory research into its governance dynamics. Thus a sequential process of deduction (elaboration), verification (checking out), and induction (theory conception) followed. According to Strauss (1987: 12), successful deduction transcends "the ability to think logically" and underscores the researcher's experience in "think[ing] effectively—and propositionally" by "making comparisons that help measurably in furthering the lines of deduction." Similarly, verification "involves knowledge about sites, events, actions, actors, also procedures and techniques (and learned skills in thinking about them). Again that knowledge is based on personal and professional experience" (ibid.: 13). Finally, the induction of the patterns of unplanning Amman and of civil society's agency emerged fortuitously through a non-linear and idiosyncratic process that can be attributed in part to the severe dearth of empirical research and writing on civil society's reactions to and engagement with formal planning in Amman. The data-driven inductive process generated these key themes by "constantly compar[ing] collections/incidents and saturating categories into each other until mature categories emerge[d]" (Creswell 1994: 156; see also: Boyatzis 1998; Fereday and Muir-Cochrane 2006; Groat and Wang 2002). According to this thematic analysis of primary and secondary data, I was able to first theorize about the notion of "unplanning" Amman once its patterns emerged, and then to categorize (and sub-categorize) the four different types of civil society's reactions and their engagements with planning and unplanning Amman, namely: coerced apathy, revolt, subversion, and innovative negotiation. According to Groat and Wang (2002: 181), "An important, distinguishing feature of grounded theory is its use of an intensive, open-ended, and iterative process that simultaneously involves data collection, coding (data analysis), and 'memoing' (theory building)." Concurrently, I embedded in my methodology several tactics to eliminate the risk of researcher bias about the end results, including multimethod data collection approaches, in situ research, understanding the dynamics through the lens of those living them by ensuring their direct input (ibid.).

Dispelling Myths: Amman's Demographic Composition and Urban Growth

The established perception that East Amman had developed around Palestinian refugee camps and, accordingly, is predominantly, if not exclusively, composed of Palestinian refugees and Palestinian–Jordanians, and that because of such a history East Amman was ignored by formal planning, is a perception that I set out to dispel as fundamentally flawed.[3] More accurately, Amman's East–West division began before the arrival of Palestinian refugees in 1948, when landowners evading eminent domain of their properties in the east and south hills of Amman on account of their designation as green spaces had encouraged rural migrants from southern Transjordan to settle in them (Munif 1996).

Since the late 1800s Jordan had received an influx of refugees fleeing political conflict, including Circassians, Levantines,[4] Palestinians, Lebanese, Iraqis, and more recently, Syrians, Libyans, and Yemeni. These refugees have been absorbed in Amman (Potter et al. 2009; Samha 1996), where their resettlement persisted within the city's socio-economic and spatial patterns, thus further heightening the disparity between an affluent northwest, a modest east, and an impoverished south. Reliable statistics about these refugees' exact numbers are absent, and there are discrepancies between the official estimates of 630,000 Syrian refugees and the unofficial claims of over 1.4 million refugees (Rogin 2015). Moreover, because Jordan is the only Arab country to allow Palestinians to become fully assimilated by acquiring Jordanian citizenship, it is impossible to give an accurate percentage of Palestinian–Jordanians in Jordan. Unofficial statistics place this figure at 65–70% of Jordan's total population, with a higher concentration in Amman of up to 80% of the city's population (Soffer 1994). Although these statistics are unofficial, it is possible to infer from the recently published official 2015 census data that of Jordan's total population of 9.5 million, 2.9 million (30.6%) are actually not Jordanian citizens. Also, nearly half of the 2.9 million non-Jordanians (49.7%) live in Amman, compared to only 38.6% of Jordanian citizens (Department of Statistics 2015; Ghazal 2016). These data indicate that Jordanian citizens (both Transjordanians and Palestinian–Jordanians) are actually a minority throughout East and West Amman. Furthermore, considering that population estimates for the primarily Transjordanian Hay el-Tafayleh people, who extend over East and South Amman, stood at 40,000 in 2001 (*Al-Sijill* 2009), and that most of the inner suburbs in East and South Amman—such as Abu

---

[3]Palestinians refers to Palestinian refugees who are not naturalized Jordanian citizens; Palestinian–Jordanians refers to Jordanian citizens of Palestinian origins, including naturalized refugees; and Transjordanians refers to East Bank Jordanians.

[4]The Levant or Greater Syria includes contemporary Syria, Lebanon, Jordan, and Palestine. These refugees arrived during the Great Arab Revolt against the Ottomans in 1916–1918 (Hamarneh 1996).

'Alanda, al-Quwaismeh, and Ajjwaideh—are home to predominantly Trans-jordanian tribes like the al-Hunaiti, al-Abbadi, al-Da'ajah, and al-Hadid, then contrary to established claims the data indicate that Transjordanians consti-tute a majority in East Amman.

Formal Planning: From Bridging Disparities to Neoliberal Marginalization

There is a common perception that municipal planners focus their efforts on West Amman at the expense of East Amman. Yet the data indicate that municipal planners throughout the twentieth century sought to bridge Amman's East–West disparities and to address its rapid expansion through four plans in 1955, 1968, 1978, and 1988. Based on the implementation outcomes—delays notwithstanding—Abu-Dayyeh (Abu-Dayyeh 2004) asserts the efficacy of these four plans. More importantly, GAM's Urban Development Department (UDD) concentrated its efforts throughout the 1980s and 1990s on improving 13 impoverished areas in East Amman through infrastructure provision, housing upgrades, and home ownership (Ababsa 2010). The UDD's efforts in the East Wahdat Upgrading Program won the Aga Khan Award for Architecture for 1990–1992 (The Aga Khan Development Network 1992). But the introduction of neoliberal policies in the 1990s, combined with political events, marked a shift in these attempts to alleviate Amman's East–West disparities and, more often than not, led to unplanning Amman.

Jordan witnessed two distinct phases of neoliberalism. The first during King Hussein's (1952–1999) last years on the throne and the second since King Abdullah II's ascension to the throne in 1999 (Abu-Hamdi 2017). The first coincided with the 1993 Oslo Accords and the 1994 Peace Treaty, amid attempts to resettle, compensate, or return Palestinian refugees (Haddad 2001). While the second corresponded with regional economic push and pull factors and, more recently, with political instability due to regional wars. The transition under King Hussein from a welfare state that subsidizes prop-erty ownership and upgrades housing to one that provides services—namely, physical infrastructure—continued under King Abdullah II, especially in East Amman, through loan-funded, multimillion dollar, road infrastructure pro-jects (European Investment Bank 2003; World Bank 2014). Moreover, King Abdullah II's economic pull policies, with their privatization and incentive programs, sought to encourage development, while concurrent economic push factors ensued from recycling Gulf petrodollars through investments in real-estate projects (Abed and Davoodi 2003). Capitalizing on these pro-cesses, King Abdullah II emphasized economic development in his letter to Omar Maani, then mayor of Amman, stating that, "It is crucial that we all continue to do our utmost to ensure that our beloved city will continue to be a magnet for pioneering development projects and a fertile ground in which innovative ideas can take root and blossom" (King Abdullah II ibn Al Hussein, in Greater Amman Municipality 2008: 10–11).

Eventually, the Amman Plan (AP) and other major urban development projects were born around 2008. The AP's authors assert that the newly proposed Amman Zoning and Building Regulations "act as an incentive to promote economic development, attract business and industry and facilitate the building of a world-class City" (Greater Amman Municipality 2008: 177). For the planning culture that sought to transform Amman into a world-class city this— rather naïvely—meant concealing aspects deemed incompatible with a world-class image, such as zoning out low-income housing developments like Sakan Karim la 'Ish Karim (decent housing for a decent life) by placing them at the urban fringes (Ababsa 2011: 224) in order to literally and figuratively remove the poor from Amman's image. It also meant glitzy, multimillion dollar, mega urban development projects that privileged market actors and that were, naturally, concentrated in West Amman—a typical act of unplanning the previous initiatives that sought to bridge Amman's East–West disparities. These projects also represented a regression in the planning processes in multiple ways.

First, GAM gave one of the very few public children's parks in Amman to a Gulf-based developer to build the high-rise development the Jordan Gate Towers (Abu-Hamdi 2017). This project started in 2005 but was been suspended mid-construction in 2009, plagued with incidents, including a fire, a collapsed crane, and outcries from local communities about the inadequacy of municipal infrastructure for such an unprecedented high-rise development in the midst of a residential neighborhood.

Second was the formation of the Abdali Public Service Commission through a partnership between the state-owned National Resources Investment and Development Corporation, dubbed MAWARED, and two private Gulf-based developers to redevelop Abdali, an established and relatively dense, mixed-use neighborhood in Amman (MAWARED 2010). MAWARED and its partners received preferential treatment from the government through direct funding, subsidies, and logistical support. Most controversial however, was GAM's use of eminent domain to appropriate private property in Abdali, including that of the Talal Abu Ghazaleh Organization, a major employer that eventually relocated the entire enterprise to Beirut after losing a contentious court battle with GAM and MAWARED (Mango 2014: 128–129). Later on, accusations of corruption and embezzlement riddled MAWARED's upper echelons of administration.

Third is the Bus Rapid Transit (BRT) project, which started in 2009 and was also suspended mid-construction in 2011 due to a public outcry against its feasibility and cost.[5] The BRT is one of the few publicly developed mega-projects but, instead of catering to those in dire need of transport, the two proposed lines of the first phase both started in the downtown sector of

---

[5]The French Development Agency loaned GAM US$152 million for the BRT project (Cornock 2015).

the city, at the edges of, but not passing through, East Amman where public transit is badly needed. Instead of servicing East Amman, the first two lines of the BRT extend exclusively through the more affluent West Amman where private automobile ownership is perceived as a status symbol, and thus public transportation is not necessary.

Finally, the construction of the Hashemite Plaza, fronting Amman's Hellenistic theater.[6] This entailed demolishing a vibrant, down-to-earth market and a small green urban park—one of very few, if not the only green park, in the downtown core.[7] Interview data reveal that the proposed plans sought to convert this part of the downtown sector into high-end boutique hotels, expensive shopping venues, and a cable car connecting the Hellenistic theater with the Citadel of Amman. Needless to say, these plans were also aborted. Currently the Hashemite Plaza is a large swath of concrete that is anything but the "social hub" claimed in the project's documents (Fig. 13.1). Collectively, the unplanning that ensued from these (and many other) planning initiatives left the general public with the perception of an incompetent GAM.

In addition to the unplanning that has occurred, a neoliberal restructuring of Amman's governance has taken place. The soft authoritarianism of Jordan's regime precluded the inclusive processes that rescaling—whether downwards, as in decentralization, and/or sideways, as in horizontal participatory governance or networked partnerships such as MAWARED—would have supposedly promoted and fostered (on governance rescaling see Jessop 2002; Swyngedouw 2005). Jordan's soft authoritarianism aligns with semi-authoritarian regimes that claim democratic legitimacy based on established conventions (e.g., representative democracy and accountability), but that simultaneously avoid the political risks of genuine democratic processes (Ottaway 2003; Schedler 2002). More than semi-authoritarianism, Jordan's soft authoritarianism constantly reconfigures governance according to the shifting political, economic, and social milieus so that, I argue, it typifies Michel Foucault's political rationality in how it discerns problems and develops pertinent strategies to maintain the delicate balance between legitimization and power (Michel Foucault in Lemke 2001). Thus, the neoliberal restructuring of urban governance in Amman is persistently manipulated to empower certain actors (the state and the market) while marginalizing, or even excluding, others (civil society). Indeed, King Hussein's era was marked by an upscaling that empowered state actors. An example of which is the 1991 merger of GAM's UDD and the housing corporation that created the national Housing and Urban Development Corporation (Ababsa 2010). Then, in 1995, GAM became semi-autonomous when it was transferred from the Ministry of Municipal and Rural Affairs directly to the Office of the Prime Minister. Under King Abdullah II, GAM's governance restructuring entailed

[6]Wrongly dubbed a Roman theatre.

[7]Details of these plans are available at: http://www.sanabel-flowers.com/landscape_public_project.html.

Fig. 13.1 The new Hashemite Plaza is a large swath of concrete

appointing 14 of the 42 city council members, electing 22 to represent each of Amman's districts, while designating six seats for a women's quota (Shami 2014)—a restructuring that rendered GAM's municipal council rather weak (see Clark 2012; Lust 2009). Additionally, and in a typical neoliberal privatization, the Amman Institute for Urban Development (Ai) was founded as a regional think-and-do-tank, drawing from nearly 50 of GAM's crème de la crème employees. Following Ai's creation, GAM controversially hired the Institute as a consultant for tasks that previously were GAM's prerogative.

Equally important is that unplanning manifests in the lost opportunities to institutionalize genuine public participation and civic engagement initiatives within the AP, or even to capitalize on the Canadian expertise in this area (Khirfan and Jaffer 2012). Instead, the state's soft authoritarianism trickled down to urban planning. Combined with the regional economic pull and push factors and with neoliberal restructuring, this soft authoritarianism empowered market actors such as Gulf-based developers, while excluding and alienating civil society actors from the planning process. Empirical findings reveal that the planning officials' claims of participatory planning in formulating the AP illustrate manipulations of uneven power. GAM selectively and unsystematically invited only influential developers and landowners to participate in consultations to regulate high-rise developments and to amalgamate fringe municipalities (Khirfan and Momani 2017). Moreover, in forming the Mayor's Round Table as a consultative body on urban affairs, then mayor Omar Maani exclusively recruited influential West Amman individuals, some of whom were architects and urban planners whose private firms were, at the time, directly involved in projects tendered by GAM (ibid.). These new governance arrangements exacerbated the exclusion and alienation of civil society, while their atypical configuration rendered their transparency and accountability questionable. Erik Swyngedouw (2005: 1993) dubs such arrangements as "a democratic deficit." Compounding these acts of unplanning, both the Mayor's Round Table and the aforementioned Ai were eventually dismantled. The dismantling of Ai followed debates in the Jordanian parliament that criticized its leadership, role, and salaries. Not surprisingly then, these unplanning dynamics intensified civil society's mistrust of GAM.

Civil Society's Reactions: From Coerced Apathy to Innovative Negotiation

The notion of civil society constitutes the distinguishing factor between government and governance (Harpham and Boateng 1997: 66). Civil society as a "social phenomena" is based on a voluntary associational life at private, civic, and political levels (Young 2000: 157). Civil society is distinct from the state and economy through "coordinated action" and the exercise of "systemic power" (ibid.). Accordingly, civil society becomes the "pivotal terrain from which social transformative and innovative action emerges and where social power relations are contested and struggled over" (Swyngedouw 2005: 1997). I distinguish four types of civil society reactions to, and engagements with, the formal planning and unplanning of Amman, namely: coerced apathy, revolt, subversion, and innovative negotiation.

Coerced Apathy

Apathy is the reaction that Middle East and North Africa (MENA) urban scholars overly highlight. Some scholars altogether discount political awareness among civil society: "There is no political awareness in Arab countries in general that would lead citizens to organize public protests if they feel that their interests are threatened" (Elsheshtawy 2008: 15). Other scholars completely overlook civil society's role in urban governance by limiting the crucial questions to "the administrative and technical functioning of the city" (Malkawi 2008: 35). More perceptive views attribute civil society's apathy to priorities where "politics with a big 'P'" dominate, such as the Palestinian issue, while "politics with a small 'p'," like urban politics, "remain outside the domains of politics and public consciousness" (Daher 2008: 64).

These urban scholars' claims and assertions contradict the theorizations on the role of civil society and the context surrounding the recent Arab Spring uprisings. More accurately, I argue that the soft authoritarianism of Jordan's monarchs that relentlessly repressed the political association of civil society actors in Amman (i.e., Young's third level) is yielding a coerced apathy. Indeed, carefully crafted restrictive legislation and laws render civil society organizations mere providers of social services rather than active participants in urban governance (on these laws, see Brynen et al. 2012; Clark 1995: 593; Wiktorowicz 2002: 610). Examples of such coerced apathy abound. For example, the aforementioned contentious planning initiatives such as the appropriation of a children's park for the Jordan Gate Towers, the use of eminent domain in Abdali, and the demolition of the Hashemite Plaza's green park, all proceeded with sporadic, even nonexistent, objections from civil society actors.

Revolt

The Arab Spring uprisings epitomized a 'moment' of significant change in the relationship between civil society, authoritarian states, and the market in MENA. Indeed, the mobilization of the Arab Spring uprisings has been partly attributed to the economic stress of the neoliberal experience (see, for example, Diwan 2014; Hanieh 2013). As the socio-economic disparities continue to amplify under neoliberal restructuring, and while MENA entrepreneurs advance their economic and political gains, the majority of civil society is bearing the brunt of neoliberalism's economic and political policies, with unemployment rampant among the youth and the educated in particular. These disparities led to the rise of "social non-movements" that, differently from (Young 2000: 157) "coordinated action," rally around "collective action" through the "fragmented but similar activities" of "non-collective actors," whose agency and activism lack formal leadership and organizational structures (Bayat 2000: 15, 20).

Recently, urban politics have garnered the attention of social non-movements around MENA. The most prominent of these social non-movements is Beirut's You Stink! campaign (since 2015), which rallied people from all political stripes in socially and politically divided Beirut around urban issues like garbage collection. Amman has its share of such non-movements in which individuals, encouraged by the events of the Arab Spring uprisings in neighboring countries, opposed urban initiatives such as municipal amalgamation and the BRT. Protests, both physical demonstrations on the streets of Amman and virtual manifestations over social media, erupted to contest these decisions. While these social non-movements challenged the long-standing apathy perspective and demonstrated the inclination of civil society actors to engage with urban politics, I nevertheless present three observations related to their consequences on planning and unplanning Amman.

First, these social non-movements emerged from those constituencies directly affected by the planning decisions, demographic makeup notwithstanding.[8] This is evidenced in how actors across a wide spectrum contested the BRT project, while the opponents of the amalgamation decision represented the tribal makeup of the affected areas. Second, these social non-movement contestations in Amman seem to be motivated by private interests rather than a collective good. Focus groups and interview participants in Muwaggar, one of the amalgamated municipalities, adamantly rejected the amalgamation, citing rising taxes and a lack of services. But further probing revealed that with aspirations for increased property values, improved urban services (minus, of course, tax increases), and increased real-estate investments, focus group participants had initially supported the amalgamation. Only when those aspired-for positive outcomes did not immediately materialize (due to the 2008 economic downturn, among other factors) did the residents of these municipalities rally against the amalgamation policy. In a typical act of unplanning, municipal residents pressured the government to rescind the policy. Finally, these social non-movement contestations of urban affairs seem impulsive and arbitrary—stemming from misinformation at best, or from ignorance at worst. For example, data reveal that the civil society outcry against the BRT project was misled by the dearth of information due to the exclusion of civil society from the planning process. Almost unanimously, interviewees confusingly assumed that 'rapid' referred to the actual speed of the bus. Even after GAM launched the BRT website, civil society actors (including news and media outlets) never questioned the BRT routes that traversed affluent West Amman while runing tangentially to impoverished East Amman. Instead, civil society protested the cost and inconvenience of construction. In a city with a dearth of transit systems, the BRT was one public transportation project that would have benefited the urban poor, its routes notwithstanding. Thus, in this case, civil society's activism essentially led to unplanning.

---

[8]There are some generalizations that these social non-movements are tribally-based (see, for example, Clark 2012).

Subversion

The farce of so-called democratic reform under neoliberalism led to a combination of marginalization, disenfranchisement, and frustration (see Jebnoun 2014; Schwedler 2012). Consequently, subversion is on the rise, including the upsurge of the dangerously radicalized who can effectively exploit instability. Most notably, the extremist ultra-conservative Islamist movements are successfully recruiting marginalized, disenfranchised, and frustrated urban youth through their social networks. Although Transjordanians have traditionally been loyal to the monarchy, interestingly they comprise most ISIS affiliates in recent terrorist subversive acts inside Jordan, such as the Karak Castle attacks in December 2016.

Subversion also manifests as unplanning though acts of vandalism and littering (Fig. 13.2) or with more positive agency through art and culture—censorship notwithstanding. Challenging the government's obstruction of Internet accessibility, its content control, and its restrictions on the press (Freedom House 2015a, b); the activist Issam al-Zaben, a retired air force pilot, founded the Jordanian Boycott Campaign on Facebook to contest the government's unjustified price hikes and fees. The campaign garnered nearly 1.5 million likes before al-Zaben was apprehended in February 2017 and his page closed (*Assabeel* 2017; Sawaleif 2017). Jordanian activists also organized concerts for the Lebanese band Mashrou'Leila, whose subversive messages in their songs tackle social and political affairs. Perturbed by these

Fig. 13.2 Litter on the streets of (West) Amman

messages, (the band's lead singer, Hamed Sinno, is openly gay), the authorities cancelled their April 2016 concert in Amman's Hellenistic theater and "unofficially informed [the band] that we will never be allowed to play again anywhere in Jordan due to our political and religious beliefs and endorsement of gender equality and sexual freedom" (BBC 2016).

Innovative Negotiation

Civil society's negotiating capacity stems from self-esteem and manifests as a non-traditional agency that either confronts or copes with Jordan's soft authoritarianism. As a technology of citizenship, self-esteem "links subjectivity and power in a way that confounds any neat separation of the 'empowered' from the powerful" by "advocat[ing] a new form of governance that cannot be critically assessed by mobilizing the separation of public from private, political from personal" (Cruikshank 1993: 340–341). Thus, in the face of the "coercion, force, or social control engineered from above" by Jordan's soft authoritarianism, self-esteem reflects an individual's "voluntary" exercise of the "technologies of selfhood" (ibid.: 340–341). In Amman's urban arena, this translates into civil society's inclination, at its three levels of associational life (i.e., private, civil, or political) to innovatively negotiate so as to counter the democratic deficit of state authoritarianism and neoliberal governance. Ironically, through decades of repressing civil society actors, Jordan's soft authoritarianism has correspondingly increased their negotiating capacity. I identify four types of such innovative negotiations that, as examples of non-traditional agency, are considered successful planning, based on indicators such as resonating with the local communities and producing both short-term and long-term positive tangible outcomes. These negotiations parallel Richard Stren's (2001: 172) informal governance mechanisms. I argue, however, that they not only supplement (as in Stren's model), but they actually supplant, the activities of formal planning by filling the gaps in social services left by the withdrawal of the traditional welfare state.

First are the informal civil society initiatives that are characterized by modest, incremental actions that take shape outside of formal planning, which, typical of unplanning, fails to engage with or capitalize on these initiatives' success. For example, in 2004 an informal neighborhood group in Jabal Amman initiated Souk Jara, a Friday market that has since been running strong. Second are the collaborations between civil society and market actors who demonstrate corporate social responsibility, such as ARAMEX's CEO Fadi Ghandour, who founded Ruwwad Al-Tanmeya in 2005 and whose services have expanded to Palestine, Lebanon, and Egypt. Ruwwad offers an array of social services that center on youth organizing, child development, community support, community-led campaigns, and micro venture funding (Ruwwad Al-Tanmeya 2015). Third are the partnerships between the public sector and civil society, such as the Islamized postal savings that provide interest-free savings tailored to religious needs. This service enables the marginalized urban poor to incrementally upgrade their housing conditions (Tobin 2017). Fourth are the innovative non-governmental organizations, such as Arini, a nonprofit private study and research institution that focuses on urbanism and architecture, whose Mapping Jabal al Natheef project documented the urban conditions of this high-density, low-income, East Amman district (Arini 2014).

CONCLUDING REMARKS

This chapter dispels established perceptions of Amman as myths: first, that Amman's East–West division is based on national origins, namely Palestinians and Palestinian–Jordanians (East Amman) versus Transjordanians (West Amman); second, and on account of first, that twentieth-century planning in Amman privileged West Amman; and, third, that it was largely ineffective. An analysis of census data reveals that Jordanian citizens (Palestinian–Jordanians and Transjordanians) are actually a minority throughout East and West Amman, while East Amman, in all likelihood, is predominantly Transjordanian. More importantly, the efficacy of twentieth-century planning is revealed, especially in terms of alleviating the disparities between an affluent West and an impoverished East Amman. In contrast, it reveals that planning initiatives during the second phase of neoliberalism epitomize an unplanning process in Amman. This unplanning not only amplifies the East–West division through an immersion in glitzy projects as part of an attempt to create a world-class city in West Amman, it also represents a regression of formal planning in general.

This was followed by an outline of civil society's reactions to the formal planning and unplanning of Amman, especially in the wake of the Arab Spring uprisings. It notes that civil society in Amman is gradually transforming its coerced apathy into other forms of non-traditional agency, including revolt, subversion, and/or innovative negotiation; and contends that whether it is the fragmented, yet collective, actions of social non-movements, the activism of coordinated groups and disconnected persons, or the innovative maneuvers of individuals and organizations, the rise of non-traditional agency in Amman (and MENA) warrants the attention of professional planners and researchers. There is a need to dissect these actors' autonomy, the nature of their social contract, and the consequences of their mistrust toward the establishment. Specifically, a thematic analysis of these nuances reveals the differences in agenda between the various civil society actors and the ensuing social non-movements, and that their demands and actions essentially constitute unplanning. More importantly, with the prevalence of regional geopolitical instability that is yielding a ceaseless influx of refugees to Amman, to the extent that refugees in Amman now outnumber the city's permanent residents, it becomes essential—rather than optional—to garner "the power to learn new practices and create new capacities" (Coaffee and Healey 2003).

This chapter argues that this power should simultaneously build on the cumulative knowledge of formal planning and unplanning, as well as on the non-traditional agency of civil society actors, both positive (informal planning) and negative (informal unplanning).

REFERENCES

Ababsa, M. (2010). *The evolution of upgrading policies in Amman. Sustainable architecture and urban development.* Paper presented at the Second International Conference on Sustainable Architecture and Urban Development, CSAAR, MPWH, University of Dundee, Amman Jordanie, Jordan.

Ababsa, M. (2011). Social disparities and public policies in Amman. In M. Ababsa & R. Daher (Eds.), *Cities, urban practices and nation building in Jordan: Villes, pratiques urbaines et construction nationale en Jordanie* (pp. 205–232). Beyrouth: Presses de l'Institut francais du Proche-Orient.

Abed, G. T., & Davoodi, H. R. (2003). *Challenges of growth and globalization in the Middle East and North Africa.* Washington, DC: International Monetary Fund.

Abu-Dayyeh, N. I. (2004). Persisting vision: Plans for a modern Arab capital, Amman, 1955–2002. *Planning Perspectives, 19,* 79–110. https://doi.org/10.1080/02665 43042000177922.

Abu-Hamdi, E. (2017). The processes of neoliberal governance and urban transformations in Amman, Jordan. In L. Khirfan (Ed.), *Order and disorder: Urban governance and the making of Middle Eastern cities* (pp. 132–154). Montreal and Kingston: McGill-Queen's University Press.

Al-Sijill. (2009, May 28). They fled to Amman in the twenties, Al-Tafayleh founded their neighbourhood in the fifties. *Al-Sijill.* Retrieved from http://www.al-sijill. com/sijill_items/sitem7524.htm.

Arini. (2014). *Mapping Jabal Al-Natheef.* Amman, Jordan: Arini.

Assabeel. (2017, February 3). *Who is Issam Al-Zaben the founder of the "boycott" page... and what is his charge?* Khirfan-PlanningandunplanningAmman-2017-08-11.docx. Accessed February 3, 2017.

Bayat, A. (2000). *Social movements, activism and social development in the Middle East* (Paper No. 3).

BBC. (2016, April 27). *Mashrou' Leila: Jordan bans Lebanese rock band with gay singer.* http://www.bbc.com/news/world-middle-east-36148343. Accessed February 28, 2017.

Boyatzis, R. (1998). *Transforming qualitative information: Thematic analysis and code development.* Thousand Oaks, CA: Sage.

Brynen, R., Moore, P. W., Salloukh, B. F., & Zahar, M.-J. (2012). *Beyond the Arab Spring: Authoritarianism & democratization in the Arab world.* Boulder: Lynn Reiner Publishers.

Clark, J. (1995). The state, popular participation, and the voluntary sector. *World Development, 23*(4), 593–601.

Clark, J. (2012). Municipalities go to market economic reform and political contestation in Jordan. *Mediterranean Politics, 17*(3), 358–375.

Coaffee, J., & Healey, P. (2003). 'My voice: My place': Tracking transformations in urban governance. *Urban Studies, 40*(10), 1979–1999.

Cornock, O. (2015, November 29). Get on the bus. *Venture Magazine: Levant Business Intelligence.*

Creswell, J. W. (1994). *Research design: Qualitative and quantitative approaches.* Thousand Oaks: Sage.

Cruikshank, B. (1993). Revolutions within: Self-government and self-esteem. *Economy and Society, 22*(3), 327–344.

Daher, R. (2008). Amman: Disguised genealogy and recent urban restructuring and neoliberal threats. In Y. Elsheshtawy (Ed.), *The evolving Arab city: Tradition, modernity & urban development* (pp. 37–68). London and New York: Routledge.

Department of Statistics. (2015). *Population and housing census 2015.* Retrieved from http://www.dos.gov.jo/dos_home_e/main/population/census2015/index.htm.

Diwan, I. (2014). Understanding revolution in the Middle East: The central role of the middle class. In I. Diwan (Ed.), *Understanding the political economy of the Arab uprisings* (pp. 29–56). Singapore: World Scientific.

Elsheshtawy, Y. (2008). The great divide: Struggling and emerging cities in the Arab world. In Y. Elsheshtawy (Ed.), *The evolving Arab city: Tradition, modernity & urban development* (pp. 1–26). London and New York: Routledge.

European Investment Bank. (2003). *Jordan: EIB EUR 26 million loan for the Amman ring road.* http://www.eib.org/infocentre/press/releases/all/2003/2003-145-eur-26-mio-loan-for-the-amman-ring-road-in-jordan.htm. Accessed February 15, 2017.

Fereday, J., & Muir-Cochrane, E. (2006). Demonstrating rigor using thematic analysis: A hybrid approach of inductive and deductive coding and theme development. *International Journal of Qualitative Methods, 5*(1), 80–92.

Freedom House. (2015a). *Freedom of the press—Jordan, country report.* Washington, DC: Freedom House.

Freedom House. (2015b). *Freedom on the net—Jordan.* Washington, DC: Freedom House.

Ghazal, M. (2016, January 30). Population stands at around 9.5 million, including 2.9 million guests. *The Jordan Times.*

Greater Amman Municipality. (2008). *Amman plan: Metropolitan growth report.* Amman: Greater Amman Municipality.

Groat, L., & Wang, D. (2002). Qualitative research. In *Architectural research methods* (pp. 173–202). New York: Wiley.

Haddad, M. (2001). Palestinian refugees in Jordan and national identity. In J. Ginat & E. J. Perkins (Eds.), *The Palestinian refugees: Old problems—New solutions* (pp. 150–168). Norman: University of Oklahoma Press.

Hamarneh, M. (1996). Amman in British travel accounts of the 19th century. In S. Shami & J. Hannoyer (Eds.), *Amman: Ville et Société* [The city and its society] (pp. 57–87). Beirut: Centre d'Études et de Recherches sur el Moyen-Orient Contemporain.

Hanieh, A. (2013). *Lineages of revolt: Issues of contemporary capitalism in the Middle East.* Chicago: Haymarket Books.

Harpham, T., & Boateng, K. A. (1997). Urban governance in relation to the operation of urban services in developing countries. *Habitat International, 21*(1), 65–77.

Jebnoun, N. (2014). Introduction: Rethinking the paradigm of "durable" and "stable" authoritarianism in the Middle East. In N. Jebnoun, M. Kia, & M. Kirk (Eds.), *Modern Middle East authoritarianism: Roots, ramifications, and crisis* (pp. 1–24). London: Routledge.

Jessop, B. (2002). Liberalism, neo-liberalism, and urban governance: A state-theoretical perspective. *Antipode, 34*(3), 452–472.

Khirfan, L., & Jaffer, Z. (2012). Canadian planning knowledge in the Middle East: Transferring Toronto to Amman and Vancouver to Abu Dhabi. *Canadian Journal of Urban Research, 21*(1), 1–28.

Khirfan, L., & Momani, B. (2017). Tracing participatory planning in Amman. In L. Khirfan (Ed.), *Order and disorder: Urban governance and the making of Middle Eastern cities* (pp. 79–102). Montreal and Kingston: McGill-Queen's University Press.

Lemke, T. (2001). 'The birth of bio-politics': Michel Foucault's lecture at the Collège de France on neo-liberal governmentality. *Economy & Society, 30*(2), 190–207.

Lust, E. (2009). Competitive clientelism in the Middle East. *Journal of Democracy, 20*(3), 122–135.

Malkawi, F. K. (2008). The new Arab metropolis: A new research agenda. In Y. Elsheshtawy (Ed.), *The evolving Arab city: Tradition, modernity & urban development* (pp. 27–36). London and New York: Routledge.

Mango, T. (2014). *The impact of real estate construction and holding companies: A case study of Beirut's Solidere and Amman's Abdali*. Ph.D., University of Exeter, Exeter.

MAWARED. (2010). *Abdali project—Overview*. http://www.mawared.jo/abdali.shtm. Accessed January 25, 2017.

Munif, A.-R. (1996). *Story of a city: A childhood in Amman* (S. Kawar, Trans.). London: Quartet Books.

Ottaway, M. (2003). *Democracy challenged: The rise of semi-authoritarianism*. Washington, DC: Carnegie Endowment for International Peace.

Potter, R. B., Darmame, K., Barham, N., & Nortcliff, S. (2009). "Ever-growing Amman", Jordan: Urban expansion, social polarisation and contemporary urban planning issues. *Habitat International, 33*, 81–92. https://doi.org/10.1016/j.habitatint.2008.05.005.

Rogan, E. L. (1996). The making of a capital: Amman, 1918–1928. In S. Shami & J. Hannoyer (Eds.), *Amman: Ville et Société* [The city and its society] (pp. 89–108). Beirut: Centre d'Études et de Recherches sur el Moyen-Orient Contemporain.

Rogin, J. (2015, October 6). U.S. and Jordan in a dispute over Syrian refugees. *Bloomberg*. Retrieved from https://www.bloomberg.com/view/articles/2015-10-06/u-s-and-jordan-in-a-dispute-over-syrian-refugees.

Ruwwad Al-Tanmeya. (2015). *Ruwwad—Our story*. http://ruwwad.net. Accessed February 28, 2017.

Samha, M. (1996). Migration trends and population growth in Amman. In S. Shami & J. Hannoyer (Eds.), *Amman: Ville et Société* [The city and its society] (pp. 191–208). Beirut: Centre d'Études et de Recherches sur el Moyen-Orient Contemporain.

Sawaleif. (2017, February 3). *How was the boycott campaign founder apprehended? Details*. Khirfan-PlanningandunplanningAmman-2017-08-11.docx. Accessed February 2, 2017.

Schedler, A. (2002). Elections without democracy: The menu of manipulation. *Journal of Democracy, 13*(2), 36–50.

Schwedler, J. (2012). The political geography of protest in neoliberal Jordan. *Middle East Critique, 21*(3), 259–270.

Shami, S. (2014). *Municipal and parliamentary elections in Jordan: Cosmetic reforms with little impact on political status quo*. Amman: Arab Reporters for Investigative Journalism (ARIJ).

Soffer, A. (1994). Jordan facing the 1990s: Location, metropolis, water. In J. Nevo & I. Pappé (Eds.), *Jordan in the Middle East: The making of a pivotal state 1948–1988* (pp. 26–55). Portland, OR: Frank Cass.

Strauss, A. L. (1987). *Qualitative analysis for social scientists* (fourteenth printing 2003 ed.). Cambridge, UK: Cambridge University Press.

Stren, R. E. (2001). Commentary—Khartoum: Canary in the mineshaft? In S. Shami (Ed.), *Capital cities: Ethnographies of urban governance in the Middle East* (pp. 169–173). Toronto: Center for Urban and Community Studies, University of Toronto.

Swyngedouw, E. (2005). Governance innovation and the citizen: The Janus face of governance-beyond-the-state. *Urban Studies, 42*(11), 1991–2006.

The Aga Khan Development Network. (1992). *The Aga Khan award for architecture: East Wahdat upgrading programme*. http://www.akdn.org/architecture/project/east-wahdat-upgrading-programme. Accessed January 16, 2017.

Tobin, S. A. (2017). Islamized postal savings: A model for risk sharing. In L. Khirfan (Ed.), *Order and disorder: Urban governance and the making of Middle Eastern cities* (pp. 189–211). Montreal and Kingston: McGill-Queen's University Press.

Wiktorowicz, Q. (2002). The political limits to non-governmental organizations in Jordan. *World Development, 30*(1), 77–93.

World Bank. (2014). *Jordan—Amman development corridor*. Washington, DC: World Bank Group.

Young, I. M. (2000). *Inclusion and democracy*. Oxford: Oxford University Press.

CHAPTER 14

Endurance, Compliance, Victory: Learning from Informal Settlements in Five Iranian Cities

Mahyar Arefi and Neda Mohsenian-Rad

INTRODUCTION

This chapter gleans some lessons from the World Bank's enablement initiative that took place in Iran from 2004 to 2009. While much research on informality has been conducted (especially in Farsi, including university theses and anecdotal evidence from reports by consulting planning firms), to date, a systematic account of the accomplishments of this joint venture between Iran and international agencies does not exist.

The history of informal settlements in Iran dates back to the post-World War II period of rapid urban development following Mohammad Reza Pahlavi's land reform policies. Cities in that era witnessed rapid growth thanks to the economic boom from the petro-dollars of the 1950s and 1960s and the rural-urban migration. As in other countries though, the reaction to the rise of informal settlements comprised coercive measures and bulldozing. This physical response to an otherwise socio-economic, cultural, and

M. Arefi (✉)
Department of Planning and Landscape Architecture,
University of Texas at Arlington, Arlington, USA

N. Mohsenian-Rad
Atlanta, GA, USA

© The Author(s) 2019
M. Arefi and C. Kickert (eds.), *The Palgrave Handbook of Bottom-Up Urbanism*, https://doi.org/10.1007/978-3-319-90131-2_14

even global phenomenon was a commonly accepted solution at the time. New experiences and dynamics gradually brought significant shifts to addressing informal urbanism.[1]

A Brief Overview

A World Bank loan to Iran in 2004 served to 'enable' informal settlements by giving poor people hopes of new opportunities for home ownership and by introducing a new discourse on policymaking. Notwithstanding, not everyone viewed this loan as a financial impetus for welfare development, many saw the loan as part of a new rhetoric that reinforced the conventional, condescending, blame-the-victim mentality of decision makers toward poor people (Nathan 1992; Lewis 1959). Despite these divergent views, a five-year loan (phase 1), with the possibility of an extension for seven more years (phase 2), was granted to physically upgrade five Iranian provincial capitals that had a high percentage of their urban population living in informal settlements (see Fig. 14.1).[2] The Iranian government discontinued the second phase and capitalized on the outcomes achieved during the first phase (Alaedini et al. 2012). Viewing as part of the city, the World Bank designated specific informal settlements in each city.

Planning and architecture consultants studied each city first and then collected community input through brainstorming sessions at local mosques or other public venues. This information provided the basis for developing upgrade plans. The following feedback from the five cities illustrates the input from participants on what they considered their key urban problems: lack of healthcare and recreational facilities in Tabriz and Kermanshah; accessibility and transportation problems, and poor street pavement quality in all five cities; drug dealing in Tabriz and Zahedan; poverty and unemployment in Tabriz, Zahedan, Kermanshah; water and sewage disposal problems in Tabriz and Bandar Abbas; high population density in Zahedan and Kermanshah; environmental and garbage collection problems in Zahedan, Bandar Abbas, and Kermanshah; safety and security issues in Tabriz and Zahedan; illegal or no utility hookups in Kermanshah; and a lack of coordination for service delivery among public agencies in Bandar Abbas.

Conceptualizing the Enablement Outcomes

Enabling informal settlements means different things to different people, but the World Bank's interpretation has three attributes: *market enabling*, providing low-income residents with access to the marketplace; *political enabling*, providing access to political capital; and *community enabling*, leveraging social

---

[1] See Dovey, Kim. 2012. Informal Urbanism and Complex Adaptive Assemblage. *International Development Planning Review* 34 (4): 349–367.

[2] Bandar Abbas (capital of Hormozgan province); Tabriz (capital of Azarbaijan Province); Kermanshah (capital of Kermanshah province); Zahedan (capital of Sistan and Baluchestan Province); Sanandaj (the capital of Kurdestan province).

Fig. 14.1 Location of the five target cities in Iran (*Source* Neda Mohsenian-Rad)

capital through grassroots efforts and shared governance (Irandoust 2010). While the distinction between enablement and empowerment is important, both concepts question the effectiveness of the state's top-down role and of the marketplace in housing provision, and emphasize good governance and bottom-up planning as an alternative (Takahashi 2009; Shatkin 2000). The market imperfection and the profit-maximizing attitude of the government toward development, allegedly intensifies the gap between the haves and the have-nots. Therefore, it makes sense to focus on governance and top-down growth models not as competitors but as complementarities. Against this backdrop, enablement aims to enhance the "democratization of decision-making systems in urban governance with citizens' views taken into consideration" (Takahashi 2009: 116).

As part of this transformative process, adaptation provides access to market forces, political and social capital, formalization/legalization, and ultimately integration into the city. Another popular interpretation of enablement in Iran comprises the improvement of settlements in four dimensions:

meaning, feeling self-motivated and self-driven by set goals; *self-fulfillment*, feeling good about successful completion of implemented projects; *self-determination*, having agency in the decision-making process; and *self-efficacy*, redressing the problems and making changes within a given period of time (Rostamzadeh 2011).

Despite good intentions and based on the implementation outcomes of the World Bank's physical upgrading projects, the enabling process in Iran has not met all the above goals. Market enabling, for example, while feasible for garbage recycling in Kermanshah, did not last because the city co-opted it; self-efficacy and self-determination also fell short. Enthusiasm in voluntarism and providing construction materials for façade improvement in Bandar Abbas fizzled out because contractors and local authorities did not act promptly. When local authorities hoped to get community input and to muster support for tangible project outcomes in Bandar Abbas, they faced community fatigue and found that meetings were useless. Before the projects even started, four fruitless community participation programs had already been launched. City officials realized that the community showed no interest in attending more meetings, indicating that the limit had been reached for community enabling and self-fulfillment.

An understanding, synthesis, and theorizing of the data from these five Iranian cities point to three separate but interrelated key themes: *models*, *policies/processes*, and *outcomes*. *Models* highlight the strategies adopted by host countries from international organizations such as the World Bank, International Monetary Fund, or the UN-Habitat (in Iran's case, enablement). *Policies/processes* comprise either top-down (i.e., the enablement processes), or bottom-up (i.e., shared governance) structures. Enabling *outcomes* describe the effectiveness of upgrading goals. These three themes can be synthesized into three outcomes that can be roughly regarded as 'stages' of enabling. While trying to continuously *adapt* informal settlements to unforeseen circumstances, policymakers typically aim to *formalize/legalize* them before *integrating* the settlements into the rest of the city.

The following illustrates these three outcomes:

1. *Adaptation* through physical, social, and cultural coping mechanisms prior to formalization/legalization and integration

2. *Formalization/legalization* through physical upgrading and granting title deeds

3. *Integration* through connectivity to the rest of the city, physically, socially, and culturally.

The data analysis revealed unpredictable dynamics among models, policies/processes, and outcomes. That is why enabling cannot result solely from top-down decrees or spontaneous bottom-up efforts. Additional provisions are required to come to grips with these highly complex circumstances.

Adaptation

As the first characteristic of upgrading informal settlements, adaptation contrasts with formalization, which typically happens by top-down decree or fiat. Mediating between formalization/legalization and integration, adaptation requires coping and adjustment to the status quo, and captures unpredictable mechanisms of survival, protection, and preservation. These coping mechanisms cannot be planned ahead of time. For example, in Sanandaj and Zahedan people adapted by chipping in partial costs of road repair, and in Kermanshah by turning garbage collection into a bottom-up enterprise.

Since informal settlements evade formal planning (Roy 2005; Dovey 2012), adaptation becomes very important. According to Dovey (2012), what distinguishes formal from informal settlements is their complex nature, characterized by redundancy (behaving in different ways), resiliency (adaptability to change), and panarchy (different parts working with other parts at higher scales or levels). These characteristics contrast planned developments that have more predictable outcomes. The home in an informal settlement can also be a place of business. Roads operate as both streets and public spaces. Informal settlements reflect "complex adaptive systems" (Dovey 2012) that are less predictable compared to formally planned systems.

In Iran, informal settlements epitomize resiliency and vibrancy despite the external forces operating against them. Unexpected spontaneous occurrences kept informal settlements resilient in all five cities, and no destruction or disintegration was reported. In formal planning paradigms significant failures or external pressures can result in vacancy and possible meltdown. Conversely, informal settlements prove to be both resilient by redundancy or change of behavior as a survival or coping mechanism, and to demonstrate panarchy.

As opposed to hierarchy, panarchy implies becoming part of a system at a higher level (Dovey 2012). For example, road repair in Sanandaj mobilized the community toward a shared purpose (community enabling). Garbage recycling in Kermanshah acted as a social glue for coalescing people around environmental and economic enablement.[3] In traditional hierarchical systems, however, actions at a lower scale generally respond to a higher level of similar functionalities. For example, street networks represent hierarchical

---

[3]This low-tech recycling process engaged a wide network of people and businesses, from shop owners who rented out carts to truck drivers distributing them among end users who collected garbage throughout the city, to people that separated and packaged garbage by type. For more information see Khatam (2002).

systems where every road belongs and connects to a higher (or lower) network, whose construction/maintenance is a public responsibility. And if, for some reason, a responsible governmental agency fails to implement a project on time, road construction comes to a complete standstill. The hierarchical structure of formally planned developments represents a complex and interwoven bureaucratic, legal, and physical structure in which not only roads follow a clear network, but specific private or public organizations are responsible for their operations before and after implementation.

Local and national governments, developers, and contractors are each responsible for the timely delivery of their projects in comprehensive plans—otherwise a project will stall or fail altogether. In informal settlements, however, residents have found ways to cope with failings and shortcomings and can even create new opportunities to mobilize their social assets for other purposes.[4] Thus, if the municipality fails to pave the roads in an informal settlement on schedule, residents would not stop their day-to-day activities just because the roads are not paved, or because trash is not collected.

This is why informal settlements do not lend themselves to straightforward and predictable planning. Subdividing the initially large lots in Dowlatabad, Kermanshah, into smaller lots shows an effective adaptation of making unaffordable large lots affordable. The efficient bottom-up garbage recycling scheme movement that took place in Jafarabad, Kermanshah, is another example of adaptation to the failure of public agencies to recycle. People's self-help in street maintenance and repair in Sanandaj, despite an adverse reaction from the local contractors, also demonstrates adaptation as a coping mechanism.

Informal settlements do not always trust conventional planning. In Tabriz, a planning consultant approached the local clergy to encourage people to attend meetings, but citizens resisted as they thought that city officials would attend to seek more votes and forget about the community residents once the local elections were over. In another case, residents of Bandar Abbas experienced obstacles to public participation and collaboration with developers, despite the fact that some residents willingly provided construction materials for façade improvements, paid for cement bags, and/or served as neighborhood mayors.[5] Understandably though, the more actively the municipality encouraged residents, the more public participation occurred in the enablement process. However, residents showed reluctance to give up part of the land they occupied for road widening, or to add sidewalks to a narrow street network (Alaedini et al. 2012). Despite the relative success in physical

---

[4] For example, in Pinar, Istanbul, people have reused discarded mattress wire mesh to separate public from private spaces and to demarcate their house boundaries.

[5] Eskandari (2008) explains how 22 local women served as neighborhood mayors and oversaw the implementation of repairing an open sewer project that ran through a neighborhood in Bandar Abbas. This project demonstrated successful trust building and cooperation between people and city officials in 2006.

upgrading, the failure to build sidewalks ultimately added to the residents' frustrations and poor perceptions of neighborhood safety, thereby increasing public mistrust. Giving up small portions of their land for the greater good (to build sidewalks), could have prevented this sense of frustration and deepening mistrust. Sometimes, residents' mistrust was justified. In Zahedan, women took leadership roles in cleaning up trash-strewn roads and providing street lighting. The project was never completed due to a contractor's personal greed, which left residents worse-off as their contributions were used for other purposes (personal interview with a local enabling project executive staff member in Zahedan 2016).

In some cases, the disconnect between residents and planners grew from differences in perceptions. In Zahedan and Bandar Abbas, for example, cultural differences between the ethnic and tribal beliefs of residents and the hierarchical mindset of planners obstructed enabling and prevented their unskilled cohort of young people from individual and social growth.

In other cases, perceptions differed between residents and experts as to the quality of the outcomes. Residents of Zahedan blamed contractors for using substandard materials in building street curbs, while experts blamed salty water in mixing concrete responsible for its short life span (in some cases less than a year). During the heated debate between protestors and contractors over the resulting poor road quality, angry people set the road construction machinery on fire (Alaedini and Bahmani Azad 2012). Similarly, when asphalt for street repair in Sanandaj wore out, residents attributed the fault to using substandard material, yet the real culprit according to experts was deferred maintenance in a cold climate with a heavy snowfall (personal interview 2015). This aura of mistrust unravels a fundamental need for cultural or social adaptation. Adaptation requires time and interaction for changing the status quo. People expect visible results where promises are delivered in a timely fashion, otherwise, in the absence of realistic expectations, unfulfilled promises do not serve as catalysts for cultural adaptation and trust building.

However, adaptation need not happen socially or culturally, or solely from the bottom-up. In Sanandaj, the government achieved public participation through a top-down process, which undercut the role of NGOs and local residents. Thus, implementation was poorly managed due to a lack of expertise, even though the planning consultants completed their work on schedule (personal interview, Moshiri and Hasan Zadeh 2015; Alaedini and Ghani 2010). Successful implementation and management would have made life much easier for residents by helping them better adapt to their settings.

Conversely, Jafarabad, Kermanashah, exemplifies bottom-up adaptation. Even though Jafarabad is not an incorporated part of the city of Kermanshah, only 1% of its residents have received formal ownership titles. This lack of enthusiasm for getting title deeds owes much to the perception that formalization has not necessarily represented a low-hanging-fruit of upgrading efforts (Khatam 2002). To people, the City Council is more concerned about macro rather than micro level problems, which explains why residents felt ambivalent about sharing the physical upgrading costs that they believed were the government's responsibility (Khatam 2002). Orangi, Pakistan, illustrates a similar situation, where residents held the government responsible for the physical upgrading costs. The local NGOs could distill such perceptions where bottom-up support for sharing part of the physical upgrading costs could pave the way for effective adaptation and eventually social integration.

Formalization/Legalization

As the second stage of transformation through physical upgrading, governments formalize or legalize informal settlements. The common perception is that if it is formalized, an informal settlement becomes officially part of the city. However, formalization could not automatically make the formal–informal divide go away. Formalization does not obviate all the problems associated with informal settlements, and may not happen at all, in which case the status quo either continues, or the settlement disintegrates. Aside from the legal discourse, formal and informal settings have obvious visual and physical differences, informal settlements are typically chaotic, unappealing, and seemingly disorganized compared to visually coherent, orderly, and harmonious formal urban fabrics (Arefi 2014).

Formalization will not immediately alter an informal townscape. For example, Ahmadabad, Tabriz, is an informal settlement that, despite being located within the city proper, is still perceived as rural. While global agencies, including the World Bank, generally promote formalization, it "is never as straightforward as simply converting informal documentation into formal titles" (Roy 2005: 152). Proponents of formalization consider it a robust way to empower the marginalized by giving them access to land markets and generating wealth and credit.[6] But even granting title deeds—as Roy suggests—does not magically obviate the cultural and social stigmas surrounding informality. A lot more work needs to be done to prepare the public for such transitions or transformations. Finally, informality offers an alternative lifestyle that caters to the needs of marginalized people without opting for title deeds and legalizing property ownership. The consulting firms that conducted the preliminary studies in the five target cities reported that residents did not welcome formalization as a step for entering the housing market. In some cases, formalization resulted in gentrification and dislocation; obviously not something residents desired.

[6]While 'formalizing the informal' presents a common urban management goal, Dowlatabad experienced a reverse trend. This originally large swath of land belonged to a wealthy local landlord who decided to subdivide and sell it all (Irandoust personal interview 2016). To avoid poor planning, a professional planner prepared a gridiron plan with plots of roughly 200 m². Knowing that large lot sizes were not affordable and would not easily sell in the future, they were further broken up into four smaller 50 m² lots. Having found these small lots affordable and more compatible with their needs, lower-income people purchased them. Unlike the organic, meandering,

To ease the transition between informal and formal settlements, urban officials can first regularize informal settlements before regulating them. Regularization characterizes a temporary relaxation of regulations that assist people to incrementally upgrade properties over time. Instead of complying with top-down zoning and construction codes in one fell swoop, squatter settlements can upgrade as their economic status gradually improves. Whereas top-down, publicly enforced regulations lie outside the control of individuals, cutting people some legal slack motivates them to upgrade their homes at their own pace.

Therefore, flexibility and spontaneity are the two key differences between regularization and the rigid, top-down nature of regulation. Regularization hence bridges the formality–informality gap. Governments that regularize rather than regulate informal settlements believe that regulation alone does not effectively solve the problems governments seek to address. Therefore, whether governments like it or not, they relax regulations. In Iran the five target cities experienced regularization, but the extent of their resulting formalization varied widely. Dowlatabad and Shirabad showcase successful formalization attempts. In the case of Shirabad this formalization was only physical, while Dowlatabad experienced both a physical and social/cultural formalization.[7] The other cities generally fit somewhere between the two. The case studies prove again that formalization alone cannot change the fate of informal settlements. In Tabriz and Zahedan, for example, social stigmas endure and outweigh the legal redlining associated with informal settlements.[8] What is needed is further integration of informal settlements into their host cities.

Integration

Adaptation, regularization/legalization and, ultimately, formalization and spatial integration do not automatically lead to cultural or social integration of informal settlements. This requires distinguishing between formalization and integration on the one hand, and between physical/legal versus social/cultural integration on the other. Shirabad and Ahmadabad show that changing the public perception and social stigmas surrounding informal settlements is much harder to accomplish than implementing physical upgrades—yet both are necessary for their ultimate integration.

and seemingly unorganized spatial structure of most informal settlements, Dowlatabad looks much like a typical formal planned neighborhood. This experience demonstrates that if there is an economic or political will, informal settlements do not have to look chaotic, and can benefit from the clarity of gridiron plans.

[7] For more information about Kermanshah's typologies of informal neighborhoods see Alaedini and Tavangar (2012).

[8] Building the Ill Goli metro station in the upscale part of Tabriz as opposed to Ahmadabad with its potentially higher ridership illustrates the public perception about enforcing redlining toward the poor.

Admittedly, physical upgrading reduces and bridges the planned–unplanned or formal–informal physical divides and promotes physical/social integration. The degree to which physical upgrading affects this integration depends on the relationship between the informal and the formal fabrics. The dominant perception of integration is that since the targeted neighborhoods in the five cities were located within and not on the urban fringe (unlike the famous Indonesian *desakotas* as urban–rural hybrids), at least physical integration occurs (Alaedini, personal interview 2016).

But, for example in Shirabad, proximity with the city has not yielded full integration. Located in the northeast of Zahedan, Shirabad grew from subdividing agricultural land, which dried up due to water shortages in the late 1970s. Shirabad comprises an organic rural fabric and a planned subdivision (Piran 2002). Increasing land and property values, rental housing in other parts of Zahedan, and the influx of Afghan refugees fleeing the Taliban's reign in Afghanistan, all contributed to Shirabad's rapid growth. Today, Shirabad is among Zahedan's largest neighborhoods. As one of the largest informal settlements in Zahedan, Shirabad benefited from two active NGOs that have played a part in the extensive physical and infrastructure upgrading over the last decade. But with a portion of the population involved in illegal drug trafficking, crime, and other urban ills, Shirabad suffers from a persisting social stigma—although the reality is not as bad as the perception of deprivation.[9] High doses of bonding social capital is an asset that keeps the residents close together, even though they are quite poor (Piran 2002). Hence, there are two different images of Shirabad: one as a close-knit introverted community with no hope for cultural integration; and one as a place that has physically integrated into the city. Based on the second image, physical adaptation can facilitate physical integration, whereas without cultural adaptation, social or cultural integration is highly unlikely. Conversely, most informal settlements in Tabriz benefit from closer proximity to more affluent and high-end neighborhoods, and are easier to integrate with the city compared to those outside city boundaries.

As a major industrial production and employment center in the northwest of Iran, Tabriz has clear advantages compared to the other four target cities, which make integration more likely by giving the residents of informal settlements easier access to amenities, infrastructure, and employment. The inhabitants of several of those neighborhoods, including Akhmeh Ghieh, do not have major subsistence problems, and already have access to public amenities and employment. However, not all informal settlements in the city are fully integrated. Despite the integration potential, overcoming the cultural and social stigmas of the settlement of Ahmadabad is much harder than overcoming its physical barriers. Many citizens believe that Ahmadabad is still

---

[9] While, according to Piran (2002), 25–30% of Shirabad's population might engage in drug-related activities, its public image is much worse.

perceived as a village even though it is a part of Tabriz.[10] This is why skeptics do not consider urban formal–informal integration a panacea that would magically make all the stigmas associated with informal settlements go away. Roy (2005), for example, critiques De Soto's call for "formalization" not so much for giving residents "property rights" per se, but as "the right to participate in property markets." In many cases, property markets do exist in informal settlements anyway, albeit not as a formal or legal transaction mechanism.[11] The key challenge is to use formalization as a way of empowering people as opposed to politicizing the informality discourse.

Dowlatabad illustrates a successful example of the informal–formal integration. Located in the west of Kermanshah, what was originally a housing redevelopment site for the survivors of the Abshouran River flood continued to grow rapidly over four decades. What differentiated this redevelopment scheme from other projects was that its owner subdivided the land parcels based on the comprehensive zoning ordinances into relatively small, 250 m^2 parcels that were far more affordable for rural-urban migrants and Iraq–Iran War exiles than other zoned sites, while maintaining sufficient land for infrastructure. Based on this innovative affordability model, the informal settlement residents further subdivided these parcels into two or four smaller plots to fit their needs. Adopting this model demonstrated a successful bottom-up effort. At the same time, top-down support by regularization (instead of a rigid adherence to regulations) helped the residents to adapt to their new conditions. Furthermore, Dowlatabad is a great example of how a particular type of physical upgrading can both provide affordable housing in an informal settlement and successfully pave the way for spatial integration. In terms of lot sizes and street networks, Dowlatabad represents an informal settlement that looks physically compatible with other formal neighborhoods. In the near future, plots may even consolidate once again to the original 250 m^2 parcels. In contrast to Shirabad, Dowlatabad has integrated both physically and socially into Sanandaj. Meanwhile, Shirabad remains socially and culturally isolated even though spatially it is part of Zahedan.

ENABLEMENT THREATS AND PROSPECTS

While enabling informal settlements involves adaptation, formalization/legalization, and integration, threats can slow down or completely stop one stage from transitioning into the next stage. Common threats were encountered in the Iranian case study during each stage of enablement. In the adaptation stage, co-optation threatens progress. When Jafarabad residents solved the city's garbage problem through innovative entrepreneurship, the city co-opted their efforts. Co-optation is the antithesis of creative

---

[10] This is what Roy calls the "rural/urban interface." See Roy (2005).
[11] See Razzaz (1998).

thinking—especially where new business incubators exist, and as bottom-up agents compete with formal powerful players. Instead of capacity building and market integration in Jafarabad, co-optation weakened this unique opportunity. On the other hand, neglect can also threaten progress, such as in Zahedan where women who actively participated in street clean-up were not properly supported. This was in some way capacity building through redundancy, which went awry and prevented full integration. In the formalization/legalization stage, regularization competes with regulation. De Soto (2000) and others have extolled the virtues of formalization as a way the marginalized population can tap into the marketplace. The tendency to regulate informal settlements closely competes with regularization, which offers settlements time to gradually catch up with zoning and building regulations.

Marginalization and social-cultural stigmas impede integration. Integration transcends mere physical connectivity, and, among other things, depends on trust and capacity building[12] (physical, social, and political), and on how well services (upgrading projects) are delivered to users. Trust building serves as the socio-political glue that binds people, plans, and officials. It also signals economic, social, and political capacity building. For example, in Sanandaj people shared the labor and costs of street maintenance to facilitate integration. However, implementation took longer than expected and the contractor's failure to meet the deadlines damaged trust building and cultural/social integration. In Zahedan, Shirabad's notoriety as a drug-ridden neighborhood overshadowed any positive social change. That said, residents in the five target cities benefited from physical upgrading projects, surprisingly as much as the public officials who sought to empower them. This is a blessing because it allowed decision makers to think out of the box.

Trust building serves as a catalyst for social and cultural integration, but it faces an uphill battle in the enablement process as residents of informal settlements often do not trust these initiatives. Mistrust persists after decades of government neglect of people's basic needs, from housing and education to health and employment. Perceived bias creates mistrust between the people who feel neglected and government officials or policies that allocate limited resources to priorities other than squatter settlements. This sense of mistrust can be challenging to overcome, and where projects are not delivered in a timely fashion, the situation is exacerbated. The urban revitalization literature has explored the nexus between unfulfilled promises in service delivery and physical upgrading and increased public mistrust in the USA and other Western countries. In his study on eight languishing Los Angeles neighborhoods (The Los Angeles Neighborhood Initiative) that received revitalization funding from the Mayor, Arefi (2004), for example, reported that lengthy project implementation heightened public mistrust. In many cases, the targeted population's mistrust arose from the perception that it was "hard for the residents to think beyond a year or two" (Arefi 2004: 15). Similar issues are at

[12] See Glickman and Servon (1998).

play in the Iranian case studies. While the aforementioned enabling initiatives increased the quality of life of the informal settlement residents, key challenges threatened the enabling process and hence generated mistrust. These challenges included delays in project delivery, co-opting bottom-up efforts, and unfulfilled promises.

In one example, the Ahmadabad focus group participants rejected the dismissive attitude of the local authorities toward informal settlements. Giving short shrift to the residents' requests and instead allocating public resources to the more affluent parts of the city reflected this attitude, which ultimately increased mistrust and delayed integration in a vicious cycle. An example of the unequal allocation of resources was the construction of a Metro station in a nearby affluent neighborhood while failing to construct a new road to the informal settlement, despite the project being promised almost five decades before. Similarly, lengthy completion or, in some cases, incompletion of the physical, social, and economic goals that did have public support caused disillusionment and dissatisfaction in Zahedan. Incomplete projects, ineffective information dissemination and publicity, and unfinished business, on the one hand, along with the poor quality of completed infrastructure upgrades on the other, caused public dissatisfaction and mistrust (delayed physical integration) (Alaedini and Bahmani Azad 2012).

Table 14.1 conceptualizes the differences between the formal and informal settlement development rationales. Based on this logic, regulation, planning, and implementation showcase the three stages of formal development, whereas adaptation, formalization/legalization, and integration form the three stages of informal development. Informal settlements do not necessarily characterize illegality. In many countries (including Iran), although equating

Table 14.1 Comparing and contrasting informal and formal planning processes

Formal development	*Informal development*
Formal/legal basis	**Adaptation**
• **Renewal**	• Redundancy
• **Regulation/rigid**	• Unpredictability
• Top-down	• Panarchy
	• Bottom-up
Planning system	**Formalization/legalization**
• Systems thinking/need assessment	• Transformation to formal planning
• Predictability/future growth	• Physical upgrading, legal upgrading
• **Hierarchy**	• Risk of gentrification
• Top-down/bottom-up	• Regularization as an alternative
	• Mix of bottom-up and top-down
Implementation	**Integration/disintegration**
• Budget allocation	• Physical/social/cultural dimensions
• **Phasing**	• Service delivery
• Maintenance/upgrading	• Capacity building
• **Top-down**	• Trust building
	• Mix of bottom-up and top-down

Source Mahyar Arefi

informality with illegality has been the norm, recent research shows that it reflects a significant part of the urban housing market rather than an army of outcasts occupying and squatting on public land. Official figures show at least one-third of the urban population live in informal settlements out of necessity.

Even though the World Bank operationalized enablement in a way that had been tested in other countries, it was not narrowly implemented in Iran. Instead, enablement fortuitously found a life of its own in each target city and signified something more powerful than what was originally intended. Portraying informal settlements in a positive light instead of demonizing them, draws from the philosophy that people can still change their own destinies if left to their own devices. This is a different mindset than associating informality with deviance and criminal behavior.

Unsanitary, unhealthy, crowded, and disorganized environments and hopeless conditions, or environments riddled with social pathologies, cannot be justified, and these settlements must be ameliorated before a wider area is spoiled. This type of thinking has persisted for decades as the top-down *modus operandi* in many countries. Over time, however, academics and policymakers alike have questioned this mantra, and have explored squatter settlements not as taken-for-granted parasites that demand immediate draconian measures, but as communities with untapped assets. This kind of thinking has prompted empowerment, enablement, adaptation, and resiliency. But, as with so many other things, there is not just one way to empower the disempowered and the five target cities in Iran revealed several unintended outcomes that help better understand enablement.

Enablement does not work in a one-size-fits-all manner, and it can appear in different guises and strengths. In some cases, it helps under-represented communities gain recognition for their accomplishments. Women who gained respect as local leaders is a case in point. This gives enablement a social charge. It also implies economic strength, as observed in Jafarabad, where those engaged in recycling were both financially and organizationally motivated and enabled. Improving people's living conditions by building better homes with better construction materials that could last longer, and reducing environmental degradation exemplify physical enablement. Enabling can also mean increasing life chances by learning new skills in locally run and managed neighborhood centers. By learning new skills residents of informal settlements not only become productive community members, but also avoid engaging in activities resulting in social pathologies that are detrimental to their futures. These are but a few positive examples of enablement gleaned from the Iranian informal settlements project.

Another way to help residents of informal settlements is to consider enablement a catalyst toward building capacity, assets, and trust. Governments learn the hard way that gaining people's trust is both politically and economically necessary. But people—especially those with a lower-income—are skeptical of government officials' promises, and gaining their trust is much harder than gaining that of more affluent people. Similarly, enablement paves the way for long-term capacity building. For example, the trash recycling plant in Jafarabad showed how bottom-up initiatives enable people not only socially and politically, but also economically and psychologically. This experience rightly showed that garbage recycling had huge enablement potential, irrespective of the age or gender of those involved. Had it not been eventually co-opted by formal city institutions, it could have positively mobilized people to keep their city clean and sanitary, generate revenue (asset building), and organize the community economically, politically, and socially. Thus, capacity building, asset building, and enablement go hand in hand.

Iran's *National Enablement Document* is another important byproduct of this initiative, which in and of itself, is advanced and forward looking. Inspired by and based on the Islamic Republic of Iran's Constitution, this document (Sarrafi 2002) seeks to offer new ways of dealing with informal settlements:

1. *Revisiting policies for low-income people:* coordinate multiple policies before adopting them
2. *Facilitating a government role for public and private sector operations:* create a vibrant housing market for all income levels, and facilitate rather than impose a top-down approach toward the private sector
3. *Mobilizing local assets and initiating self-help:* view informal settlements not just physically, but also socially and culturally; leverage residents' strong social networks to cope with hardship and uncertainties
4. *Right to the city and safety along with undertaking urban and civic responsibility:* ensure informal settlement residents' citizenship and safety rights; recognize their social contracts with the city by initiating social contact
5. *Enhancing socio-economic foundations of family by housing and employment support:* informal settlement residents deserve government assistance to secure housing and employment opportunities and to integrate socially
6. *Initiating a comprehensive and forward-looking approach:* Physical upgrading, socio-economic development, empowerment, and capacity building help the urban poor integrate
7. *Enhancing local leadership roles in the enabling process and organization of informal settlements:* Promote bottom-up initiatives by delegating responsibilities to the lower tiers of governance.

As a malleable, open-ended, community engaging endeavor, design illustrates another important aspect of enablement. The common perception is that especially those who live in informal settlements are incompetent when it comes to design and that they need expert knowledge. But just because experts can help, does not mean that local knowledge cannot be part of a collaborative urban design discourse. The cyclical and incremental nature of urban design, when faced by different bureaucratic or administrative hurdles such as budget deficits, is not solely within the realm of architects and planners. Instead, urban design is quite similar to the gradual steps taken by local residents in some informal settlements to complete unfinished projects or to seek support from different sources. This similar approach demonstrates that urban designers can learn a great deal from informal settlements.

CONCLUSION

When reviewing the implementation of the Iranian enabling initiative, the World Bank (2010) criticized it as "ambitious, complex and over-optimistic, without long-term vision and manageability of the inputs and outcomes" (p. 9). This conclusion was drawn for three reasons: the subjectivity of the outcomes assessment; the vagueness on post-implementation status beyond pilot projects in each target city; and, the fact that 90% of the loan was earmarked for urban upgrading, with only 3% on housing reform financing to create more options for housing demand. These assessments suggest that the loan did not promote long-term capacity building in the housing market as was originally intended. Policymakers should tap the enablement momentum by addressing these potential areas of improvement. Our five case studies concur that the enablement record of accomplishment has an average record in Iran. Yet, despite its challenges, positive signals from each of the five target cities suggest policymaking and urban management potentials. The government gets credit for adopting or modifying macro-level policies relative to housing and employment. While the government performs quite well in enhancing local leadership in some of the five target cities (especially with respect to women's roles), the symbiosis between enablement, capacity building, and physical upgrading needs more work. Ensuring informal dwellers' right to the city also needs more work; there are still government officials who view residents of informal settlements as criminals and parasites, rather than considering informality as a lifestyle.

Table 14.2 ranks the multifaceted nature of enablement in the Iranian case studies and the physical, social, political, and legal challenges to achieve positive outcomes. As shown, capacity building has fared more poorly compared to physical upgrading, as the latter is much quicker and easier to achieve in the case of, for example, road repairs or relaxing land use and zoning regulations. The empirical research from all the case studies shows that a lot more work on governance and self-help is necessary to achieve higher marks on capacity building. Physical upgrading in all target cities also trumped leveraging the informal social networks of trust or, simply put, investing in shared governance. Had the government mobilized these resources to a greater extent, entrepreneurial examples, including the garbage recycling operations in Jafarabad, would have been more successful as opposed to being co-opted.

Table 14.2 Ranking the accomplishment of the enablement goals

Enablement	*Degree accomplished*
Adaptation	
• Service delivery	+
• Governance	✓
• Leadership	✳
Formalization	
• Title deeds	✓
• Regularization	+
• Physical upgrading	✳
Integration	
• Capacity building	+
• Trust building	✓
• Asset building	✳
✓ Low-Medium	
+ Medium-Good	
✳ Good-Excellent	

Regularization has been more successful than issuing title deeds, as it is a means toward integration and not an end in itself. This is another indication that while formalization continues, issuing title deeds alone has not helped the target populations much, because, with or without the deeds, residents still make transactions in the housing market. Probably the least successful outcome of the Iranian enablement initiative was trust building in some of the targeted neighborhoods. The two correlating indicators of service delivery and partnership (shared governance) also fared poorly. In reality, trust building happens in an atmosphere of mutual interaction and facilitation. If promises are not delivered in a timely fashion, trust building is unlikely. This has been witnessed in all five target cities. The enablement track record is mixed, even though government efforts have positively affected the informal settlements. To sum up the three stages of enablement in the Iranian case: adaptation fosters *endurance*, formalization *compliance*, integration *victory*. Endurance is exactly what adaptation entails; compliance more often than not brings forth formalization or legalization and, of course, if informal settlements happen to integrate spatially, socially, and culturally (which is awfully difficult for various reasons), then both people and policymakers can call it a true victory.

REFERENCES

Alaedini, P., & Bahmani Azad, B. (2012). The target population's evaluation on upgrading and empowerment activities of informal settlements in Zahedan: Case studies of Karimabad, Siksouzi and Shirabad. *Research in Social Welfare, 1,* 29–47 (in Farsi).

Alaedini, P., & Ghani, A. (2010). The empowerment and upgrading experience of informal settlements in Sanandaj: The stakeholders' views. *Haft Shahr, 33 & 34,* 11–21.

Alaedini, P., Poorshad, M. M., & Jalali-Mousavi, A. (2012). Promoting the welfare state of informal settlements in Iran. *Social Welfare, 41,* 69–91 (in Farsi).

Alaedini, P., & Tavangar, F. (2012). Improving safety and security of informal settlements through urban upgrading and enabling activities: The experience of dowlatabad neighborhood of Kermanshah, Iran. *Haft Shahr, 37 & 38,* 94–104 (in Farsi).

Arefi, M. (2004). Neighborhood jump-starting: Los Angeles neighborhood initiative. *Cityscape, 7*(1), 5–22.

———. (2014). Order in informal settlements: A case study of Pinar, Istanbul. *Built Environment, 37*(1), 42–56.

De Soto, H. (2000). *The mystery of capital: Why capitalism triumphs in the West and fails everywhere else.* New York: Basic Books.

Dovey, K. (2012). Informal urbanism and complex adaptive assemblage. *International Development Planning Review, 34*(4), 349–367.

Eskandari, Z. (2008). Teacher-Mayor: Managing the city like a classroom: Revisiting an experience from Bandar Abbas. *Haftshahr, 23 & 24,* 106–113 (in Farsi).

Glickman, N., & Servon, L. (1998). More than bricks and sticks: Five components of community development corporation capacity. *Housing Policy Debate, 9,* 497–539.

Irandoust, K. (2010). A brief overview of the informal settlements' enabling experience: The case study of Kermanshah. *Geography & Development, 20,* 59–78 (in Farsi).

Khatam, A. (2002). People's share, government's share in neighborhood physical upgrading & enabling projects: The Jafarabad, Kermanshah experience. *Haftshahr, 9 & 10,* 33–42 (in Farsi).

Lewis, O. (1959). *Five families: Mexican case studies in the culture of poverty.* New York: Basic Books.

Nathan, R. P. (1992). *A new agenda for cities.* Columbus, OH: National League of Cities.

Piran, P. (2002). On informal settlements again: A case study of Shirabad, Zahedan. *Haftshahr, 9 & 10,* 8–24 (in Farsi).

Razzaz, O. (1998). Land disputes in the absence of ownership rights: Insights from Jordan. In E. Fernandes & A. Varley (Eds.), *Illegal cities: Law and urban change in developing countries* (pp. 69–89). New York: Zed Books.

Rostamzadeh, Y. (2011). Organizing and empowering informal settlements in urban areas with the attitude of local residents: Case study: Tohid area, Bandar Abbas. *Urban Management, 28*(Autumn & Winter), 321–336 (in Farsi).

Roy, A. (2005). Urban informality: Towards and epistemology of planning. *Journal of the American Planning Association, 71*(2), 147–158.

Sarrafi, M. (2002). Recommendations for organizing and empowering the informal settlements in Iran. *The Journal of Social Welfare, 11,* 111–122 (in Farsi).

Shatkin, G. (2000). Obstacles to empowerent: Local politics and civil society in Metropolitan Manila, the Philippines. *Urban Studies, 37*(12), 2357–2375.

Takahashi, K. (2009). Assessing NGO empowerment in housing development frameworks: Discourse and practice in the Philippines. *International Journal of Japanese Sociology, 18,* 112–127.

The World Bank: Implementation Completion and Results Report (IBRD-47390). (2010). *Report No: ICR00001412.*

CHAPTER 15

Space, Time, and Agency on the Indian Street

Vikas Mehta

In our rapidly urbanizing world, one-third of urban dwellers live in informal or self-made settlements and habitats (Neuwirth 2005; Davis 2006; UN-Habitat 2008) that are inextricably linked to the social, political, and economic processes of the other two-thirds of urban world, as well as the remaining non-urbanized population. In the twenty-first century, urban informal processes dominate overall urbanization in the developing world (Roy 2005), particularly in India (Harriss-White 2003) where the majority of retail sector is informal (Naik 2009). Unfortunately, the discourse of formal-informal duality overrides the glaringly evident interconnectedness of the two and undermines the value of informality (Roy 2005; Portes 1983). Because of the numerous marked differences between the two, informality is largely portrayed, simplistically, as being an unusual and provisional response to human needs and is generally perceived as detrimental to places, particularly in terms of place image (Roy 2004, 2011). These overarching conceptual constructions of informality that classify it as dysfunctional and an inferior makeshift effort (AlSayyad 2004), hinder an understanding of the complex processes that eventually support social and economic sustainability for numerous places, phenomena, and actors. By viewing informality as the 'other' and through the lens of the formal, we cannot fathom and appreciate these processes and the benefits they entail. A closer look at any part of the urbanized world shows the undeniable link between institutionally supported and self-made processes (Bromley 1978; Ward 2004). This chapter chronicles the street in India as a case study to demonstrate unclear boundaries and the interconnectedness between informal and formal sectors and processes (Schindler 2014). The observations and findings are based on two

V. Mehta (✉)
School of Planning, University of Cincinnati, Cincinnati, OH, USA

© The Author(s) 2019
M. Arefi and C. Kickert (eds.), *The Palgrave Handbook of Bottom-Up Urbanism*, https://doi.org/10.1007/978-3-319-90131-2_15

neighborhood commercial streets in New Delhi, studied as both a research project and as a participant observer, employing systematic observations throughout the day, interviews with shopkeepers and vendors, and extensive field notes. By understanding social and economic processes through empirically studying space and place, the chapter contributes to debates that reveal gaps in urban theory and argues moving beyond the predetermined colonial frameworks that have relegated self-made processes (Roy 2005).

THE STREET IN CONTEMPORARY INDIA

The image of the Indian street, captured in films, novels, other media, or a visit in person, is characterized by a hyperintense and overstimulated experience. This image is shaped by the juxtaposition of eclectic forms that create a richly patterned and textured enclosure of Indian streets. The sensory overload is further pronounced by the aural, olfactory, and tactile qualities of the space. In this image, the street is crowded and chaotic and human sounds are intermingled with sounds of: traffic; of music playing; of vendors calling out their goods and services; of sounds associated with manufacturing and repairing objects; and of birds and animals. The smells also have an immense range: the soothing scents of flowers and incense; the aroma of spices and oils; and the smell of foods; sometimes intermingled with the stench of garbage and rain water collected in ditches. In this image of the Indian street, path and place compete for the space on the street, gathering and lingering behaviors occur simultaneously with the act of movement, often negotiating and compromising for space. The linearity of the street is constantly challenged. The Indian street is characterized by a multiplicity of use and meaning produced through the agency of many actors of varying social classes, reinforcing an image of a lack of a rational order and control over the street. Through the actors' astute use of space and time, the street performs as a vibrant and representative space for various groups in the Indian city. The commercial street in India demonstrates how the use of space, accompanied by the temporal patterns of use, creates affordances and agency for diverse groups in the city, and offers interactions in public space that are an important building block of urban social order.

The ubiquitous commercial street in India has existed over a centuries-long continuum—not only as a place of local commerce, but also one of the most vibrant urban social spaces in the city. The street—as a complex and wide-ranging communal space—is at the core of the social and economic workings of the neighborhood and the city. It is a place for daily activities, rituals, survival, socializing, and recreation; a place for walking, sitting, standing, sleeping, grooming, cooking, eating, smoking, washing, cleaning, preaching, praying, panhandling, and playing; and certainly also a place for commerce. The Western trend of shopping malls has undoubtedly penetrated the development pattern of towns and cities in India. Yet, there is continued demand for local shopping, particularly given the increasing urban population and the limited mobility of numerous groups—women, children, elderly, and the poor. A detailed analysis of behavior and interactions on an Indian street, over the time of day and week, reveals its workings as a negotiated and fluid space of access, commerce, leisure, sociability, and survival. Operating under complex relationships among the institutional structures, colonial frameworks, and class divisions, the formal and the informal sectors cohabit the street to serve the many needs of various social and economic classes in society.

The Inherent Duality: A Shared Space of Cohabitation

Delhi, as many other Indian cities, was a clearly divided, dual city in pre-independence India. The centuries-old native city, contained in the walled space also known as Old Delhi, is a space of labyrinthine streets and uneven blocks, with compact urban forms housing tightly knit communities. The other city, in contrast to the native city and aptly named New Delhi, was constructed in the early twentieth century by the ruling British class. New Delhi comprised an expansive government political space, planned on the principles of the City Beautiful movement, which displayed the power of the imposed political structure and the ruling class. This political city space was surrounded by colonial residential quarters for the ruling class and high-ranking Indian government employees, also planned on the principles of the Garden City movement (Irving 1982). The duality between the two old and new cities was clearly evident. The native city was compact, crowded, and intense—a complex and hybrid space. The colonial city was expansive and open, with ample sun, light, air, and greenery—a rational space—designed to express control and power through a formal and legible spatial order. Post-independence planning in Delhi, in the name of modernity, continued the colonial city narrative of development. Evident in the layout and form of the rapidly growing city is an orderly colonial conception visible in the rational, formal, gridded layouts of streets and blocks, also introduced in the everyday rules of conduct translating into a disciplined and orderly behavior for the citizenry (Escobar 1995; Gidwani 2008). For example, doormen, watchmen, and sentries, present at the entrance of many public and private buildings, are expected to greet or salute not only their seniors but everyone above their economic strata. Social superiority is granted through economic status. It must be recognized by addressing people above one's economic class as sir or madam and by bodily gestures such as getting up if seated and bowing to the passing superior. The bourgeois middle class that emerged in post-independence India adopted this colonial order in everyday life, and the new urban immigrants, often the rural poor and illiterate populations, had to obey and adapt to these norms. However, numerous other social- and culture-based systems and norms coexist in Indian society and the juxtaposition of many is evident in any given space. As a result, what is often visible in the Indian city is a disciplined and ordered colonial form and social norms overlaid with a localized use of space based on native social and cultural practices and customs (see, for example, Mehrotra 1992).

In the twenty-first century Delhi the street as public space displays an increasing disparity between the haves and the have-nots, as the middle class stakes out its social space. As a result of the economic reforms of the final years of the previous century, twenty-first century India is a place for this growing middle class, with a new social identity based on mass consumption (Baviskar and Ray 2011; Srivastava 2009; Varma 1998). In contemporary urban India there is a constant tension between the aspirational twenty-first century "world class" city that is a sanitized and secure space of consumption and the everyday post-independence city of the twentieth century (Dupont 2011). This tension is most clearly visible as a dichotomy between the burgeoning middle class and the service underclass of the city—vendors, shop helpers, domestic servants, local janitors and cleaners, security guards, panhandlers, local priests, delivery aids, and so on—and their conflicts over space (Rao 2010). Much of the city space displays a geography of separation. On the one hand, affluent and middle-class neighborhoods have controlled access—particularly restricting the poor underclass—to their residential streets and commercial malls. On the other, the affluent and middle class seldom set foot in the informal settlements that are home to the underclass. Yet, in contrast to the gated neighborhoods and malled spaces of contemporary urban India, the commercial street remains a porous territory, resulting in a space that facilitates a coexistence between the multiple identities and social practices of many social classes. Unlike in the gated malled spaces of the bourgeois city, the underclass, the poor, and the underprivileged are participants in the production of public space on the commercial street. The service class depends on the middle class for economic survival and the middle class depends on the service class to maintain their lifestyle. The interrelationship between the classes is neither hostile nor friendly, but accommodating (Schindler 2017). The service class requires access to urban space, particularly the contemporary urban space of mass consumption, for providing services to their clientele, the middle class. Simultaneously, the service class is essential because their labor enables the middle-class practices of consumption and leisure, and the transformation of urban space to a "world class" space (Brosius 2010). Thus, in this diverse milieu, people of all classes end up claiming space on the street, albeit temporarily. The very act of using the street becomes a way for numerous social groups, particularly marginalized ones, to belong to society and express their right to space in the city (Lefebvre 1996).

There is another important cultural agreement between the middle and under classes that permits numerous forms of bottom-up and self-supporting forms of urban social practices. In traditional Indian society each class has a responsibility toward the weaker being. For example, the underprivileged and deprived, whether due to disease, poor health, disability, or other circumstance, make a living through panhandling. Culturally this is acceptable because it is intricately connected to religion. Ascetics and other religious persons in a state of renunciation depend on society for charity and alms for survival. This form of panhandling is considered a dignified way to survive. In fact, providing alms to the ascetics and the deprived helps legitimize or even elevate the social status of the donor. Again, unlike the malls and other gated commercial spaces, the commercial street remains an open public space that ascetics and the deprived may occupy. Thus, such remnants from the native city are often visible as layers of social practices in the streets of the contemporary, world class, city and are largely tolerated and supported by the middle class.

CLAIMS TO SPACE

Examining the physical patterns of occupancy and use of space helps understand the social communications, exchanges, and interactions on the Indian street. Along the street may be found a range of immobile commercial businesses, such as grocers, hardware stores, pharmacies, electronics stores, cleaners, apparel stores, barber shops, restaurants, teashops, and a myriad other uses, along with religious buildings, offices, small industries, and apartments above. As expected, these businesses in the buildings lining the street clearly establish their presence on the street by using the street space as an extension of the interior with displays, storage, and even sales, spilling out onto the street. Their open shop fronts further blur the boundary between the street and the shop interior. Through their legitimate existence, as part of institutionally recognized forms, these business owners establish a hegemony over the street. Yet, the street itself is occupied and claimed by numerous other independent actors and self-made processes—a bottom-up urbanism. The Indian street is seldom complete without the presence of food vendors, small household goods and other services, makers and menders of goods, including locksmiths, cobblers, potters, and mechanics, joined by panhandlers, traveling entertainers, soothsayers, and ascetics. Vendors, most of whom belong to the underclass, augment the official businesses and the range of goods and services and make the street a useful and meaningful place for many more, increasing the diversity and difference on the street (Fig. 15.1). By creating nodes that often extend from sidewalks into the roadway to anchor stationary activities, vendors spatially expand the pedestrian reach on the street. For example, vendors may create a place to sit for a cup of tea or a snack, to gather for a smoke, listen to a commentary on a game of cricket, or to get a haircut and talk. Through this pattern of occupancy that spreads beyond the edges of the street, vendors create a field of space that expands not only across the width of the sidewalk but often into the vehicular space (Figs. 15.2, 15.3). This results in a larger set of opportunities for interaction, encounter, and social cohesion among merchants and customers of varying backgrounds and class.

Fig. 15.1 Vendors make the street a useful and meaningful place for many more by increasing the range of goods and services, thus also increasing the diversity and difference on the street. Photograph by Prem Mehta

This use of the street is possible due to the codependency among shop owners and vendors—the top-down and bottom-up actors. Shop owners (mostly lower-middle or middle class) understand that vendors (mostly the underclass) near their shops expand the diversity of goods and services and thus attract more potential customers to the street, as long as they do not duplicate goods and services. Knowing that most vendors do not yield much power, and are deemed illegal in many locations, shop owners establish their hegemony on the street edge and may push vendors away to the periphery of the sidewalk or to other less trafficked areas of the street if they deem it necessary. All this territorialization, adjacencies, and overlaps of space on the street occur through an elaborate act of dialogue, negotiation, and compromise between the occupants, even as some are more powerful than others (Fig. 15.4). The expansive range of use generates numerous activities in space and sets up a stage for overlap, and thus interactions that result in a myriad conditions for social interaction and agency. Through their own understanding of space and an empirically developed sense of demand for their products and services, the numerous actors gauge the extent of claims they can make on the street. For example, a food vendor starts by claiming a small space on the threshold of the sidewalk selling snacks from a basket that he props up

Fig. 15.2 Vendors create an inhabitable field of space that expands the occupancy on the street beyond the edges. Photograph by Shilpa Mehta

Fig. 15.3 Vendors negotiate and occupy space, in this case an area for parking, in ways that challenge the linearity of the street. Photograph by Shilpa Mehta

Fig. 15.4 The complex pattern of territorialization, adjacencies and overlaps on the street occurs through an elaborate act of dialogue, negotiation and compromise among the occupants. Photograph by Shilpa Mehta

on a folding wooden stand. But finding a location is not easy. He is asked to move away by other vendors, a nearby shopkeeper complains about the trash his business will generate, and the employee of a nearby shop berates him for being in his way as he sweeps the sidewalk. He tries out a few locations before settling down at a spot not far from a bus stop, where he seems to generate enough sales. Over time, as demand grows, he acquires a larger basket that is strung to the back of a bicycle. The larger basket and the other bags strung on the bicycle allow him to carry more goods more easily. The bicycle also allows him to claim more space and serve a larger offering of snacks while still being mobile. The snack vendor creates an amenity for his customers while making a living through this bottom-up process by using his own agency—his ability to negotiate, be resilient, and take risks—and without creating much burden on the existing physical or commercial structure of the street—an Indian rendition of Jacobs' Manhattan sidewalk ballet (Jacobs 1961).

Each street tenant—the shop owners, workers, vendors, entertainers, ascetics, and others—attracts a clientele that crosses boundaries of class. The wide-ranging customers from various economic and social classes interact with the occupants, many of whom are from the poor underclass. Thus, the very structure of the transactions plays a critical role by creating a framework for

Fig. 15.5 The street displays various levels of contact between the classes. Photograph by Shilpa Mehta

social interaction among buyers and sellers from different classes (Anderson 2011). For example, a group of young middle-class men use a vendor for an evening snack. They chat and laugh but also engage with the vendor in the local gossip. In another location, three well-to-do middle-class girls use the services of a vendor to decorate their hands with henna. In this case, contact among the classes engages all the senses, including tactile (Fig. 15.5). Such seemingly minor interactions of diverse populations, produced through the bodily experiences of others (Lefebvre 1991), are significant for social cohesion in a society that is traditionally deeply divided by class.

TIME MAKES SPACE ELASTIC

As a prime place of commerce, the limited space on the commercial street in India is in high demand. At the same time, this ubiquitous public space is actively contested, socially and politically. Often due to the lack of space, there is extensive use of time to differentiate the use and purpose of space (Kim 2015). Thus, space is made elastic by segmenting it into multiple uses over time; and as uses of space change over time, so do its meanings and its regulations. By manipulating the space over the time of day and week, street space is expanded to become a viable place for numerous social practices, transactions, and survival as it performs as a vibrant and representative space for many groups in the Indian city. Multiple uses of space over the course of the day result in overlaps that permit interactions among different, and sometimes disparate, users and groups. This creates a complex web of interconnected activities and phenomena, eventually resulting in a space that is contested but is simultaneously a place of sharing and empathy. Studying the street over the passage of the day clearly reveals this. For example, the space in front of a corner shop is a site of people sleeping on the threshold and on a wooden cot laid out in front. At daybreak, the threshold and the cot are vacated, the bedding rolled up and packed away. The wooden cot is now used as a platform for the milkseller to set up shop for about an hour. In the few hours from then and the time the shop opens, the space in front of the shop transforms first into a place of gathering and resting by a group of retired men who return from their morning walk to stop, sit, and chat. The same spot is then used as a school bus stop, with children occupying the corner, using the wooden bed to sit on, stand next to, and play around. Next, the shop owner arrives with an employee and opens the shop. The employee moves the wooden cot close to the shop to display goods on it. He sweeps and occasionally washes the sidewalk in front of the shop and brings out a wooden bench, an electric generator, and trays on stands to display goods—items that will occupy the threshold of the sidewalk

Fig. 15.6 Multiple social practices of active and passive interaction and transaction make the street space dynamic and fluid. Photograph by Shilpa Mehta

and the shop. After the shop closes in the evening, the space is occupied by men who gather near the corner. Some sit on the wooden cot while others stand nearby, smoking and chatting. In a couple of hours, the wooden cot is again transformed into a bed and the cycle continues. By using the same space and attributes differently through time, the street is flexible and adaptable. A variety of activities occur in various parts of the street at any given time, and numerous activities occur in the same place over the hours of the day. The multiple social practices, thus, make the street space dynamic and fluid in nature—one that resists the normative static taxonomies typically associated with the street in the West (Fig. 15.6). The street itself is understood less as a physical public space between private and public properties and buildings and more as a place realized by social and economic processes over time.

Path Is Also a Place

Urban streets in Delhi have a similar morphology, including the dimensions of enclosure, to the street space in the urban areas in the West: the majority of pre-independence inner city streets are narrow and labyrinthine, while the post-independence street layout is a mix of wide, automobile-oriented avenues, boulevards, and streets with narrow alleys. However, the Delhi urban street is articulated and used in ways that question its geometry. Path and place compete for the space on the street; gathering and lingering behavior occurs simultaneously with the act of movement, often negotiating and compromising for space. Movement seldom has hegemony on the street; various speeds and rhythms coexist, making moving cumbersome whether in a vehicle or on foot. Moving translates into a negotiated labyrinthine path that pushes, nudges, and squeezes through the myriad territories and ongoing activities. The image of the street as a space of flows competes with that of a place of interactions, activities, and rest, reflecting the coexistence of self-made, self-supported and top-down processes.

Merchants typically encroach and appropriate substantial space immediately outside their businesses, creating zones for use of the street space—and also ones that hinder movement. Vendors and panhandlers further punctuate the linear space of the street. The linearity and unidirectionality of the street is constantly challenged. These social and economic practices fracture the structure of the path due to the inevitable occurrence of encounters. For example, moving along the street to get from one end to another one first encounters the street corner where the shop owner has encroached upon most of the sidewalk with makeshift furniture to prepare and serve snacks. This leaves only a few feet of the otherwise generous sidewalk to get by, because a car is parked half on the sidewalk. Walking by one cannot ignore the smell of the snacks and the conversations of the customers. Moving along one notices the varied shapes and sizes of encroachments through personalized displays outside each shop punctuated with regular projections of bench-like furniture that camouflages the gas-powered electricity generators outside of most shops. These displays include food items and spices, garments, inexpensive jewelry, electronic accessories, small household goods, and so on. Further, the walking space is squeezed again as a family of potters has displayed a range of pots and earthenware outside their shop, across the sidewalk, and into the roadway. Motorcycles and scooters parked just a few feet away from the shops leave barely enough space for people to walk by but they also provide places to sit on the street outside the shops. Walking from one end of the street to another often requires one to leave the sidewalk and walk in the roadway thus confronting oncoming vehicles that do not travel in designated lanes. Further up the street, one encounters a cow that has decided to sit on the pavement. No one will bother the cow, and some will even feed it as it is considered sacred by Hindus. Next, one has to carefully walk by so as not to step on the *sarees* that are drying over the sidewalk. As one travels one has to stop, zig-zag, nudge, ask for space, and negotiate to move on.

Whether by choice or not, the slow-moving body on the Indian street acts like a sponge, taking in stimuli and interactions as it moves through the street. The street becomes a place of "social immersion" (Edensor 1998: 209) and the body, actively or passively, participates in the street. Through such social practices of communication and negotiation, the body becomes both a consumer and a producer of stimuli, interactions, and meanings. In its rich milieu of social practices, the Indian street is characterized by difference and diversity and the street is both a path and a place. Not all streets are equally crowded, of course. The labyrinthine streets of the historical inner city may be the most arduous to get through, but even commercial streets display the path-place overlap that intermingles commerce, dwellings, and transportation, making travel cumbersome but also engaging and an active way to participate in the city.

Negotiation Is Key to Survival

The coexistence of formal and informal practices and processes that result in the street as a dynamic space is a result of elaborate negotiations among its numerous actors—landlords, tenants, shop keepers, vendors, municipal authorities, police, residents, customers, street cleaners, and many more. The Indian street is a distinctive environment of complexity and contradiction, a diversity of use, and a place of overstimulated sensory experience. The location of activities and the use of space on the street is always negotiable, morphing to fit the needs of the users.

Order and operation on the Indian street is determined more by the limits of adaptability of the space and the negotiation between individuals and local institutions than by predetermined laws executed by public authorities. Although there are formal rules set in place by jurisdictions, corrupt practices by those in power—politicians, police, landlords, and local gang members—often break and make new rules to their advantage. However, the most important rules and regulations are enacted by the negotiations among the users of the street. For example, a shop owner may allow an employee or a vendor to sleep overnight on the entrance threshold of the shop, providing shelter for the individual and free security for the shop owner. In another common scenario, local police officers may harass vendors and shopkeepers—the vendors for setting up shop on the street, and shopkeepers for encroaching on the sidewalk with goods and other items. But the police negotiate with each to continue their business for a small fee for the week or month, when they return to demand their ransom. This practice is particularly punishing for vendors since this bribe amounts to a substantial portion of their revenue. Nevertheless, this localized negotiation allows shopkeepers and vendors to continue to sell, even if the regulations prohibit their activities. In another case of negotiation, even as shopkeepers dominate the street, they may also protect the vendors from authorities during municipal raids by providing them with shelter and hiding their goods in their shops, sometimes for a small fee.

Numerous such symbiotic relationships exist on the Indian street. This adaptability also requires interactions among the many different users of the space—those that occupy the street by following institutionalized norms and ones that do so by self-made rules based on myriad social and economic needs. These interactions lead to more social compacts among the actors.

CONCLUSIONS

Examining the city through the lens of bottom-up urbanism has provided an opportunity to understand the city as a malleable space, especially for the underrepresented groups that claim the right to the city in multiple ways. The street, as a quintessential public space, has always been a contested space across the world. As a result, the claim to the city via a claim to the street has been exercised by numerous groups, both through top-down and bottom-up structures, and the many hybrids in between. This multiplicity of actors that claim the street, the coexistence of top-down formal and bottom-up self-made urbanisms, is not a new phenomenon on the Indian street—many Western cities hosted similar street scenes for centuries and many other developing countries still do today (Kim 2015).

As discussed earlier, the formal and informal use and appropriation of space on the Indian street is inherently a display of the juxtaposition of native pre-colonial cultural practices and colonial orders. In fact, reading the Indian street exposes the limits of the use of informality as a way to describe the diverse and complex processes in the city. What may seem informal economic and social activities at first sight are actually a result of multi-layered relationships, power struggles, and negotiations among formal institutions, local populations from different classes, and the actors who are engaged in these activities (Schindler 2017). Empirically studying spaces without predetermined categories, and focusing on the agency of numerous actors, helps us understand the richness, complexity, and fluid nature of places and phenomena (see, for example, Duneier 1999; Low 2000; Mehta 2013).

Examining the Indian street provides many lessons that can help further our understanding of bottom-up processes and, perhaps more importantly, the inextricable link between bottom-up and top-down structures in creating equitable and sustainable places. First, it is clearly evident that bottom-up, informal activities and processes are inherently linked to the economy of cities. They provide a means of income for a substantial populace (Sanyal and Bhattacharya 2009; Bhowmik 2001), and also form an important basis for the formal economy. Second, although conflicts frequently occur, and most actors engaged in a bottom-up process have to fight for their right to public space, this bottom-up urbanism is not primarily characterized by insurgency. Rather, it operates through complex modes of negotiation. Third, the Indian street beckons us to rethink the street as a place to stay, to question its unidirectionality as a space of movement. Fourth, it shows us that planning and design must accommodate multiple uses across the day and week, inspired by the latent needs of the neighborhood and city, to use the space to its full potential. Finally, examining the street demonstrates how various, seemingly incongruent, groups and activities have the ability to cohabit public space. In doing so, the Indian street teaches us how the city can thrive on the coexistence of diverse groups, activities, forms, objects, and modes of control and negotiation, as it operates as a social, cultural, economic, and political space.

References

AlSayyad, N. (2004). Urban informality as a "new" way of life. In A. Roy & N. Alsayyad (Eds.), *Urban informality: Transnational perspectives from the Middle East, Latin America and South Asia* (pp. 7–30). New York: Lexington Books.

Anderson, E. (2011). *The cosmopolitan canopy: Race and civility in everyday life.* New York: W. W. Norton.

Baviskar, A., & Ray, R. (2011). *Elite and everyman: Cultural politics of the Indian middle classes.* New Delhi: Routledge.

Bhowmik, S. K. (2001). *Hawkers and the urban informal sector—A study of street vending in seven cities.* Patna: National Association of Street Vendors of India.

Bromley, R. (1978). The urban informal sector: Why is it worth discussing? *World Development, 6*(9/10), 1033–1039.

Brosius, C. (2010). *India's middle class: Forms of urban leisure, consumption and prosperity.* New Delhi: Routledge.

Davis, M. (2006). *Planet of slums.* London: Verso.

Duneier, M. (1999). *Sidewalk.* New York: Farrar, Straus and Giroux.

Dupont, V. (2011). The dream of Delhi as a global city. *International Journal of Urban and Regional Research, 35*(3), 533–554.

Edensor, T. (1998). The culture of the Indian street. In N. Fyfe (Ed.), *Images of the street.* London: Routledge.

Escobar, A. (1995). *Encountering development: The making and unmaking of the third world.* Princeton: Princeton University Press.

Gidwani, V. (2008). *Capital interrupted: Agrarian development and the politics of work in India*. Minneapolis: Minnesota University Press.

Harriss-White, B. (2003). *India working: Essays on society and economy*. Cambridge: Cambridge University Press.

Irving, R. (1982). *Indian summer: Lutyens, Baker and Imperial Delhi*. New Haven: Yale University Press.

Jacobs, J. (1961). *The death and life of great American cities*. New York: Vintage Books.

Kim, A. (2015). *Sidewalk city: Remapping public space in Ho Chi Minh City*. Chicago: University of Chicago Press.

Lefebvre, H. (1991). *The production of space* (D. Nicholson-Smith, Trans.). Oxford: Blackwell.

Lefebvre, H. (1996). *Writings on cities* (E. Kofman & E. Lebas, Trans.). Cambridge, MA: Blackwell.

Low, S. (2000). *On the plaza: The politics of public space and culture*. Austin: University of Texas Press.

Mehta, V. (2013). *The street: A quintessential social public space*. London: Routledge.

Mehrotra, R. (1992). Bazaars in Victorian Arcades. *Places, 8*(1), 24–31.

Naik, A. K. (2009). *Informal sector and informal workers in India*. Paper Prepared for the Special IARIW-SAIM Conference on "Measuring the Informal Economy in Developing Countries", Kathmandu, Nepal.

Neuwirth, R. (2005). *Shadow cities: A billion squatters, a new urban world*. New York: Routledge.

Portes, A. (1983). The informal sector: Definition, controversy, and relation to national development. *Review (Fernand Braudel Center), 7*(1), 151–174.

Rao, U. (2010). Making the global city: Urban citizenship at the margins of Delhi. *Ethnos, 75*(4), 402–424.

Roy, A. (2004). Transnational trespassings. In A. Roy & N. AlSayyad (Eds.), *Urban informality: Transnational perspectives from the Middle East, Latin America, and South Asia* (pp. 289–319). New York: Lexington Books.

Roy, A. (2005). Urban informality: Toward an epistemology of planning. *Journal of the American Planning Association, 71*(2), 147–158.

Roy, A. (2011). Slumdog cities: Rethinking subaltern urbanism. *International Journal of Urban and Regional Research, 35*(2), 223–238.

Sanyal, K., & Bhattacharya, R. (2009). Beyond the factory: Globalisation, informalisation of production and the new locations of labour. *Economic and Political Weekly, 44*(22), 35–44.

Schindler, S. (2014). Producing and contesting the formal/informal divide: Regulating street hawking in Delhi, India. *Urban Studies, 51*(12), 2596–2612.

Schindler, S. (2017). Beyond a state-centric approach to urban informality: Interactions between Delhi's middle class and the informal service sector. *Current Sociology Monograph, 65*(2), 248–259.

Srivastava, S. (2009). Urban spaces, Disney-divinity and moral middle classes in Delhi. *Economic and Political Weekly, 44*, 338–345.

UN-Habitat. (2008). *The state of the world's cities report 2008/2009: Harmonious cities*. Nairobi: United Nations Human Settlements Programme

Varma, P. K. (1998). *The great Indian middle class*. New Delhi: Penguin Books.

Ward, P. (2004). Introduction and overview: Marginality then and now. *Latin American Research Review, 39*(1), 183–187.

CHAPTER 16

Bottom-Up Urbanism in China Urban Villages and City Development

Stefan Al

INTRODUCTION

China has witnessed the world's largest urbanization drive, which has added 450 million people to its cities in only 25 years (United Nations 2014). Most of the new urban areas are car-oriented, mono-use 'superblocks,' without human scale—planning mistakes the national government is presently aiming to correct through urban design guideline reform. But, surprisingly, an alternative type of urbanization has emerged from within China's cities: urban villages, former agricultural villages that have become 'urban' because of an influx of a large migrant population. These informal settlements exhibit pedestrian-oriented, mixed use, urban conditions that the formal city, restrained by dated zoning regulations that advantages vehicular transportation and mono use, could not achieve.

The emergence and proliferation of these urban villages is a unique phenomenon of urban China that occurs in many regions and in different forms. They happen as a consequence of the rapid urbanization that has resulted from land reform, the dual urban and rural land ownership and management system, and the large influx of an underprivileged migrant population. Literally villages within the city, or *chengzhongcun*, these are previously agricultural villages that have been engulfed by the city and have become urban, but in their own way. Buildings stand so close to each other they are dubbed 'handshake houses'—you can literally reach out from one building and shake hands with your neighbor. Although it is easy to misperceive these villages as slums,

S. Al (✉)
University of Pennsylvania, Philadelphia, PA, USA

© The Author(s) 2019
M. Arefi and C. Kickert (eds.), *The Palgrave Handbook of Bottom-Up Urbanism*, https://doi.org/10.1007/978-3-319-90131-2_16

a closer look reveals that they offer a vital, mixed use, spatially diverse, and pedestrian alternative to the prevailing car-oriented, monotonous, modernist, planning paradigm in China.

However, these countless villages wedged within urban areas are now being 'redeveloped' in what is arguably the world's largest urban demolition. This chapter argues for the value of urban villages as places and for their incorporation into cities rather than their demolition, and suggests that there are lessons in the urban villages for the formation of China's new urban design guidelines. Although much academic research has pointed to their role in providing affordable housing for a migrant population, they have been approached insufficiently from an urban design perspective. It is this urban design argument that could potentially help persuade city governments to integrate villages, rather than to demolish them. Chinese modern cities, planned rigidly from the top, could benefit from the unique qualities of urban villages that grew from the bottom-up.

The Rise and Fall of Urban Villages

In 2011, bulldozers tore down nearly the entire village of Dachong, destroying over ten million square feet of village housing and evicting more than 70,000 residents, many of them migrants (*Shenzhen Daily* 2011). In what was called one of the key urban 'upgrades' of the decade, a vibrant community had been turned into a rubble-ridden demolition site. Only a few old trees, historic temples, and ancient wells stood on the site, to be preserved, further accentuating the bleak new hole that had formed amid the skyscrapers of Shenzhen.

Located inside Shenzhen Special Economic Zone, Dachong Village had become a prime real-estate location when it was engulfed by the explosive development of the surrounding city. Developers and government officials saw the village's adjacency to a new high-tech industrial zone as both a major nuisance and a business opportunity. Following the familiar *tabula rasa* approach to planning, the village would be subsumed in the anonymity of the surrounding city only after it was razed. Billboards with images of corporate office towers, a five-star hotel, and a colossal mall, already visualized the future of the village on the demolition site. Banners celebrated the "scientific urban planning" and the "collective transformation" of what was to be the largest redevelopment project of its kind in the Pearl River Delta, aspiring to become a national model for upgrading older urban areas.

The local press chose not to question these empty slogans, nor the eviction of the countless politically disadvantaged migrants, but rather to frame the redevelopment as a conflict between the real-estate company and the village households who owned land. A few families had refused to transfer their property rights, but after the district government approved eminent domain, the remaining homes were razed. Those who agreed were given more than 30 billion RMB compensation to sell their properties, propelling the former farmers into the ranks of the *nouveau riche:* 400 villagers became multimillionaires and ten made it to the ranks of RMB billionaires (*Global Times* 2010).

Dachong Village is just one of countless villages wedged within urban areas and now being eliminated. But what local people call a 'village' is, in reality, an urbanized version of a village: an 'urban village.' They no longer consist of the picturesque farms of rural China, but of high-rises so close to one another that they create dark claustrophobic alleys—jammed with dripping air-conditioning units, hanging clothes, caged balconies and bundles of buzzing electrical wires—crowned with a small strip of daylight, which locals call "thin line sky" (see Fig. 16.1).

On the cusp of the economic reform in China in the 1980s, in a process that continues today, municipal governments gradually converted rural to urban land. Before land reform, government-owned land was allocated to users without charge. While the reform privatized land-use rights, not land, it effectively created a de facto land market since land-use rights could be sold. It marked the end of free land use in China, opened up the market to private investors and real-estate development, and kick-started rapid urbanization.

However, the municipal governments could only achieve partial land acquisition in the countryside since it was too costly to compensate and relocate villagers, so they could only transfer the farmland surrounding the villages into industrial areas and housing. The now landless villagers had to find another source of revenue and went from growing vegetables to leasing out apartments to a steady stream of migrant workers, who sought employment in the nearby newly built factories. Since their collectively owned villages, officially categorized as rural by the household registration system, or *hukou*—the system that designates all Chinese citizens as either rural or urban—they were unconstrained by the city building laws and regulations of the new cities that surrounded them. Villagers were able to add story after story to their homes, leading to the literal extrusion of the village's narrow building lots from low rise to high rise. As a result, disproportionally narrow streets delineate the new high-rise version of the village.

Initially, the resulting cramped 'kissing buildings' worked out for all stakeholders. They helped the government to transfer large portions of collective land to urban property ownership, while supplying the villagers with a new livelihood now that they could no longer farm. The villagers were promoted to landlords, and many chose to enjoy their new affluence in more opulent parts of the city, moving out of the village. In their place came migrants from all over China, in search of cheap rent. Largely excluded from the general housing distribution system and from home ownership because of their limited rights and low incomes, their housing options were limited. Often, they preferred living in urban villages to the monotonous dormitories in the factory compounds, since the villages, in close proximity to the factories, offered many services, including different types of shops and restaurants. To them, urban villages provided a suitable place to live and, with their burgeoning economies, to work.

Fig. 16.1 'Thin Line Sky' in Xiasha urban village, located in Shenzhen. Caged balconies, hanging clothes and dripping air-conditioning units adorn the narrow streets, with only a small strip of daylight seeping in. Credit: Stefan Al

The success of these villages seems to be short-lived, however, as they are being wiped out almost as soon as they pop up. The case of Dachong Village, one of the largest urban villages in south China until the bulldozers rolled in, is paradigmatic of the demolition of urban villages all over China. More than a 1000 village redevelopment plans exist all over China, affecting the homes of millions of people. For instance, a 2000 plan for Guangzhou mandated the destruction of all 138 urban villages in the city's central districts alone (Crawford and Wu 2014). It is the largest urban demolition in the world's history, but has thus far been given little attention outside of China.

As urban villages have become valuable real estate in now urban locations, city governments aspire to deal with the 'problem' of the villages permanently, eager to transfer the collective, village-held, land-use rights, back to the state. The village redevelopment benefits the government and developers above all, who make a fortune by developing large swaths of land in prime locations, and only those native villagers who successfully negotiate their transfer of land-use right for housing, moving costs, and loss of livelihood. Moreover, unless villagers have been able to negotiate for an urban *hukou*, the household registration system that privileges urban over rural residents, they could end up having no access to social security or health care. Despite their organization in village collectives, the government can exercise eminent domain at any time. Migrants end up losing most, left with little or no alternative to affordable housing. They are also the last to be considered, as they suffer under a rural *hukou*, an inferior form of citizenship. Urbanization in China is certainly wealth creation, but it is unequally distributed and forced through land evictions and the maintenance of the political inequalities systematically produced by the *hukou* system.

Furthermore, the city as a whole loses its unique history and places to the relentless repetition of cookie-cutter office blocks and residential enclaves. It is poignant that Chinese city planners, desperately pondering ways to infuse identity into their newly built homogeneous cities, overlook the urban village. Their unique urbanisms, histories, spatial experiences, culture, and cosmopolitanism could bring a more diverse texture to the future of the city.

A Respite from the Standardization of Formal Planning

Chinese urban planning post-economic reform has been a smashing success in building infrastructure and promoting economic growth at rapid speed. Strong top-down planning that favored standard urban forms helped urbanization and the provision of housing on a large scale. But it failed to make places with a human scale. Urban villages, exempt from most formal planning restrictions, were one of the few places that offered a pedestrian-friendly alternative to modern Chinese planning.

Before discussing urban design and planning problems in China, it is important to remember that the urbanization of China, since Deng Xiaoping began the reform of the economic system in 1978, has been an astonishingly successful achievement. A complete new national road system, a new national air system, a new rail network (including high speed trains and rail rapid transit in almost every major city), the annual average growth of 9.4% GDP, the construction of more than a 100 million units of housing, and the elevation of more than half a billion people out of poverty and into the middle—or even the upper—class, has been unparalleled in the history of national development. Nearly 54% of the country's 1.35 billion people now live in cities, up from less than 18% in 1978. China added 400 million people to cities in a period of only 35 years. This is roughly the populations of the USA and Germany combined. Moreover, China's 2014–2020 Urbanization Plan called for the country to be 60% urbanized by 2020, adding another 90 million people to cities (National Congress 2014). Both the scale and the speed of this urbanization process are unprecedented. London, for instance, during the nineteenth century, transformed into the world's largest city, from one million in 1800 to 6.7 million a century later—a 1.9% average annual growth. The Shenzhen population, on the other hand, increased 40-fold in 35 years from 280,000 to an estimated 11 million residents—a massive 11% annual compound growth rate (United Nations 2016).

China's first urbanization drive triggered by economic reform of the early 1980s was largely based on the repetition of urban forms, in particular the mass reproduction of standard footprints for housing estates and factory towns (Fig. 16.2). To facilitate rapid urbanization the government created a rule book for designing and building cities, including standards for all kinds

Fig. 16.2 Shanghai's typical modern housing areas are devoid of the fine-grained urbanism of its urban villages. Credit: Lawrence Wang, Flickr (licensed under Creative Commons)

of street system, a standard regional development plan, a required system for making land-use plans, population standards for residential areas as a hierarchy of three different kinds of housing organization, and clear and strict standards for the housing itself—including detailed rules for how much sunlight has to reach every dwelling unit, leading to vast and lackluster spaces between buildings, or 'towers in the park.'

The paradox is that the world's largest urbanization drive ever has been undertaken with a very small number of planners. The per capita number of city planning professionals and architects is much lower than in other countries. The Beijing planning department, for instance, only has a few hundred people in charge of an area with a population of 21 million. That same department is also in charge of building the massive number of 200,000 affordable homes, a year—which could house the entire population of Washington, DC. This explains why the emphasis lies less on innovation and customization, and more on standardization. Typically, existing urban models are copied, and building codes are followed rigidly—for instance the Urban Residential Code GB 50180-93, which mandates standard requirements for residential estates, including for sunlight (item 5.0.2), building setbacks, and open space (Ministry of Construction 2002). While this strategy succeeded in setting a minimum quality standard for a large volume of buildings and open spaces, when it was followed overly rigidly, it led to a large number of monotonous urban environments lacking the spatial specifics of a particular geography, culture, or climate.

Not surprisingly, much of the new development in China has ended up following a pattern which is similar from city to city, and from region to region. This standardized urbanization has spread over agricultural land and scenic landscapes, with little regard for environmental considerations. Following this powerful set of rules, written almost 40 years ago, is no longer in keeping with the complex and sophisticated society that China has become, and the lack of consideration for compact development, cultural heritage, and pedestrian spaces. The Central Committee of the Communist Party of China and the State Council have recognized these problems in one recently published guidance document: *Opinions on Further Strengthening Urban Planning and Construction Management* in February of 2016. The document kick-starts an urban design guideline reform that aims to correct the mistakes of ubiquitous gated superblocks in cities, and aspires to mixed-use, pedestrian-friendly environments. However, there is no word on urban villages, let alone their urban design qualities.

Although, usually, no urban design professionals have been involved in their design, the urban design merits of the urban village are plentiful. Their densely grouped, compact footprints are highly efficient, with much higher population densities than in the surrounding city. Since most of the ground level of the village has a commercial function, urban villages are truly mixed use, giving residents the convenient proximity to neighborhood stores,

Fig. 16.3   Visitors and villagers in Daxin Village play pool outdoors. Photo by Stefan Al

restaurants, and places to work. This also contributes to an active street life with plenty of "eyes on the street" (Jacobs 1961).

In an urban China dominated by droves of generic skyscrapers, traveling through urban villages presents an alternative vision of modernity that evokes Marco Polo's journeys in *Invisible Cities* (Calvino 1974). Typically, you can only access them by going through a gateway, which doubles as a security gate since most villages have their own private police force. Once inside, there is an air of cosmopolitanism, with dialects heard from all over China, and restaurants with cuisines from many regions. The narrow and populated market streets in some villages, with their open display of exotic products, remind one more of *souks* then they do of *hutongs*. These lead to unexpected open spaces, with children running around outside, or to ancient temples where the elderly are playing mahjong (Fig. 16.3).

Often, a village is dominated by a special industry, whether information technology, shoe manufacturing, ceramic production, replica products, or massage parlors. Some villages are well-known backdrops for filming Chinese operas, while others have a more infamous status as a red-light district. Some villages display wealth and even build their own plaza and museum to celebrate the village's history and future, while others are less prosperous, with unpaved streets littered with rubble and trees filled with drying clothes. The small shops in mixed-use villages provide services to the local population and tourists, opportunities for small business owners and entrepreneurs, and work for residents.

The fine grained urban fabric provides more intimate and human-scale urban spaces than most modern Chinese development. The streets are usually too narrow to accommodate cars, and the small blocks provide a denser network of pedestrian connections than the oversized, modernist, mega-blocks outside of the villages. For these reasons alone, people in the urban village mostly travel by foot or use public transit rather than a car. Walking through their undulating streets gives an interesting experience, enhanced by the variability of village buildings. As much as urban villages have an important role in providing an affordable housing option to a disadvantaged migrant population, they can also offer a mixed-use, human scale, and pedestrian alternative to the prevailing car-oriented superblocks.

The Stigma of Urban Informality

It is easy to misperceive the urban village as a slum, and many Chinese do exactly that. Dickensian nightmares portrayed by the local press often describe the urban village in medical terms, as an "eye sore," "scar," "ill," or even "cancer" of the city. They further stigmatize the migrant residents as filthy, as burglars, drug users, or even murderers. The reporters' quotes of unsanitary conditions and crime rates help governments justify their destruction, who perceive them as a messy threat to their more sterile vision of modernity.

Although many villages have dirty alleys and dilapidated buildings with poor lighting and ventilation conditions, and buildings of up to 15-stories that may lack an elevator and are topped with roof shacks, the people living in these homes are not the urban poor; they are productive, if politically disadvantaged, citizens with jobs (Wu et al. 2010: 74). Many urban villagers have television sets, refrigerators, and occasionally even cars. For them, the place is not a slum but an important, affordable, and well-located entry point to the city where they can become a full urban citizen after a few years of a steady job, so they can get decent health care, social benefits, and send their children to proper schools (Song et al. 2008). Furthermore, even white-collar workers or college students frequent the urban village to enjoy its many services and shops, and sometimes they even prefer to live there.

From this perspective, the emergence of urban villages in China fits the worldwide trend of "urban informality" (Roy and AlSayyad 2004). Much of the world's urbanization occurs in the informal sector, outside of institutional structures, such as building regulations, zoning laws, or land tenure. Hundreds of millions of people around the world are excluded from formal housing, which explains the existence of the *favelas* in Rio de Janerio, the *barrios* in Mexico, and the shantytowns in India (Fernandes and Varley 1998; Hernandez et al. 2010). Developed nations, too, have their forms of extralegal and unplanned communities, for instance, the *colonias* border settlements in Texas. As research shows, these communities are not marginal, but are fully embedded in the economy (Perlman 1976).

The study of Chinese urban villages can contribute to this scholarship, particularly because urban villages are not synonymous with the urban poor. Urban villages are anything but marginal; they are integral to an economy that relies on low paid, value-added labor, created by the state's inability to provide adequate housing to millions of blue-collar workers who are playing an important part in the economic development of China.

Counting the number of stories of buildings in urban villages almost literally enumerates the different level of urbanization of the enclosing cities. The ones in Shenzhen and Guangzhou are tallest, since these cities were urbanized first. The height of the urban village is thus a barometer of urbanization, a marker of the lack of affordable housing, and sadly, also an indicator of impending demolition; the higher they get, the more they signal the value of the land, and the more likely they are to become the prey of developers and governments.

Alternatives to Demolition

Total demolition, the default option of the state, is problematic for more than a lack of proper substitutes. Not only does it erase the unique historical and cultural traces of a village, the redevelopment can put pressure on the surrounding infrastructure, and is also expensive. In addition, demolition eventually forces migrants to resettle into suburban areas that have potential for social trouble. China's Twelfth Five-Year Plan announced the building of 36 million affordable housing units by 2015, but most of them are located on the outskirts of the city (Central People's Government 2015). These are lesser alternatives to the urban villages, since they are a long commute from work, while their isolation from the city and lack of social diversity gives them the gloomy prospect of turning into ghettos, much like the *banlieues* in Paris. Instead of urban villages, cities would be better to redirect their anxieties to even more threatening disruptions of their vision of modernity: the newly constructed ghost towns and malls—the empty and high-end antithesis of the urban village—such as Ordos City in Mongolia and the South China Mall in Dongguan, the largest mall in the world and also the emptiest, with a 70% vacancy rate.

The upgrading of urban villages makes sense from a housing perspective. China's current social housing policy is insufficient. While living conditions in urban villages are currently substandard, tested village upgrading strategies to overcome persistent village ailments offer a more viable alternative to outright demolition. Urban planners could be guided by important counter-paradigms to the Dachong redevelopment, for instance, Dafen and Huanggang Villages.

Dafen Village in Shenzhen is internationally infamous for its cultural production of 'fake' paintings ranging from Da Vinci to Warhol, which are exhibited in the countless alleys that make the village a popular tourist attraction. Dafen represents hope for urban villages in general, not only because of its economic success, but also because the village is recognized as a model by city officials, who agreed to feature Dafen in the Shenzhen Pavilion of the 2010 World Expo in Shanghai. The 2017 Shenzhen Biennale for Architecture and Urbanism even chose another urban village, Nantou Old Town, as the site for the government-funded progressive architecture exhibition. It shows that a new generation of planners and officials, thanks to the efforts of several architects and activists, may soon offer more progressive ways to deal with the villages.

Huanggang Village, also in Shenzhen, has been inventively and independently redesigned by its own village shareholding corporation. The village managed to redevelop, walking a fine line between respect for its rural past and aspirations for an urban future that includes 40-story towers and even the ubiquitous closed circuit television. The village became, according to one city planner, "even more urbanized than the city," which led one researcher to reverse the understanding of Huanggang from "village in the city" to "city in the village" (Smith 2014). Huanggang challenges the urban-rural dichotomy, and shows that villages, like cities, can be key actors of urbanization, as they have been in the past in, for instance, Ancient Greece. In a process called *synoikism*, urbanization started with villages growing, encroaching on one another, and coalescing into towns, such as the city of Nicopolis (Sève 1988; Kostof 1991).

Now that the first urbanization wave has passed, and major cities such as Beijing, Shanghai, and Guangzhou are seeking to contain further urban growth by setting growth boundaries, the new focus no longer lies on urban expansion, but on urban regeneration. The moment is right to implement a smarter, more flexible form of urbanization that departs from the rigidity of the current urban development system. Reinstating the focus on the village could lead to a richer, more varied pathway to urbanization.

But the poor condition of individual buildings in an urban village does not justify the eradication of entire village areas. As in any city, buildings come and go, but streets, open spaces, and everything else that gives a long-term identity to a place can be sustained and even integrated into the future of a city. Villages could be treated like the older historical villages that some Western cities have been smart enough to incorporate in their greater urban fabric—places like Gràcia in Barcelona, or the Village in New York City. Their irregular and small-grained urban fabric provides a welcome variety to the larger, homogeneous grid of the formal city, whereas the small lots provide opportunities to accommodate smaller businesses and dwellings.

Instead of what the *Shenzhen Daily* claims to be a "gloomy picture" of the urban village, with "rampant burglary, drug abuse and trafficking, prostitution, organized crime and even murder" (Guangqiang 2011), planners should paint a picture that celebrates the urban village's uniqueness, pedestrian friendliness, human scale, accessibility, vibrancy, and spontaneity—in short, all the elements that make up a good city.

Refeferences

Calvino, I. (1974). *Invisible cities.* London: Vintage. Original in Italian, (1972). *Le città invisibili.* Turin: Giulio Einaudi Editore.

Central Committee of the Communist Party of China. (2016). *Opinions on further strengthening urban planning and construction management.* State Council Gazette Issue No. 7, Serial No. 1546.

Crawford, M., & Wu, J. (2014). The beginning of the end: Planning the destruction of Guangzhou's urban villages. In S. Al (Ed.), *Villages in the city: A guide to South China's informal settlements.* Hong Kong: University of Hong Kong Press.

Fernandes, E., & Varley, A. (Eds.). (1998). *Illegal cities: Law and urban change in developing countries.* London and New York: Zed Books.

Guangqiang, W. (2011, April 18). Urban villages, an ill of the city. *Shenzhen Daily.*

Hernandez, F., Kellett, P., & Allan, L. (Eds.). (2010). *Rethinking the informal city: Critical perspectives from Latin America.* Oxford: Berghahn Books.

Jacobs, J. (1961). *The death and life of great American cities.* New York: Vintage Press.

Kostof, S. (1991). *The city shaped: Urban patterns and meanings through history.* London: Thames and Hudson.

Ministry of Construction. (2002). *Urban residential code GB50180-93.* Beijing: Architecture and Building Press.

National Congress of the Communist Party of China. (2014, March 17). National New Urbanization Programme 2014–2020. *Announcement by Xinhua News Agency.* http://politics.people.com.cn/n/2014/0317/c1001-24649809.html. Accessed July 1, 2017.

Perlman, J. (1976). *The myth of marginality.* Berkeley: The University of California Press.

Roy, A., & AlSayyad, N. (Eds.). (2004). *Transnational perspectives from the Middle East, Latin America, and South Asia.* Lanham, MD: Lexington Books.

Sève, M. (1988). Colonies et fondations urbaines dans la Grèce romaine. In J.-L. Huot (Ed.), *La Ville neuve, une idée de l'Antiquité?* (pp. 185–201). Paris: Errance.

Smith, N. (2014). City-in-the-village: Huanggang and China's urban renewal. In S. Al (Ed.), *Villages in the city: A guide to South China's informal settlements*. Hong Kong: University of Hong Kong Press.

Song, Y., Zenou, Y., & Ding, C. (2008). Let's not throw the baby out with the bath water: The role of urban villages in housing rural migrants in China. *Urban Studies, 45*(2), 313–330.

United Nations. (2014). World Urbanization Prospects: The 2014 Revision, Highlights. Department of Economic and Social Affairs. *Population Division, United Nations.*

United Nations, Department of Economic and Social Affairs, Population Division. (2016). The world's cities in 2016—Data Booklet (ST/ESA/ SER.A/392).

Villagers to get billions from Shenzhen demolition. *Global Times*, January 25, 2010.

Works starts on Dachong renovation. *Shenzhen Daily*, December 22, 2011.

Wu, F., Webster, C., He, S., & Liu, Y. (2010). *Urban poverty in China*. Cheltenham: Edward Elgar.

Zhonghua Renmin Gongheguo Zhongyang Renmin Zhengfu (Central People's Government). (2015). *Guomin jingji he shehui fazhan dishier ge wunian guihua gangyao* [People's Economy and Social Development 12th FYP Outline]. http://www.gov.cn/2011lh/content_1825838.htm.

PART IV

Theory Versus Practice

CHAPTER 17

Bottom-Down Urbanism

Nezar AlSayyad and Sujin Eom

In the classic film *Metropolis* (1927), Fritz Lang depicted a city divided between an upper-class world of nightclubs, outdoor sports, and heavenly gardens and a subterranean realm where workers lived in dingy, dark slums. By showing a modern future where levels in space correspond to social class, *Metropolis* presented a vision of place where top and bottom urban practices commingle, and utopia and dystopia intertwine.

To understand the underbelly of cities, one must recognize that these places do not simply grow on their own. Indeed, cities result from the actions of many individuals, communities, and institutions over time; they are also physical manifestations of conflict and collaboration among competing ideologies and values. Within the spectrum of cities, however, those of the, so-called, Third World are often seen as 'informal' because they are a product of subcultures at the lower end of economic and political power. They are thus sometimes seen as reflecting 'bottom-up urbanism,' or urban informality.

The recent portrayal of bottom-up urbanism within the disciplines of geography, sociology, and urban studies has come to include strategies variously labeled as do-it-yourself, creative, guerilla, vernacular, hybrid, pop-up, and tactical. These are all typically seen as noble endeavors. However, many have also critiqued their association with a particular geography (the Third World) or a particular class (the urban poor). Instead, informality must be understood as a practice at all economic and social levels (Roy and AlSayyad 2004).

N. AlSayyad (✉)
University of California Berkeley College of Environmental Design,
Berkeley, CA, USA

S. Eom
Dartmouth College, Hanover, NH, USA

© The Author(s) 2019
M. Arefi and C. Kickert (eds.), *The Palgrave Handbook of Bottom-Up Urbanism*, https://doi.org/10.1007/978-3-319-90131-2_17

271

Within the First World and with regard to elite groups, the term can thus refer to activities outside the purview of established laws and regulations (Mukhija and Loukaitou-Sideris 2014).

But, bottom-up from where? And top-down from what? If it is top-down planning against which bottom-up urbanism is pitted, 'top' encompasses countless faces of hegemonic power. In this polemical piece, we propose to interrogate the ground line, or *datum*, from which decisions emerge, thus allowing us to track the nature of top and bottom. By ground line, we refer to absolute basic needs or minimal conditions for proper housing. In some sense, then, we seek to relocalize these levels of intervention and their directionality. Only then is it possible to discuss top or bottom urbanism in a balanced manner.

On Top-Down Planning

While planning, in crude terms, refers to "an activity that precedes both decisions and actions" (Friedmann 1987: 39), top-down planning has long remained a standard mode of building and managing settlements. Historic examples range from Baron Haussmann's transformation of Paris in the mid-nineteenth century to the design and construction of modernist Brasília as a new capital for Brazil. According to James Scott (1998), the top-down approach can be understood as an attempt by the state to make society "legible," so that various modern practices, from taxation and conscription to the prevention of rebellion, can be more easily accomplished. A planner, within this paradigm, retains control over a complete process, with a clear vision of the public interest in, and self-confidence about, scientific and technological progress. Goethe's Faust was such a figure, producing totalizing plans and visions for a new society (Berman 1988). From Le Corbusier to Robert Moses, top-down planners have, however, also been known for their abhorrence of messy urban forms and conditions, whether slums or squatter settlements. On the one hand, this attitude may reflect a professional aversion to disorder; but it may likewise express class bias toward these types of settlements as a threat to order and authority.

Top-down planning has worked well in creating vibrant environments, but it is often achieved only at tremendous cost, including the demolitions and evictions that typically accompany urban renewal. One is reminded of the famous debate between Jane Jacobs and Lewis Mumford in the 1960s. While Mumford saw the potential of top-down planning (Mumford 1962), Jacobs focused on the street spaces of ordinary people and rejected the large-scale public interventions involved in master planning.

In many parts of the Third World, top-down planning has often been the norm for spatial governance—with presidents, ministers, and governors making direct decisions without a consultative process. It was not until issues of equity, race, and ethnicity came into focus in the mid-twentieth century that challenges emerged to such practices.

On Bottom-Up Planning

From its beginnings, bottom-up planning has been about empowering users, particularly the residents of neighborhoods and cities. In the USA it originated with figures such as Paul Davidoff (1965), who highlighted the role of advocacy and pluralism and called for community action from the "bottom-up." The role of planners in this case was to advocate for citizen input. Jacobs was an important figure in this movement, heralding the capacity of ordinary people to understand and plan for their own needs. Against Moses's top-down approach, exemplified by large-scale projects of urban renewal, she emphasized the importance of streets, neighborhoods, and small blocks— which might appear to top-down planners as chaotic, but which were in fact "complex systems of functional order" (Jacobs 1961: 376). "Participation" soon became a mantra in this type of planning, signifying a strategy by which citizens could obtain "full managerial power" (Arnstein 1969: 217).

It was principally in the First World that advocacy garnered attention as a form of resistance to top-down planning. Bottom-up planning did not appear in the same way in the Third World. Instead, a range of external experts— architects, planners, and economists—intervened, highlighting 'local capacity' as a way to address the 'spontaneous settlements' that had emerged following the rapid urbanization of the late 1950s and early 1960s. Not only were national governments concerned with housing the urban poor, but international agencies, scholars, and architects became deeply engaged in this challenge in a variety of ways depending on profession and class interest.

With regard to Third World housing problems, the most well-known approach became the self-help strategy of the British architect John F.C. Turner. In *Housing by People* (1977) he critiqued the incompetence of professional architects and planners, who he said were incapable of deciding what was best for actual residents. In general terms, Turner drew a contrast between centrally administered systems (heteronomy) with self-governing systems (autonomy), arguing that housing and services must be "autonomous." Centralized technologies, hierarchical bureaucracy, and the large-scale production and distribution of heteronomous systems resulted in serious diseconomies, he wrote, including the extra expense of administrative procedures and factory overheads. In contrast, the small-scale, low-energy operations and labor-intensive technologies of autonomous production could reduce housing prices substantially.

Self-help housing soon extended beyond the Third World to be regarded as a general strategy for acquiring land and building houses for the poor. Paul Baross (1990) later captured the contrast between planned and unplanned development by conceiving of orthodox development as involving four phases. The first, planning, involves obtaining administrative approvals. This is followed by, second, servicing and, third, building, which allow the construction of houses on legally subdivided plots. The final phase, occupation, then involves renting or selling houses to users. In many cities in the Third

World, he pointed out, every phase of the planning, servicing, building, occupation (PSBO) process is beyond the reach of the poor. What actually takes place is a reversed sequence of occupation, building, servicing, planning (OBSP). Because planned development requires high up-front investment and overheads, unplanned development (which may appear as urban informality or bottom-up urbanism) may be the only way for the urban poor to obtain serviced land and housing.

Building on the work of Turner, recognition of these conditions soon led to new policy recommendations, most notably those of the Peruvian economist Hernando De Soto. In *The Other Path* (1989), he identified four phases of informal urban development, and argued that as an "extralegal" process, these could be improved, and made more "legal," through deregulation, debureaucratization, and privatization. In *The Mystery of Capital* (2000), De Soto subsequently extended these recommendations, calling for the "integration" of the poor into the formal sector by providing title to extralegal real estate. In essence, De Soto argued that the poor are wealthier than people realize. The main problem is that they are incapable of converting informal assets (dead capital) into live capital. His solution was to formalize their holdings, and so make them visible within state-sanctioned legal systems.

Even before it was picked up and eulogized by neoliberal consultants like De Soto, the self-help approach had encountered considerable criticism as a basis for housing policy. In particular, Rod Burgess (1982) maintained that its primary flaw was that it uncritically assumed the free, unprotected labor of the poor. He also argued that heteronomy and autonomy (or institutional and self-help housing) were "falsely polarized categories" (1982: 67). These were not different systems at all, he pointed out. Rather, he argued, they were different parts of capitalism—one representing the formal valorization of landed capital, and the other the petty-commodity production of housing as entwined with the general conditions of capitalism. By tracing the structural context in which self-help housing emerged, Hans Harms (1982) further contended that followers of the self-help credo confounded "freedom to build" with "necesssity to survive." He thus differentiated "self-help," as initiated and controlled by residents, from the similar processes mandated by the state. And he concluded that while the former represented "class struggle from below," the latter embodied "class struggle from above"—an attempt to "increase integration into the existing social order and to perpetuate capitalist accumulation and domination" (1982: 20).

Critics at the time also pointed out that self-help as official policy shifted responsibility on to the urban poor to such an extent that the state could extract itself from its proper roles and responsibilities. Depoliticizing the housing question thus helped depolicitize the state. Even though Turner had identified hierarchy, bureaucracy, and scale as the source of the housing problem, his critics argued that he had done so in a way that treated the state in merely technical terms. He, and De Soto after him, had thus failed to properly account for its constitutive role in the capitalist mode of production.

As this brief review indicates, it is not that self-help housing—or, by extension, bottom-up planning—is itself problematic. It is more a matter of who practices and advocates it, and for what purposes. It is therefore necessary to reflect upon what bottom-up urbanism means within capitalist regimes today.

THE DATUM FOR ACTION

So, again: Bottom-up or top-down from where, and in what direction? Top-down and bottom-up are concepts that assume actions exist in a zone whose top is endowed with authority and power and whose bottom has little or no power. And while top and bottom indicate opposite locations, up and down suggest opposite directionality. Hence, it seems that the key to answering the main question is to understand and redefine the *datum*, or ground line, as a frame of reference from which power is enacted.

Planning as a process, whether top-down or bottom-up, is restrained by the institutional hierarchies through which it operates. In top-down planning those with power take direct action; while in bottom-up planning institutional structures establish spaces in which those without power may act. In a sense, the idea of what is top and bottom thus creates an overall space that defines the directionality of decision-making and the processes by which wealthy or poor, powerful or powerless, may become engaged. In top-down planning, we assume an arrow coming down from the power above. In bottom-up planning, we imagine an arrow rising up from the powerless below. Such a conceptualization, however, assumes a datum, or limit of action, on both top and bottom. But what really interests us is the capacities these create in the zone between.

One may argue that the relationship between top and bottom, and the directionality of up and down, has changed over time. The early urban revolutions contributed to the primacy of the top-down dynamic. As a result, the *political city* of the classical age established the citadel as the top-central institution of life. By medieval times, however, the development of the *mercantile city* allowed space to be organized as a network of independent city-states, and thus expand capital and political power. While top-down processes shaped many cities at the time, a process similar to bottom-up urbanism also existed and influenced the shape of such settlements. It was only with the arrival of the *industrial and post-industrial city* that the urban crisis necessitated the involvement of professionals, especially in the Global North, who followed a top-down approach and viewed their work entirely from above.

Professional planning also appeared in cities of the Global South, but their development continued to be dominated by bottom-up practices. Urban conditions in the Global South also continued to worsen in a manner that complicated the structures of power and the directionality of decision-making. Both top-up urbanism, represented by gated communities, and bottom-down urbanism, represented by massive squatter settlements, stand witness today to this phenomenon. While different in every respect, both types of communities in places such as Latin America, South Asia, and the Middle East, share one common trait: they exist outside land use and building laws.

It is within this context that neoliberal consultants such as De Soto have advocated the formalization of the informal sector. However, the financial crises of the present century already provide ample evidence that this approach does not benefit the poor. Instead, these events show how the formal sector, from banks to governments, is itself prone to acting informally—by exploiting legal loopholes to benefit the powerful. Indeed, numerous recent cases of financial fraud reveal how success in capitalist societies relies not only on deregulation and free markets, but on the "manipulation of the mass media to divert attention," "predatory lawsuits to silence opposition," and "money to purchase the support of key politicians" (Bromley 2004: 276). Along the same lines, the formalization of self-help housing can be seen as "an inexpensive policy for housing provision without changes in resource allocation or structural changes" (Harms 1982: 23). It encourages the integration of the poor into structures of power as taxpayers and consumers of privatized goods and services without offering them much in return.

Today, the promise of urban informality to deliver effective bottom-up urbanism in the form of self-help housing and other interventions seems as far away as ever. The issue is not only that what is celebrated as bottom-up urbanism is often divorced from reality; it is also that the bottom-up approach seems doomed to create a downward spiral—or what we could term bottom-down urbanism. For instance, bottom-strata communities have fought for decades to obtain land and property titles from the state, without much success. Ultimately, they have also built their own settlements using vernacular practices, informal strategies, and often illegal means. But these achievements are more likely to reflect the dedication of local organizations; alliances with political parties and leaders; or negotiations with the army, the business elite, and, in some cases, drug cartels. Thus, during election periods in Latin America and South Asia, or political unrest in the Middle East, such enclaves may suffer radically different fates—with some expanding rapidly and others being totally demolished.

By contrast, gated communities are an instantiation of top-up urbanism, through which only the already wealthy prosper. Here, if buildings are built in violation of codes, state authorities find ways to condone them. "Elite informalities" are thus made formal by governments, where other informalities are simply criminalized (Roy 2011: 233). The metropolitan fringe has become a particular location for such activities. The former homes of rural–urban migrants and their auto-constructed settlements are now being reclaimed under state authority for gated communities, the construction and maintenance of which is then subsidized by state infrastructure spending. Informality is thus not only supported by the state, but is also "produced by the state" (Roy 2011: 149).

The development of gated communities in the Third World indicates a form of planning that does not need to seek government legitimacy. This is why we label it top-up; because it often exists beyond the reach of planning authorities. Bottom-down urbanism, by contrast, may refer to conditions in which the poor fail to acquire legitimacy, despite their efforts to achieve it. Top-up communities can live without legitimacy; bottom-down ones cannot acquire it, however hard they try.

The Promises of Urban Informality

Self-help settlements and auto-construction created within the sphere of urban informality may be conceptualized as bottom-down urbanism—or bottom-down planning if professionals are involved—if they fail to deliver on the promise of bottom-up urbanism and simply perpetuate conditions of poverty and powerlessness. What follows are a series of examples from different countries of the Global South.

The Rocinha *favela* in Rio de Janeiro, one of the largest in Brazil, has been regarded as a "squatter success story" (Neuwirth 2007). Contrary to conditions in the 1970s, when residents lived in wood or mud shacks without tap water or electricity, by 2007 its population of 150,000 had full access to such services. The change was largely driven by the Favela-Barrio program, initiated in 1994 with financial help from the Inter-American Development Bank. Its goal was to physically upgrade slums, to "transform squatter settlements (*favelas*) into officially recognized neighborhoods (*barrios*)" (Roy 2005: 150). The program also reflected a shift in public policy in Brazil—from eradicating *favelas* to attempting to integrate them into the city.

In a follow-up to her 1970s ethnographic investigation of *favelas* in Rio de Janeiro, however, Janice Perlman discovered that physical upgrading had made little difference to the residents of Rocinha. Where the language of poverty had once been used by parents to discipline children ("If you don't stay in school and study, you'll end up being a garbage collector"), three decades later even applying for a garbage collector job required a diploma. Many residents also told Perlman that they had been "harmed" by the government or by international agencies such as the World Bank, IMF, and Inter-American Development Bank. As she observed: "With too many candidates courting the favela vote, and too many promises that go unfulfilled, political corruption has become too visible, and cynicism has set in" (Perlman 2004: 137). Nowadays Rocinha is also a site for *favela* tourism. Thus, on the website Brazil Expedition, a tourist can arrange a visit for US$30, during which a local guide will serve as an informant on "life in the favela and what it means to the people living here."[1] Business magazines tout such activities as an

---

[1] http://www.brazilexpedition.com/tours-in-rio/favela-walking-tour/. Accessed March 16, 2017.

"entrepreneurial opportunity" for *favela* residents.[2] But the reality is that residents of Rocinha must endure this voyeuristic practice while remaining at the bottom of the economic ladder, with little prospect of moving up.

Villa El Salvador in Lima, Peru, is often cited as an example of successful state-sanctioned informal urbanism. It was started as a semi-spontaneous invasion of land in an area named Pamplona in 1971. During this contested action, one leader, Salvador Saldivar, was killed, and many prospective settlers were injured. With the intervention of the Catholic Church, however, the invaders were able to create considerable publicity and gather strength by framing the government's reaction as evidence of a prevailing lack of social justice. General Armando Artola Azcárate, the minister of the interior, was subsequently forced to resign, and the invaders were allowed to name their settlement after the deceased leader (De Soto 1989). By 1978, its population had expanded to 300,000 (AlSayyad 1993). And by 1983, it had become an independent urban district with a population of more than half a million people today.

Fifteen years later, Villa El Salvador also became central to the policies of Hernando De Soto. Between 1996 and 2005, 650,000 properties were legalized in Lima alone (Calderón 2005, as cited in Plöger 2012). However, legal recognition and formalization paradoxically led to a decline in community organizations (Plöger 2012). Meanwhile, the neoliberal policies of the Alberto Fujimori regime (1990–2000), allowed small-scale enterprises in Villa El Salvador to be damaged by global competition; while health-threatening environmental conditions persisted, with 40% of its residents still having no access to safe drinking water or sewerage by the late 1990s (Hordijk 2005). Since 1999, the mayors of Villa El Salvador have launched a range of programs to develop participatory governance and projects to improve physical conditions. But, as a result of poor funding, a shortage of professional expertise, and sheer mistrust, most of these have failed to produce positive outcomes. The fragmentation of public participation has further undermined residents' capacity to voice their opinions. Continuing poverty has only worsened the situation. Thus, despite the installation of water pipes, many residents cannot use them because they cannot afford the connection fee (Ioris 2012).

An example from the Philippines offers another interesting case. Manila, the capital of the Philippines, has been a city of squatters since World War II. Their number in the city and its suburbs increased from 46,000 in 1946, to 98,000 in 1956, and to 283,000 in 1963.[3] In 1991, the documentary *On Borrowed Land* portrayed the dismal conditions endured by many of these

---

[2] "Favela Tourism Provides Entrepreneurial Opportunities in Rio," *Forbes*, May 16, 2011. Accessed March 16, 2017.

[3] http://philrights.org/from-squatters-into-informal-settlers/. Accessed April 2, 2017.

rural migrants. More than 50,000, for example, lived in the informal settlement of Reclamation alongside Manila Bay, established after the overthrow of Ferdinand Marcos in 1986. The site eventually became a city within a city, built with scavenged materials and with its own transportation network and water distribution system. But its residents were never allowed to acquire security of tenure. Eventually the Philippine government began selling the area to foreign developers, one of whom envisioned a project named Asiaworld City. Such speculative urbanism provides a stark contrast to the conditions of the urban poor in Manila, who have been largely "neglected" and "forgotten" (Shatkin 2004). Although every government post-Marcos has professed concern for poverty alleviation, a range of globalization concerns has led them to pursue neoliberal economic policies since the 1990s. Recently, following pressure from international aid agencies to cut national government spending, the responsibility for the poor has been delegated to local governments, which lack the resources to address their concerns. In part this explains the chronic housing shortage in Manila, where, as of 2010, there were an estimated 2.9 million squatters, representing 556,526 families.[4]

Egypt too has had its share. *Zabaleen* is an Arabic term that literally means garbage people. In Cairo, it also refers to a district on the city's eastern fringe where its garbage collectors live. Mostly Christians (a religious minority in Egypt), they collect food waste, paper, boxes, bottles, and plastic from the rest of the city and transport it to Zabaleen, where they clean and process it, and then resell usable items. The district functions as its own city, with grocery stores, butchers, coffee shops, and a school. Several attempts have been made to "formalize" waste collection in Cairo (Assaad 1996), the most notable of which have involved international waste- management companies. In 1981, the Zabaleen Environmental Development Program was launched to improve living conditions with funds from the World Bank and, by the 1990s, the area's makeshift shacks had largely been replaced by more solid structures built with concrete and brick (Fahmi 2005). Yet, despite the improvement in housing quality, a survey in 2004 indicated that residents still feared being evicted as a result of official upgrading and private speculation. This is partly because projects initiated by the Cairo Governorate to relocate Zabaleen residents to the city's outskirts, would result in enhanced land values and benefit developers and investors who wished to "beautify" the area (Fahmi 2005). Moreover, access to clean water and sanitation was still inadequate, especially given the site's historic use.

Several destructive rockslides have also raised the profile of efforts to move the Garbage City (Fahmi and Sutton 2010). The first, in 1993, killed 70 people; in 2008, residents again awoke to the sound of boulders crashing down on them. Nearly 100 brick buildings were destroyed, and more than 100 people killed. Residents vented their anger at the government, which

[4]Ibid.

Fig. 17.1 Zabaleen after the collapse

they claimed had neglected the warning signs—including cracks in a retaining wall—and failed to prevent a repeat of the 1993 disaster. While searching for bodies of family members "buried alive," residents lamented that "no one had listened to them because they were seen as poor, powerless and less than human" (Amnesty International 2009: 6) (Fig. 17.1).

Kibera, the largest slum in Kenya's capital city, Nairobi, houses 600,000 people in an area the size of New York's Central Park. It has received national and international aid simply on account of its size. One such massive effort was the Kenyan Slum Upgrading Programme (KENSUP), initiated by the Kenyan government and UN Habitat in 2000. But Kibera has also witnessed design interventions by smaller nonprofit organizations. One of these, the Kounkuey Design Initiative (KDI), won the Renewal Award from *The Atlantic* magazine and the Allstate insurance company in 2017 for its efforts to transform unused land into productive community space. In Kibera, the KDI (a nonprofit founded by graduates of the Harvard School of Design combining the efforts of architects, engineers, and planners) has tried to create spaces, from public gardens to community centers, that will allow residents to generate additional income by selling vegetables, water, or basic services such as toilets and showers. Beyond designing the spaces, its goal is to encourage residents to profit from their own "talents and resources."[5]

[5] https://www.citylab.com/design/2017/03/public-space-projects-kounkuey-design-initiative/521346/. Accessed March 16, 2017.

However sincere the intentions of such a project may be, the reality of poverty in Kibera is far graver than such design activism can address. And it has only been compounded by slum upgrading projects such as the Kibera High Rise project and the Pumwani-Majengo slum redevelopment. Although these efforts may ultimately improve the lives of the relatively better-off, they displace the poor to new slums in less convenient locations (Huchzermeyer 2008). While a distorted housing market is among the primary causes for the failure of such interventions, they also critically fail to acknowledge the inner workings of slums. The reality is that some residents have an economic stake in housing and service delivery that is overlooked by dysfunctional local governments. Thus, more often than not, slum upgrading only benefits 'slumlords,' perpetuates the status quo, and worsens already poor conditions (Marx et al. 2013).

The above condition may be particularly relevant when the commodification of basic services, such as water, sanitation, or garbage collection, functions as "a barrier or disincentive" to service provision by the public sector (Huchzermeyer 2008: 31). In this sense, one must ask what the real priority should be when addressing conditions of poverty. Should it be to increase income-generating activity by turning basic services into privately sold commodities? Or should it be to interrogate the role of the state, while protecting housing stocks from market invasion, as a resource for the poor?

What residents in all the above examples have needed most was institutional and economic *legitimacy*. What they have typically acquired was, at best, minor forms of *legalization*, which do not necessarily provide the same power. Legitimacy refers to the recognition and acceptance of people, structures, and authorities as valid and proper. While, in theory, legitimacy is bound up with being legal, in actuality something can be legitimate without being legal, or it can be legal without being legitimate. In this case, legalization is what the datum—the symbolic line indicative of the institutional power structure—can offer the poor. Legitimacy is much more complex; it is something the residents desire in order to upgrade their physical environment and to improve the social conditions therein. For instance, settlements such as Villa El Salvador have undergone some legalization, and now receive better services than before. But there has been little improvement in the socioeconomic condition of residents.

Should these not then be considered examples of bottom-down urbanism? Certainly, the bottom-up, self-help strategies used in these settlements have not brought a substantial improvement in the lives of residents. But this is not because physical conditions have worsened; rather, it is because they have never reached beyond the social datum from which they started. This condition is engrained in the process of their genesis. When upscale gated communities get established, developers and residents are not concerned about government involvement, because they are legitimate by virtue of who they are and what they own. The poor, however, no matter how hard they try, can never penetrate the datum. The datum is always assumed to be a line but, in

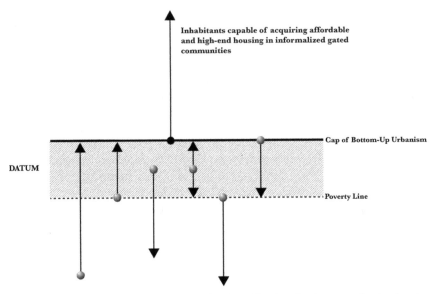

Fig. 17.2 The datum is always assumed to be a line but in the case of urbanism is a zone

the case of urbanism it can be argued that the datum is a zone (see Fig. 17.2). The urban poor start from that zone; the best they can achieve is stay in it or climb up to a certain extent, whereas they may also slip below the datum zone. The rich who live in equally informalized gated communities, by contrast, start from above the datum and can only go upward.

A Bottom-Down Urbanism

Economist Peter Edward (2006) has argued that the paradigm of development should be shifted away from the "under-development" of the Third World to the "over-development" of the First World. His argument is that underdeveloped countries should not have to "catch-up" with developed ones, but that over-developed ones should "catch-down" by lowering their levels of consumption. His call to revisit the directionality of up and down in thinking about development and global poverty corresponds with what we have explored here through the language of bottom-down urbanism.

In many instances of bottom-up urbanism, architects and planners conclude that squatters and slum dwellers are not competent to use the resources and materials at their disposal. Like the early colonial administrators, these metropolitan experts tend to see their expertise as helping the urban poor by providing architectural or design solutions. But what they fail to recognize, based on their own ethnic and class bias, is that informality as a process cannot be applied to conditions of poverty in isolation from global capitalism. In effect, doing so may relegate responsibility for poverty itself to the poor by erasing the role of the state.

But not only does this attitude toward bottom-up action fail to improve conditions, the architects and planners who advocate it may actually be damaging the built environment. For instance, the Favela-Barrio program, initiated by the city of Rio de Janeiro and Jorge Mario Jáuregui Architects in the 1990s, was meant to engage *favela* residents in "a participatory design process with architects and other technical experts" (Architecture for Humanity 2006: 216–220). Its design interventions, from underground power lines, to brightly colored stair rails, and gateways made from brick and steel I-beams, were subsequently celebrated as "the arrival of basic services and the favelas' assimilation into the formal city." And the installation of facilities such as day-care centers, communal laundries, and salsa halls was heralded as an attempt to "invite and encourage the community to embark on its own journey of self-improvement." Meanwhile, streets were built for the first time to allow "police and sanitation departments crucial access to once-inaccessible parts of the city." The real problem here, however, was not that the language reflected a view of *favelas* as isolated, inaccessible, and unassimilated urban entities that needed the touch of metropolitan experts. It was rather that, to build an infrastructure of integration that would assimilate the poor into the formal city, the designers had to remove and demolish many of the existing houses.

Informality can help reduce vulnerabilities and increase options for the economically and politically disadvantaged, as we have seen in the example of Zabaleen. It would be a mistake to conflate this, however, with the failure or success of the urban poor. Instead, we suggest that a structural failure of contemporary neoliberal capitalism in recognizing, let alone addressing, the problems of the urban poor is to blame. Architects and planners alone can neither address issues of urban poverty nor solve urban questions that are deeply rooted in structural economic inequality. Self-help approaches to housing, such as those initiated through the Favela-Barrio program, are philanthropic endeavors at best; but they are also nothing less than policy solutions that isolate the housing question from the structural conditions of capitalist development. At stake is neither design itself nor the intentions of architects and planners. What architects and planners are striving to deal with is often far beyond the capacity of their profession. Chronic poverty is far more than a lack of public space, which architects and planners can address with design solutions. Bottom-up planning can never be comprehensive or complete. It may at best remedy situations that are already degraded; it cannot eradicate the root causes of poverty.

What we have problematized here is a zone in which top and bottom continue to move or shift in each direction. The movement at the top takes place under exceptional circumstances, where undemocratic processes and outcomes remain invisible due to the complicit relationship between capital and authority. Even though a change may occur within the bottom of the zone, it has very little significance for the urbanism of the poor. Although the poor continue to move up and down within the zone, their quality of life remains a matter of sheer survival. Recognizing the datum as a zone thus requires a rethinking of top and bottom from the perspective of planning theory and action. Bottom-down urbanism, in this regard, can be understood as the datum which is formed and constructed by the glass ceiling of the neoliberal capitalist order.

REFERENCES

AlSayyad, N. (1993). Squatting and culture: A comparative analysis of informal developments in Latin America and the Middle East. *Habitat International, 17*(1), 33–44.

Amnesty International. (2009). *Buried alive: Trapped by poverty and neglect in Cairo's informal settlements.* London: Amnesty International Publications.

Architecture for Humanity. (2006). *Design like you give a damn: Architectural responses to humanitarian crises.* New York: Metropolis Books.

Arnstein, S. R. (1969). A ladder of citizen participation. *Journal of the American Institute of Planners, 35*(4), 216–224.

Assaad, R. (1996). Formalizing the informal? The transformation of Cairo's refuse collection system. *Journal of Planning Education and Research, 16,* 115–126.

Baross, P. (1990). Sequencing land develompent: The price implications of legal and illegal settlement growth. In P. Baross & J. J. van der Linden (Eds.), *The transformation of land supply systems in third world cities* (pp. 57–80). Aldershot, Hants; Brookfield, VT: Avebury.

Berman, M. (1988). *All that is solid melts into air: The experience of modernity.* New York: Penguin.

Bromley, R. (2004). Power, property, and poverty: Why De Soto's 'mystery of capital' cannot be solved. In A. Roy & N. AlSayyad (Eds.), *Urban informality: Transnational perspectives from the Middle East, Latin America, and South Asia* (pp. 271–288). Lanham: Lexington Books.

Burgess, R. (1982). Self-help housing advocacy: A curious from of radicalism. In P. M. Ward (Ed.), *Self-help housing: A critique* (pp. 55–97). London: Mansell.

Davidoff, P. (1965). Advocacy and pluralism in planning. *Journal of the American Institute of Planners, 31*(4), 331–338.

De Soto, H. (1989). Informal housing. In *The other path,* by Hernando De Soto (pp. 17–55). New York: Harper & Row.

———. (2000). *The mystery of capital: Why capitalism triumphs in the West and fails everywhere else.* New York: Basic Books.

Edward, P. (2006). The ethical poverty line: A moral quantification of absolute poverty. *Third World Quarterly, 27*(2), 377–393.

Fahmi, W. S. (2005). The impact of privatization of solid waste management on the Zabaleen garbage collectors of Cairo. *Environment and Urbanization, 17*(2), 155–170.

Fahmi, W., & Sutton, K. (2010). Cairo's contested garbage: Sustainable solid waste management and the Zabaleen's right to the city. *Sustainability, 2,* 1765–1783.

Friedmann, J. (1987). *Planning in the public domain: From knowledge to action.* Princeton: Princeton University Press.

Harms, H. (1982). Historical perspectives on the practice and purpose of self-help housing. In P. M. Ward (Ed.), *Self-help housing: A critique* (pp. 17–53). London: Mansell.

Hordijk, M. (2005). Participatory governance in Peru: Exercising citizenship. *Environment & Urbanization, 17*(1), 219–236.

Huchzermeyer, M. (2008). Slum upgrading in Nairobi within the housing and basic services market. *Journal of Asian and African Studies, 43*(1), 19–39.

Ioris, Antonio, A. R. (2012). The geography of multiple scarcities: Urban development and water problems in Lima, Peru. *Geoforum, 43*, 612–622.

Jacobs, J. (1961). *The death and life of great American cities.* New York: Random House.

Marx, B., Stoker, T., & Suri, T. (2013). The economics of slums in the developing world. *Journal of Economic Perspectives, 27*(4), 187–210.

Mukhija, V., & Loukaitou-Sideris, A. 2014. Introduction. In V. Mukhija & A. Loukaitou-Sideris (Eds.), *The informal American city: Beyond taco trucks and day labor* (pp. 1–17). Cambridge, MA: The MIT Press.

Mumford, L. (1962). Home remedies for urban cancer. *The New Yorker*, pp. 132–139.

Neuwirth, R. (2007). Squatters and the cities of tomorrow. *City, 11*(1), 71–80.

Perlman, J. (2004). Marginality: From myth to reality in the favelas of Rio de Janeiro, 1969–2002. In *Urban informality: Transnational perspectives from the Middle East, Latin America, and South Asia* (pp. 105–146). Lanham: Lexington Books.

Plöger, J. (2012). Gated barriadas: Responses to urban insecurity in marginal settlements in Lima, Peru. *Singapore Journal of Tropical Geography, 33*, 212–225.

Roy, A. (2011). Slumdog cities: Rethinking subaltern urbanism. *International Journal of Urban and Regional Research, 35*(2), 223–238.

Roy, A. (2005). Urban informality: Toward an epistemology of planning. *Journal of the American Planning Association, 71*(2), 147–158.

Roy, A., & AlSayyad, N. (2004). *Urban informality: Transnational perspectives from the Middle East, Latin America, and South Asia.* Lanham: Lexington Books.

Scott, J. C. (1998). *Seeing like a state: How certain schemes to improve the human condition have failed.* New Haven: Yale University Press.

Shatkin, G. (2004). Planning to forget: Informal settlements as 'forgotten places' in globalising Metro Manila. *Urban Studies, 41*(12), 2469–2484.

Turner, John, F. C. (1977). *Housing by people: Towards autonomy in building environments.* New York: Pantheon Books.

CHAPTER 18

Property, Planning and Bottom-Up Pedestrian Spaces in Toronto's Post-war Suburbs

Paul M. Hess

Before I built a wall I'd ask to know
What I was walling in or walling out,
And to whom I was like to give offence.
Robert Frost "Mending Wall" from which the better-known line is "Good fences make good neighbors." (Frost 1917)

INTRODUCTION

Various bottom-up urbanisms, including "everyday" (Chase et al. 2008), "tactical" (Lydon and Garcia 2015), "insurgent" (Hou 2010a), and "do-it-yourself" (Talen 2015), contrast localized activities to transform urban spaces with more formal, top-down, institutionalized, larger-scale processes. Drawing on Certeau (1984), authors like Crawford make the distinction between strategic organization and tactical behavior, where people find the gaps and spaces to contravene strategically created structures (Crawford 2008b). These activities redefine "both 'public' and 'space' through lived experience … restructuring urban space, opening new political arenas, and producing new forms of insurgent citizenship" (Crawford 2008a: 22). Likewise, Hou refers to "alternative spaces, activities, expressions, and relationships … that open up new possibilities of public space and public realm in support of a more diverse, just, democratic society" (Hou 2010b: 12). Most of the literature in this vein focuses on case studies in which organized groups

P. M. Hess (✉)
Department of Geography & Planning, University of Toronto, Toronto, Canada

© The Author(s) 2019
M. Arefi and C. Kickert (eds.), *The Palgrave Handbook of Bottom-Up Urbanism*, https://doi.org/10.1007/978-3-319-90131-2_18

institute micro-level physical changes to the urban environment, or stage temporary activities that contravene the designed, institutionalized, and 'master-planned' urbanism that shapes formal public spaces. Very few authors, however, articulate how the top-down, formal, and institutional processes do more than passively set the context for more informal, bottom-up ones. These top-down planning processes structure the geographic and socio-legal patterns of public and private property, and thus the location and nature of the public and collective spaces that everyday or insurgent urbanisms may violate. Thus, essays on various bottom-up urbanisms often refer to the broad public space literature on the increasing privatization and regulation of public space (Banerjee 2001; Blomley 2007; Smith 1996; Loughran 2014), but often do not lay out the specific and strategic or top-down ways this space is structured.

This chapter focuses on post-war North American suburbs where there are distinctive and highly planned patterns of public and private space. These landscapes are often critiqued as places where conventional public spaces have been replaced with private spaces—as places where shopping malls have replaced the public space of the sidewalk and square (Kohn 2004), and where gated communities with private streets increasingly comprise residential landscapes (Grant and Mittelsteadt 2004). Rather than address these common tropes of privatization, this chapter addresses the ways that public space has been created in geographically and socially uneven ways as part of core suburban planning models. This has shaped and even necessitated bottom-up practices by many residents, especially those living in the large apartment areas that are an integral and planned part of these landscapes.

More specifically, the chapter focuses on Toronto's inner suburbs, a location that shares similarities with many other US and Canadian metropolitan areas. First shaped by nineteenth-century land survey practices, and then transformed in the post-war period using planning models such as the traffic arterial and the neighborhood unit, Toronto's inner suburbs incorporate the core ideas of mid-century North American planning. Like US suburbs, these models include large areas of multifamily housing (Moudon and Hess 2000; Larco 2010). The chapter then turns to the ways contemporary residents of these apartments encounter these landscapes as pedestrians, via the tactical behavior referred to in Certeau's famous essay (1984). However, rather than Certeau's abstract and theoretical vision of the streets of Manhattan, the chapter examines the more concrete and everyday spaces common to suburban pedestrian environments, including large urban arterials designed and regulated for vehicle movement, and various types of routes across private property used to access schools, shopping, transit, and other daily activities. Pedestrians are thoroughly disciplined by the strategic spaces of arterial streets, which minimize the possibilities for bottom-up practices; at the same time, property owners often contest the use of informal routes across private land, which nevertheless persist because they are an almost necessary part of carrying out everyday activities. Indeed, these 'transgressive' practices are deeply enmeshed in the very planned patterns of public and property being defended. Residents work to carve out new spaces of collective, if not public, use, and organize to make community claims for the control and improvement of these spaces. However, in contrast to the tenor of the bottom-up, insurgent, or everyday urbanism literature, these claims rarely fundamentally challenge property relations.

Top-Down Construction of Suburban Property and Neighborhood Units

A central process of top-down planning is the way it structures the geography and nature of property and development rights through plans, zoning, subdivisions, and other regulatory tools. In this context, planning's foundational act in North America was the colonial imposition of cadastral surveys as a key part of the process of transforming native land into non-native property. These surveys also established the initial spatial organization of public and private land that still underlies modern, planned development. In Ontario, colonial surveys organized property as a grid of quadrilateral 'concession blocks' bounded by public roadway allowances in a manner similar to, if not as systematic as, the Jeffersonian public land survey used west of the Ohio River in the United States. In the Toronto region most roadway allowances framed either 500 or 1000 acre blocks, establishing some of the only land that was not converted into private property (Weaver 1968; MacGregor 1981). These public rights-of-way that were one survey chain wide (66 feet or 20 meters), represented a mere 2–4% of the area of a concession block. Thus, long before the development of modern "traffic logic" (Blomley 2008), public space across much of the colonial property system was created for the sole purpose of circulating people and goods.

In the post-war period, new planning authorities and institutions were established by the Province of Ontario that almost fully implemented the central ideas of post-war North American suburbanism on top of this older property framework (Sorensen and Hess 2015). Rules for developing private property were spelled out in provincial legislation, such as the 1946 Ontario Planning Act, authorizing municipalities to establish official plans, to enact zoning, and to regulate subdivisions consistent with the plans. In addition, a new regional government, the Metropolitan Municipality of Toronto (Metro), provided additional planning oversight and funded large-scale infrastructure projects to allow development to proceed across its territory (White 2016). With this robust system, public planners helped shape the private development of neighborhood units for small, detached houses, each organized around public schools and parks, with local public streets that deterred through traffic. Thus, planned suburbanization actually created new types of public space within the old pre-war patterns of private land in the interior of concession blocks (see Fig. 18.1).

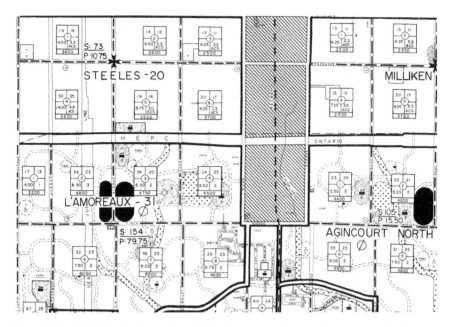

Fig. 18.1 Neighborhood units within the concession grid as part of Scarborough Township's official plan, 1966. Copyright: In public domain, crown copyright expired, Scarborough Planning Board, Ontario

These new public spaces were intended to foster community, but they were created almost as an extension of the private space of the detached house and its idealized associations with family domesticity (Hayden 1984). The fine scale and geographic interconnectedness of subdivisions also created a new network of property interests in which resident-owners consistently seek to protect their neighborhoods from social and physical change. Zoning and other planning policies work to maintain these zones as 'stable areas' and to prevent 'traffic infiltration' into the local public streets of subdivisions. Thus, the new public streets, schools, and parks within neighborhood units are prioritized for use by their (largely landowning) residents, a process that was integral to suburban planning development long before the current neoliberal era.

THE STRATEGIC REALM OF THE SUBURBAN ARTERIAL AND TRAFFIC LOGIC

On the exterior of neighborhood units, post-war planners and engineers also refashioned the old road allowances to accommodate new suburban arterials designed to move motor vehicles across the city. Large parcels along arterials were left for shopping plazas, gas stations, garages, and other commercial infrastructures that support automobility (Scheer 2015), along with sites for rental apartments. Where the neighborhood unit was intended to create a controlled realm for domestic community, arterial streets were their corollary: movement corridors stripped of social life. Thus, the subdivision is largely in the regulatory domain of the community planner and the arterial is firmly in that of the traffic engineer (Hebbert 2005). For the engineer the purpose of the arterial is clear. As the Toronto Road Classification System states, arterials are "intended to serve primarily a traffic movement function" (City of Toronto Transportation Services 2013).

Although it is publicly owned, such movement space is not public space as typically theorized. Blomley (2008: 71) argues that "traffic logic departs from the humanist view of the street." Instead, the freedom and autonomy offered by the car—at least when traffic is moving—allows drivers to pass through the city in an environment that literally protects the self within the private vehicle's shell of steel and glass. Here, the public space potential for interaction and thus for 'insurgent' or bottom-up activities is radically reduced. As Urry puts it, "communities of people become anonymized flows of faceless ghostly machines" (Urry 2004: 30). While Blomley (2010) writes of traffic logic at the scale of the sidewalk, the suburban arterial maximizes flows and organizes the entire territory of the post-war city, connecting neighborhood units to places of employment, shopping, services, and recreation, where the intersections, traffic signals, driveways, and other connections to the local environment are minimized.

In their idealized state, there are no pedestrians along arterials, but pedestrians must use them to access transit and services, and they are provided with sidewalks and other minimal infrastructure. Pedestrian movement is, however, highly disciplined by both design and regulation. Formalized crossings for pedestrians are fit into the overall traffic logic, with signals located only every two kilometers at arterial intersections along the old survey lines. When traffic signals partially halt through-traffic, pedestrians are afforded a short opportunity to cross the roadway. The Ontario Highway Traffic Act states that pedestrians must cross with the light and within marked crosswalks, where they are provided (Province of Ontario 1990, c. H.8, s. 144). Local municipalities, however, further discipline pedestrian behavior. For example, Toronto by-laws make it illegal for pedestrians to interfere with traffic. As long as the driver is operating a vehicle safely and within the law, it is illegal for a pedestrian to force a car to slow down or stop. Although drivers have an obligation to avoid striking a pedestrian, the by-laws suggest that if a legally operated vehicle hits and kills a pedestrian, the pedestrian, by definition, has interfered with traffic (Reid 2007). Thus, while pedestrians are allowed to enter the traffic realm of the arterial, unlike the car driver, their bodies are thoroughly vulnerable both physically and under the law. Pedestrians do legally engage in minor 'tactical' behaviors by crossing arterials at non-official locations, but this does not challenge the designed and regulated environment of control they are navigating.

Pedestrians travel along, as well as across, arterials, but on the minimal sidewalks provided there is little need to regulate pedestrian behavior. Loukaitou-Sideris and Ehrenfeucht (2009) are interested in the ways sidewalks are linked to abutting properties where all kinds of competing social claims and informal and bottom-up activities may occur, but this relationship has been broken on suburban arterials where buildings are pushed well back from the public street. If activity and claims on space extend beyond the envelope of buildings, it is mostly into parking lots and does not reach the sidewalk or street. Indeed, there is little reason to be on these sidewalks, except to travel or to wait for the bus. Theorists such as Kohn (2004) are interested in the spaces of the suburban shopping mall as replacing the public space of traditional main streets, but they have not yet focused on the ways that planners and engineers have disciplined actual streets in subdivisions and on arterials to minimize their function as a public space and thus their potential for protest and insurgent or tactical behaviors.

Pedestrians, and Suburban Apartments as Strategic Landscapes

Pedestrians are an integral part of these post-war landscapes due to the unexpected ways that planning models shaped their social geography. As described above, Toronto's post-war suburbs contain the hallmark features of many of their American analogues with arterial roads, subdivisions, strip malls, shopping centers, and industrial parks. Multifamily housing has also been part of North American post-war suburban development and is even integral to the idea of the neighborhood unit. However, apartments in the model suburb are deliberately pushed to the streets dedicated to traffic and located away from the domestic space of the single-family house and the public space of the neighborhood school and community park (Hess 2005, 2009).

Although commonplace in suburban landscapes across the USA and Canada, Toronto's apartment development took place in greater numbers and on a larger scale compared to US cities. Toronto Metro planners saw apartments as a primary tool to achieved density targets and generate tax revenues for large, new infrastructure systems for transportation, sewers, and water. Apartments were also attractive to local municipalities because it was thought that they would come with few children and would therefore generate more in taxes than they cost in services. In other words, apartments were used to finance the infrastructure and services for neighborhood units, and by the mid-1960s more than seven in ten housing starts in Metro were multifamily buildings. Currently, the city is estimated to have almost 1200 apartment towers of eight stories or more, containing about 280,000 housing units, mostly privately developed in the 1960s and 1970s, and employing modernist designs and site layouts (Stewart and Thorne 2010). The majority of these are located not in the older, central city, but in what were, in the period of their construction, new suburbs and are now identified as inner suburbs.

Although suburban apartments were developed adjacent to areas of single-family houses around parks and schools, planners followed the logic of the neighborhood unit model by physically separating them. In a 1967 Metro apartment control policy, for example, planners wrote that "some form of separation must be sought ... in order to protect the privacy of the low-density dwellings" (Metropolitan Toronto Planning Board 1967). Thus, apartments were located in their own zones with open space buffers, they were not allowed to be located on or connected to subdivision streets, and planners required fences between apartments and single-family housing. The need for this separation was always discussed in terms of planning impacts—shadowing, traffic, and 'privacy'—but the underlying logic should be seen as part of a basic social unease with apartment living, especially for children and families (Dennis 1998). In this context, high-rise apartments were seen as a way to ease a post-war housing crisis and to be transitional housing for people without children (Metropolitan Toronto Planning Board 1959). It was also assumed that most people could afford a private vehicle and would have few restrictions on their physical mobility.

Over the decades, however, the demographic profile of the city in general, and suburban apartment towers in particular, has changed. Although there are many middle-class, well-maintained apartments, many suburban apartment towers are also filled with people with limited incomes, new immigrants, large multigenerational families, people with children, people with complicated travel needs, and people who do not own a car or who only have access to a car part of the time (Hulchanski 2010; MacDonnell 2011). Thus, these landscapes are now inhabited in ways that do not conform to the mid-century top-down social imaginings for which they were designed. For example, single-family subdivisions were planned and developed so that children could walk to school—this is a key idea of the neighborhood unit—but with generational and social change, some single-family areas now have relatively few children. The apartments, however, were consciously separated from schools, but many are now full of children. Often these children must now find ways to walk to school through spaces that were not designed for pedestrians. In general, areas that were designed for cars now house people that must rely on walking and transit to carry out their lives.

In a project on walkability and to understand how residents get around their areas in Toronto's high-rise neighborhoods we conducted surveys, focus groups, and mapping exercises in eight areas of modernist apartments. Although the research was exploratory, not representative, licensing and auto-ownership rates suggest a large population is walking in these areas.[1]

---

[1] These areas are underrepresented in Toronto's large regional transportation survey, as recruitment is difficult. Our strategy was to work alongside neighborhood and community organizations. We encountered 35 languages in the eight communities we worked with, giving a sense of the difficulty. A more complete description of the study and methodology can be found in Hess and Farrow (2011).

Of the approximately 250, mostly low-income, residents who participated, 56% did not have a driver's license, with women being far less likely to be licensed than men. Likewise, 42% of respondents reported living in house-holds without a vehicle. Another 43% came from households with only one car, with only 18% of households having as many or more cars as they had adults living in them.

Unsurprisingly then, walking was one of the most important modes for grocery shopping, for doing other kinds of errands, and for bringing children to school. For grocery shopping for example, 32% of respondents reported walking as their normal way to get to and from the store. This was equaled by the combined total of those who drove (25%) or were driven by a family member or friend (7%). It does not include the 21% of people who used mul-tiple modes, mostly walking to the store in one direction and taking transit or a taxi on the return trip with their load of groceries. For work trips too, residents were highly dependent on walking. Sixteen per cent of participants reported walking to work as their principal mode, with another 41% using transit, which, of course, includes a walk to and from the bus stop. In other words, many residents rely on walking to carry out their daily lives.

Bottom-Up Routes Across Private Land

Pedestrians rely on walking along and across arterial streets in these places but, based on maps made by study participants, about 24% of the total length of the routes they use are, on average, across private property. These routes were created through bottom-up practices as residents forced connections across property boundaries and through and around fences to get to basic destinations such as the local school, the grocery store, or the bus stop. This was necessary because apartments were developed on large parcels of land, without the planning requirements for new infrastructure or public space that are institutionalized parts of the subdivision process in single-family areas. Even though many of these bottom-up routes were of very poor quality, local pedestrians were dependent on them. These routes were also vulnerable to being shut down. These issues are illustrated using two short case studies drawn from the many examples residents described in the focus groups.

The Peanut

The Peanut is a large neighborhood planned by the municipality of North York in the late 1960s.[2] It is named for a distinctively shaped piece of land at its center, the location of community services including a middle school, park, and shopping plaza. More than a dozen high-rise apartment towers surround

---

[2] The formal name for the area is Don Valley Village. North York is now part of the city of Toronto.

Fig. 18.2 The Peanut as planned in 1965, right, and built, left, with modernist slabs arrayed around a community space and adjoining neighborhood units (North York Planning Board. 1965. District 12 Plan. North York, Ontario. Air photo courtesy of Lockwood Surveying Corporation)

these facilities, which in turn are surrounded by neighborhood units of single-family houses carefully planned around parks and schools (see Fig. 18.2). In many ways, the Peanut can be characterized as a hybrid suburban model. The area was comprehensively planned by the municipality, with the central area resembling the mass-housing estates that were being contemporaneously developed in Europe, but the area was developed parcel by parcel as part of a private-property regime more akin to the suburban development of post-war USA. Therefore, what was envisioned as an open, modernist landscape has been fragmented and divided along property lines.

Residents generally expressed a high degree of satisfaction with the overall design of the community. As one resident noted, "I like my neighborhood as I am [close] to the mall, library, parks and groceries, and community center." Walking to these facilities, however, is highly problematic because property owners have closed walkways that were originally designed to lead through the superblocks on which the apartments are located. As one resident explained:

> In '69 there was a path, and they put [in] a fence and I was bloody mad. I had to walk all the way up Goodview to take my kids to the pool. And then they built a path, and we used to scramble up the path.

Far from being 'built,' this latter path is simply the trodden ground formed by the bottom-up practice of constant use. It leads up a short, steep hill to a hole torn in the fences between the apartment properties and out to the main arterial. Asked who uses the route, some focus group participants were clear: "We all do, of course." Beyond issues of night-time security, or mud and snow during winter, some residents also felt uncomfortable crossing

Fig. 18.3 Common routes across private property used by Peanut focus group participants (left) and the 'Goodview' informal path through a fence cut (right). Photo: Katherine H. Childs

private property: "It's wrong—we shouldn't be using [it] that way to get to the other side." Rather than claiming a right to the city, some residents were reluctantly transgressing property relations (see Fig. 18.3).

Scarborough Village

Scarborough Village is a neighborhood in the former municipality of Scarborough, now the eastern borough of the city of Toronto. The area was not master-planned like the Peanut, but still confirms to the basic neighborhood unit model. There is a series of apartment towers strung along six-lane arterials and interspersed with strip malls and a shopping plaza. Separated from the arterial and the apartments, there are areas of single-family housing organized around parks and schools.

Focus groups were held in an area of four large towers organized around Cougar Court, a public cul-de-sac that connects to a main arterial. However, more than using this public connection, pedestrians used several routes across private property to more directly connect to many of the area's basic services (see Fig. 18.4). One such destination was the local public school, originally designed to be accessible from surrounding single-family houses. With shifting demographics, however, the apartments are full of children, but their walk to school that used the formal, public street system was roughly four times as long as the most direct, informal route. The most direct route to the school passed behind the apartments, through a property with a vacant house, and then connected to a local subdivision street, just a short walk from the school. There was some anxiety about this shortcut, as illustrated in this conversation from the workshop:

> *Young Girl*: People think [the house] is haunted... A witch comes out at night. The witch walked outside the house and put paint on the path, to light it up ...

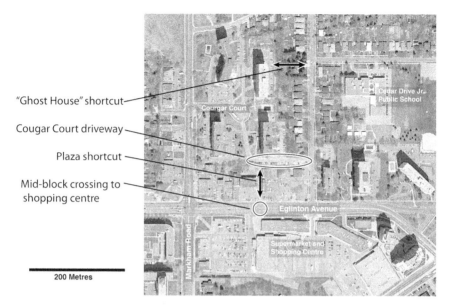

Fig. 18.4 Important informal Scarborough Village routes. Figure by author

Adult: And the house is abandoned, so that probably makes people shaky.

Facilitator: Do all the kids go there to go to school?

Adult: Yeah. I don't want my son to go there. I took the long way... [but most kids] go through the shortcut.

Adult: It's so lonely. Nobody lives there. One day they said ... don't pass by here, because it's my property. They're blocking the fence.

Adult: Sometimes they refix the fence.

Facilitator: So that's a big issue?

Adults: Yeah.

As suggested by this conversation, this route was not always open, with the fence frequently repaired to dissuade use. Users then broke it open again or found other ways across it. Residents had mixed feelings about the route, with some recognizing it as an important route for many children, but others being uncomfortable that it was isolated and went across private property (Fig. 18.5).[3]

As an alternative route, many residents traveled through parking lots and out a long, active driveway that carried much of the traffic in and out of the

---

[3]As of May, 2017, the property was under redevelopment and the route seems to be permanently closed.

Fig. 18.5 Opened and closed conditions of Ghost House shortcut fence. Photo: Katherine H. Childs

apartment area. Residents had many complaints about the driveway, reporting that the surface was poor, with potholes, that rainwater did not drain well, and that it was slushy and icy in winter. The most common complaint was that there was an almost complete absence of lighting. One resident commented:

> This whole road needs more lighting. Where the houses are on the next street from here—those lights, perfect. But here .... it can be better lighting. And then all the cars that park ... they cast a shadow across. So it's jet black.

The comment notes the differences in the quality of the infrastructure on adjoining public subdivision streets compared to what is referred to as a 'road,' but is really just a private apartment driveway. It does not recognize, however, that these differences relate to their public or private status, nor that this difference was a planned, integral part of post-war development—with public infrastructure built for the public streets associated with single-family houses, but not for the internal areas of apartment sites, even though they may be used by more people. Residents experienced these differences as part of their everyday, bottom-up practices as they walked their children to school and went to other destinations, but they did not necessarily understand or consciously challenge their 'strategic' origins.

Bottom-Up Urbanism and Public Versus Private Property

In some instances, however, residents do make conscious, collective efforts at improving their walking conditions and infrastructure. In discussing public space in a liberal property regime, Blackmar (2006: 76) mirrors some of the everyday and insurgent urbanism literature writing that, "Locally, public spaces continue to offer arenas of assembly and can even prompt meaningful political fights on behalf of access, fairness, accountability, or redistribution." In limited ways, we encountered these claims in our work with residents, although they were mostly claims of basic access rather than the types of broader principles to which Blackmar refers. Still, they go further than the simple forging and reforging of routes across private property boundaries; residents sometimes also organize to improve them. Because of the strength of the status of public and private property, however, forging these claims requires enormous amounts of effort for relatively little effect, but still, they more closely resemble claims of a right to the city described in the everyday or insurgent urbanism literatures.

The local residents' association that includes Cougar Court, for example, had been successful at involving the city in minor improvements for a route across several private parcels. The association did annual community clean-ups, clearing out garbage on an important informal pedestrian route. The route led through a rear service area behind two strip malls and out to a location on the busy arterial roadway without a traffic light. Residents crossed there to get to a shopping center located on the other side. The route was used frequently, although it had no lighting, poor sightlines, and rough surfaces. However, in 2010 the residents' association made a sustained effort to improve it, engaging adjoining property owners, garnering the support of the city's Tower Renewal office and its police anti-gang unit, getting some trees cut to improve visibility, and removing some minor barriers.[4] After which a muddy area of the path was paved and some small planter boxes installed. Still, according to a member of the residents' association whom we interviewed, the strip mall owners and police initially suggested fencing off the route rather than improving it. He argued that a new fence would just be broken down like the old one and convinced them that making improvements was a better option.

These types of efforts, however, are severely constrained by property issues. In interviews, the planner involved in the Cougar Court improvements described how they had to be done "off-budget" because the city could not legally spend public monies on private property. Instead, the basic planning strategy to get private land owners to make infrastructure improvements was

---

[4] Cougar Court was then a pilot site for the City's Tower Renewal Program to reinvest in postwar apartment blocks. The program continued to evolve and worked with the public housing authority to test upgrading strategies on nine buildings. The privately owned buildings at Cougar Court, however, were never upgraded under this program.

to offer them additional development rights. In 2009, the Tower Renewal planners even briefly discussed allowing Cougar Court property owners planning permission for new high-rise condominiums in exchange for a new public through street on the development. To my knowledge, the strategy was not seriously explored, as it would be technically difficult, highly political, require the cooperation of several land owners, and it assumed that there was sufficient development pressure in a socially stigmatized community to develop new housing.

More realistically, because the owner wanted a zoning variance to allow for the conversion of former community meeting rooms into more rent-generating apartments, as a condition of the permit, planners sought to require investments in a walkway and lighting along the driveway commonly used by school children. It is but a small irony that in order to create the infrastructure that would simply be assumed on public streets, residents would have sacrificed some of their former collective space within their building. This strategy, however, was also unsuccessful; the driveway was eventually repaved, but no other improvements took place. Likewise, the local councilor in the Peanut area was unsuccessful at getting apartment property owners to reopen the old walkways across their sites. Instead, more effective 'pedestrian proof' fencing was installed to prevent trespassing.

These difficulties arise because the collective, everyday space residents rely on is usually private property. This status is built into strategic structures of planning and governance that are little threatened by small tactical violations. Residents' general reliance on private property for collective use leaves them not just politically vulnerable, but it also creates literal barriers to them carrying out basic activities to sustain their daily lives. For the residents of the Peanut, property owners closed walkways that were initially built to connect people to public spaces, institutions, and community services. Residents without cars had to either take very long, indirect walks to get to bus stops or to go shopping, or they had to take rough pathways forced through fences and trespass across private property. In Scarborough Village, there have never been public routes to directly connect apartments to local public schools. Residents battled to keep routes across private land open, or they coped with long walks along a private driveway system without walkways or lighting.

CONCLUSIONS

While the privatization of suburban space as represented by the shopping mall and the gated community has long been recognized, this chapter argues that planned suburbanization also created other kinds of public space within neighborhoods in which apartment dwellers who do not own vehicles or property are largely excluded. In other words, uneven distribution and access to public streets, public space, and the provision of public infrastructure that goes with it, is planned into these places. The original cadastral surveys used to implement a colonial property regime only carved out enough public land for basic circulation. With the advent of post-war suburbanization, municipal planners helped reshape the arrangements of public and private property. Within the land area of private concession blocks, new public space was recreated in the form of local streets, parks, and schools that were carved out with development, but largely as extensions of the domestic space of detached housing. Zones of apartments were included in suburban planning, but they were intentionally separated from single-family housing, and were associated with the realm of commerce along suburban arterials. Planners also helped reshape the old, rural road allowances into these arterials, but they barely qualify as public space for those without a vehicle. For pedestrians, of course, this is a matter of life and death. The weight and speed of moving vehicles compared to the vulnerable bodies of pedestrians means the balance of power is uncontested, but this is also well codified in the public management of the street which helps discipline pedestrian behavior.

Thus, pedestrian behaviors that forge routes across fence lines, behind strip malls, use private driveway systems as major walking routes, and cross major arterial roadways where there are not traffic signals or crosswalks—the kinds of tactical behaviors alluded to by Certeau and more contemporary authors discussing bottom-up urbanism—must be understood within these planned, institutionalized, strategic structures. Apartment residents, especially those without cars, dwell between the subdivision and the arterial, in a privatized realm without streets or basic pedestrian infrastructure. Blomley discusses "the right to pass freely" as used by the state to control behavior in public space, but pedestrians living in suburban apartments do not have even this right. Indeed, the logic of unimpeded movement breaks down when there are no streets. To discipline behaviors in these spaces does not require laws like those enacted to discipline begging or other 'anti-social' behavior on public streets. Rather, the existing property regime is simply enforced, and even apartment residents that rely on these bottom-up and legally illicit routes often accept this as non-controversial. Once these property regimes are established through planning institutions and authorities, planners have relatively little power to change them, except, perhaps, by creating new development rights that accrue to property owners.

Terms such as bottom-up urbanism, guerrilla urbanism, everyday urbanism, which are current in the urban design and planning literature, capture some of the ways that suburban apartment dwellers cope with their surroundings by modifying them, individually and collectively, sometimes in transgressive ways, and sometimes more formally by working with city officials. These bottom-up efforts should be acknowledged as important, but the connotations of political resistance and social change often associated with bottom-up urbanisms insufficiently highlight how they are structured by top-down strategic, planned structures such as the institutionalized logics of property. Of course, only a few selected examples have been described, and there is a much wider range of bottom-up activities that make claims on everyday and public space. Langregger's (2015) rich work on Latino uses of public space in the context of gentrifying areas of Denver comes to mind, illustrating a very different scale and politics of expropriation. Context is important though. As Certeau (1984: 37) notes, tactics take place within "a terrain imposed" on them and are "the art of the weak." Indeed, for most residents in Toronto's suburban apartment areas, they are simply trying to carry out their lives in environments that were planned for an idealized social world that no longer exists. To make these areas better places to live will take a new vision and both strategic and tactical change.

REFERENCES

Banerjee, T. (2001). The future of public space: Beyond invented streets and reinvented places. *Journal of the American Planning Association, 67*(1), 9–24.

Blackmar, E. (2006). Appropriating "the Commons": The tragedy of property rights discourse. In S. Low & N. Smith (Eds.), *The politics of public space* (pp. 49–80). New York: Taylor & Francis.

Blomley, N. (2007). How to turn a beggar into a bus stop: Law, traffic and the "function of the place". *Urban Studies, 44*(9), 1697–1712.

Blomley, N. (2008). Civil rights meet civil engineering: Urban public space and traffic logic. *Canadian Journal of Law and Society, 22*(2), 55–72.

Blomley, N. (2010). *Rights of passage: Sidewalks and the regulation of public flow.* New York: Routledge. Retrieved from http://myaccess.library.utoronto.ca/login?url=http://www.tandfebooks.com/isbn/9780203840405.

Certeau, M. (1984). *The practice of everyday life.* Berkeley, CA: University of California Press.

Chase, J., Crawford, M., & Kaliski, J. (Eds.). (2008). *Everyday urbanism.* New York: Monacelli Press.

City of Toronto Transportation Services. (2013). *Toronto road classification system: Summary document* (p. 40). Toronto: City of Toronto.

Crawford, M. (2008a). Blurring the boundaries: Public space and private life. In *Everyday urbanism* (pp. 22–35). New York: Monacelli Press.

Crawford, M. (2008b). Introduction. In *Everyday urbanism* (pp. 6–11). New York: Monacelli Press.

Dennis, R. (1998). Apartment housing in Canadian cities, 1900–1940. *Urban History Review, 26*(2), 17.

Frost, R. (1917). Mending wall. In A. Lowell, E. A. Robinson, R. Frost, E. L. Masters, C. H. D. Sandburg, & J. G. Fletcher (Eds.). *Tendencies in modern American poetry* (pp. 92–93). New York: Macmillan.

Grant, J., & Mittelsteadt, L. (2004). Types of gated communities. *Environment and Planning B: Planning and Design, 31*(6), 913–930. https://doi.org/10.1068/b3165.

Hayden, D. (1984). *Redesigning the American dream: The future of housing, work, and family life.* New York: W. W. Norton.

Hebbert, M. (2005). Engineering, urbanism and the struggle for street design. *Journal of Urban Design, 10*(1), 39–59. https://doi.org/10.1080/13574800500062361.

Hess, P. M. (2005). Neighborhoods apart: Site/Non-sight and suburban apartments. In A. Kahn & C. Burns (Eds.), *Site matters* (pp. 223–247). New York: Routledge.

Hess, P. M. (2009). Avenues or arterials: The struggle to change street building practices in Toronto, Canada. *Journal of Urban Design, 14*(1), 1–28. https://doi.org/10.1080/13574800802451049.

Hess, P., & Farrow, J. (2011). *Walkability in Toronto's high rise neighbourhoods.* Cities Centre: University of Toronto. Retrieved from http://janeswalk.org/files/8114/5331/6149/Walkability_Full_Report.pdf.

Hou, J. (Ed.). (2010a). *Insurgent public space: Guerrilla urbanism and the remaking of the contemporary cities.* New York: Routledge.

Hou, J. (Ed.). (2010b). (Not) your everyday public space. In *Insurgent public space: Guerrilla urbanism and the remaking of the contemporary cities* (pp. 1–17). New York: Routledge.

Hulchanski, J. D. (2010). *The three cities of Toronto: Income polarization among Toronto's neighbourhoods, 1970–2005* (No. Bulletin 41, p. 32). Toronto: Cities Centre, University of Toronto. Retrieved from http://www.urbancentre.utoronto.ca/pdfs/curp/tnrn/Three-Cities-Within-Toronto-2010-Final.pdf.

Kohn, M. (2004). *Brave new neighborhoods: The privatization of public space.* New York: Routledge. Retrieved from http://link.library.utoronto.ca/eir/EIRdetail.cfm?Resources__ID=767842&T=F.

Larco, N. (2010). Suburbia shifted: Overlooked trends and opportunities in suburban multifamily housing. *Journal of Architectural & Planning Research, 27*(1), 69–87.

Langregger, S. (2015). Curbing cruising: Lowriding and the domestication of Denver's Northside. In S. Zavestoski & J. Agyeman (Eds.), *Incomplete streets: Processes, practices and possibilities* (pp. 119–138). New York: Routledge.

Loughran, K. (2014). Parks for profit: The high line, growth machines, and the uneven development of urban public spaces. *City & Community, 13*(1), 49–68.

Loukaitou-Sideris, A., & Ehrenfeucht, R. (2009). *Sidewalks: Conflict and negotiation over public space.* Cambridge, MA: MIT Press.

Lydon, M., & Garcia, A. (2015). *Tactical urbanism: Short-term action for long-term change.* Washington, DC: Island Press.

MacDonnell, S. (2011). *Poverty by postal code 2: Vertical poverty: Declining income, housing quality and community life in Toronto's inner suburban high-rise apartments.* United Way Toronto. Retrieved from http://myaccess.library.utoronto.ca/login?url=http://site.ebrary.com/lib/utoronto/Top?id=10443475.

MacGregor, J. G. (1981). *Vision of an ordered land: The story of the dominion land survey.* Saskatoon: Western Producer Prairie Books.

Metropolitan Toronto Planning Board. (1959). *Official plan of the Metropolitan Toronto planning area.* Toronto: Metropolitan Toronto Planning Board.

Metropolitan Toronto Planning Board. (1967). *Metro Toronto (Planning Board) (1967) Report of the proposed metropolitan apartment development control policy.* Toronto: Metropolitan Toronto Plannning Board.

Moudon, A. V., & Hess, P. M. (2000). Suburban clusters: The nucleation of multifamily housing in suburban areas of the Central Puget Sound. *American Planning Association. Journal of the American Planning Association, 66*(3), 243–264.

Province of Ontario. Highway Traffic Act, RSO 1990, § c. H.8, s. 144 (1990). Retrieved from https://www.ontario.ca/laws/view.

Reid, D. (2007, November 20). *Pedestrians crossing mid-block in Toronto: The definitive guide.* Retrieved May 11, 2017, from http://spacing.ca/toronto/2007/11/20/pedestrians-crossing-mid-block-in-toronto-the-definitive-guide/.

Scheer, B. C. (2015). Strip development and how to read it. In E. Talen (Ed.), *Retrofitting sprawl: Addressing seventy years of failed urban form* (pp. 31–56). Athens: University of Georgia Press.

Smith, N. (1996). *The new urban frontier: Gentrification and the revanchist city.* New York: Routledge.

Sorensen, A., & Hess, P. (2015). Building suburbs, Toronto-style: Land development regimes, institutions, critical junctures and path dependence. *The Town Planning Review, 86*(4), 411.

Stewart, G., & Thorne, J. (2010). *Tower neighbourhood renewal in the greater golden horseshoe an analysis of high-rise apartment tower neighbourhoods developed in the post-war boom (1945–1984)* (p. 150). Toronto: E.R.A. Architects, planning Alliance, and the Cities Centre at the University of Toronto for the Ontario Growth Secretariat Ministry of Infrastructure.

Talen, E. (2015). Do-it-yourself urbanism: A history. *Journal of Planning History, 14*(2), 135–148. https://doi.org/10.1177/1538513214549325.

Urry, J. (2004). The 'system' of automobility. *Theory, Culture & Society, 21*(4–5), 25–39.

Weaver, W. F. (1968). *Crown surveys in Ontario.* Toronto: Department of Lands and Forests.

White, R. (2016). *Planning Toronto: The planners, the plans, their legacies, 1940–80.* Vancouver: UBC Press.

CHAPTER 19

Privilege and Participation: On the Democratic Implications and Social Contradictions of Bottom-Up Urbanisms

Gordon C. C. Douglas

From city-sponsored creative placemaking, tactical urbanism, and participatory planning efforts to wholly unauthorized acts of improvised improvement, guerrilla urbanism, or DIY urban design, bottom-up urbanisms suggest new ways of making and remaking streets and public spaces. Crucially, they suggest ways for democratizing the process. These things are rightly gaining attention not only for the creativity, practicality, and site-specific applicability inherent in many of them, but also for their grassroots and popular spirit; at their best, a form of participatory citizenship open to anyone and everyone. It is tempting, in other words, to view bottom-up urbanism as a medium for direct popular participation in urban planning and placemaking, and to view the results as reflective of local values and priorities. Yet this raises questions of what benefit these interventions actually provide for the communities they mean to improve, and whose values and priorities they really reflect. Because many things that we gather under the umbrella of bottom-up urbanism can be problematic in a number of ways and are, in fact, anything but democratic.

This chapter works to complicate such assumptions about the social meaning and local democratic value of bottom-up urbanism. It does so by looking to a number of examples across American cities, some rather extreme, that have much to tell us about the persistence of social inequality in bottom-up urbanism. It argues that this is visible in who participates in the creation of these sorts of would-be improvements and in the types of altered places that result.

G. C. C. Douglas (✉)
San José State University, San José, CA, USA

© The Author(s) 2019
M. Arefi and C. Kickert (eds.), *The Palgrave Handbook of Bottom-Up Urbanism*, https://doi.org/10.1007/978-3-319-90131-2_19

Different concerns arise depending on which sorts of interventions and what sense of bottom-up urbanism we are talking about. The term is used broadly here, to encompass a variety of non-traditional, quasi-informal place-making activities, so some brief definitions and typologies are in order. Without spilling too much ink on the effort, we can note first simply that bottom-up urbanisms can be divided into those actions that are unsanctioned (perhaps even illegal) and those that may be citizen- or community-driven but operate with some approval or through formalized processes (or are even created by professional actors but still in the 'spirit' of bottom-up, tactical urban space interventions).[1] In addition to just 'unauthorized' versus 'authorized,' the first group is referred to here as do-it-yourself (DIY) urbanism or urban design (Iveson 2013; Douglas 2014; Finn 2014), and the second as participatory placemaking (an imperfect nod to the open but not illegal or insurgent nature of more formally sanctioned efforts. I might have chosen from numerous terms for either of these categories[2]). This simple but important point of distinction between the unauthorized and the authorized in bottom-up urbanism serves to divide this analysis into two parts, each raising different social questions.

In both cases, the focus here is on *physical* design interventions in the urban landscape that are ostensibly intended to functionally improve or mod-ify it much as a traditional (official planning, streetscape, or architectural) design element might, however different they might be in terms of particular goal, process, or aesthetic. (So many less physical interventions, such as data-driven 'open city' projects online, community planning charrettes, and citizen science initiatives, are excluded, even though these are parts of the broader bottom-up urbanism trend. Nor are less functional improvements, such as many acts of street art, graffiti, or public art, for instance, of particular inter-est here.) For a handful of reasons, discussed further below, the discussion is also confined to cases in major American cities—a very particular slice of the bottom-up urbanism pie, to be sure.

There are numerous other criteria across which we might categorize acts of informal or bottom-up urbanism. There are broad categories that reflect the focus and intention of the project (streetscape improvements, public services,

---

[1] In this sense, "bottom-up" is not a perfect umbrella term, but it is convenient enough for consistency and in the context of the wider discussion here. Projects on both sides of this unsanc-tioned/ sanctioned divide are still more bottom-up (grassroots, participatory, citizen- or commu-nity-based) than top-down.

[2] The other obvious term for the second group here would be creative placemaking itself, which is a well-established name for a mainstream planning trend that absolutely does encompass many things like parklets, in-street plazas, pop-up shops, and other creative interventions, but may be a little too broad and is defined more by the activation of the arts in placemaking than any bottom-up community-based participation *necessarily* (see, e.g., Markusen and Gadwa 2010; Project for Public Spaces 2015). The name tactical urbanism likewise applies to most of these projects, but that term is more about technique, process, and intention than who in particular participates or whether they do so officially or not, and so it cuts across both unauthorized and sanctioned urban space interventions (see Lydon and Garcia 2015).

efforts to declare or change a certain policy), or more specific breakdowns that might describe the basic form of an intervention itself: seating, signage, planting, street or traffic markings, structural installations, functional modifications, events and occupations, publications, apps or web projects, and more. Others might look at scale, from minor fixes or one-off prototypes to hugely impactful placemaking efforts or city-wide campaigns.[3]

Aesthetics—or at least attention to aesthetics—seem important too, if necessarily somewhat subjective. For instance, outside of graffiti and street art, most popular attention in the North American context has been given to interventions that bring innovative and standout design to urban improvements (see Merker 2010; Ho 2012). A contrasting second group therefore is functional but the opposite of flashy—simple interventions like homemade street signs, improvised places to sit, guerrilla bike lanes, even DIY speed bumps, some of which hope not even to be noticed at all. Interventions of the first group can be unauthorized or not, but many ultimately operate with some permission; interventions in the second group seem rather more likely to be informal. (Interestingly, the latter may nonetheless try harder to *look* official—as part of their essence, and so as not to be removed—while the former are often happy to stand out.)

All of these differences are under consideration here, though I try to maintain a sense of how the issues raised by any particular intervention have implications for how we think about bottom-up urbanism in general. The chapter draws upon existing qualitative research on bottom-up urbanism and related trends, including my own, as well as media coverage of these things, and a database I have compiled of many interventions in the United States. My own empirical work comprised five years of interviews and ethnographic fieldwork with the people who take part in many forms of unauthorized urban intervention, including a total of more than 100 interviews, as well as participant observation and analysis of media coverage, providing information on 80 DIY urban design interventions in 17 cities (see Douglas 2018 for a lengthy description of this research). The database was built from a

---

[3]A wide variety of typologies can be found in different accounts, with varying relevance to the present discussion. Visconti et al. (2010) propose six categories of "place marking" ranging from "pure resistance" to "beautification." Coreil-Allen (2010) identifies dozens of sites, components, and qualities to build his "typology of new public sites." The 2012 exhibition Spontaneous Interventions (see Ho 2012) organized its projects by how much they prioritized information sharing, accessibility, community involvement, sustainability, economic issues, or pleasure and enjoyment. A related New York-based precursor to this (Guiney and Crain 2011) identified five such trends (accessibility, beauty, connectivity, enjoyment, and equity), as well as 16 focus issues by which hundreds of citizen planning ideas could be grouped. I have previously considered DIY urban design projects in terms of whether or not they make use of professional and scholarly knowledge (Douglas 2016), and elsewhere divided them into broad categories of streetscaping, greening or reusing, and aspirational ideas (Douglas 2014); see also Douglas (2018: 189–198) for an accounting of 80 DIY interventions across these three categories and six other criteria from year and location to aesthetic style and the attitude motivating them.

combination of the US cases in that empirical study (comprising 65 projects, all unauthorized examples) and additional US projects tracked in the research (41 more, a mix), plus the relevant projects from an existing database of submissions considered for inclusion in an architecture exhibition on bottom-up urbanism (111 projects, mostly sanctioned but some DIY).[4] The database contains a total of 217 projects—128 unsanctioned and 89 sanctioned.

This chapter turns first to unauthorized DIY interventions, where we see how one person's improvement may be another's unwelcome intrusion, and that such things are inherently individualistic. Next, it looks at officially sanctioned participatory placemaking, which shows how the tactics and aesthetics of bottom-up urbanisms as a design trend and placemaking strategy can have unintended consequences, from privileging certain uses and users to potentially playing a role in broader gentrification processes. Each section describes both positive and problematic examples, and analyzes the projects on two levels: the people who participate in the practices of bottom-up urbanism, and by extension the issue of those who do not; and the interventions they create, whether broadly civic-minded and locally appreciated or more selfish, unwelcome, or exclusionary. Of course, these things are somewhat subjective: even the most well-intentioned intervention may be poorly received; a project of great meaning and benefit to one group may not appeal at all to another. The politics and cultural meanings of urban spaces, especially changes to them, are always somewhat murky. But in wading into them we can see how bottom-up urbanism, in practice and in principle, is not quite the democratizing force of popular participation and active citizenship it is often associated with in progressive planning and creative urbanist circles. Despite its great potential, the trend itself ultimately embodies some of the most problematic aspects of contemporary neoliberal planning and development processes and can reproduce the same deficiencies and inequalities.

Vigilante Vigilante: Individualized Responses to Civic Concerns

It is possible that the single most common motivating issue for simple DIY urbanism interventions is traffic in urban neighborhoods. More than 15% of the DIY projects tracked are about traffic calming to some degree or another, which does not even account for things like homemade roadside signs that we know occur at rates nearly impossible to accurately count in cities across the

---

[4]The design exhibition from which the data comes, the US Pavilion at the 2012 Venice Architecture Biennale (Ho 2012) both held an open submission process and identified/recruited some participants, resulting in a preliminary list of 218 and a cleaned/consolidated list—removing submissions that did not qualify and grouping some redundant ones—of 171 projects. (Fewer still were ultimately selected for exhibition.) After then removing a handful of projects for redundancy with my own list and a few others that did not fit the definition of either category of bottom-up-urbanism proposed here, this resulted in 74 sanctioned and 37 unsanctioned projects to add to the larger database.

country. Other examples include hand-painted crosswalks and intersections, fake potholes, plantings *in* potholes, and guerrilla speed bumps. The latter have been documented in New York, Detroit, Los Angeles, Oakland, and (curiously enough) numerous parts of Colorado, among other places.

DIY speed bumps present a particularly stark illustration of what it can mean to have a sensible concern such as traffic calming addressed informally by everyday citizens, because, not surprisingly, not everyone views them as a benefit. They can damage cars and bicycles (some are far from standard shape or size), or simply irritate drivers. And yet, unlike official speed bumps, there is not necessarily any process behind their placement and perhaps nobody to hold accountable for them. (Although, in some cases, community members have been fined for creating them; see CBS4 Denver 2011.) However sympathetic to the creators of these actions many urbanists and fellow community members might be, there are others who disagree.

This last observation can be applied to any alteration of urban space, official or unofficial. But it is particularly significant to understanding unauthorized urbanism, because such interventions have all the impact with none of the official legitimacy or accountability. Even those that feel more unquestionably positive, such as DIY benches and other seating placed where needed in public, can introduce problems if not designed well or not maintained. For instance, as proud as they were of their local improvements, several DIY bench builders in Los Angeles expressed concern about "someone getting hurt," a sentiment echoed by city officials. A DIY bench installed by community members in one South Los Angeles neighborhood I visited had to be removed after attracting vandals and unwelcome loiterers. Guerrilla bike lanes, striped by night along dangerous roadways, may sound like beneficial or even necessary safety enhancements to some, but are dangerous impositions on roadways from the perspective of others, not to mention lawsuits waiting to happen in the eyes of city departments of transportation. One such lane in Brooklyn, installed by cyclists to replace an official one recently removed by authorities at the behest of the local community, provoked aggressive responses on the street (and ultimately even fines) as it became part of a wider cultural conflict (Idov 2010; Douglas 2018).

We begin to see then how one person's well-meaning improvements could be another's unwelcome intrusion. DIY interventions are, almost by definition, the imposition of one person's or group's values onto public space. A noted guerrilla greening effort in St. Louis involved design students cutting gashes into a couple of parking spaces that were deep and wide enough for plants to grow in them (though parking was still possible). Some even go further, pushing the limits of 'improvement' at the expense of others, including activities such as: illegal advertising removal or 'adbusting;' anti-graffiti vigilantism; and provocations like the bright orange "gated community viewing platforms" installed by the group Heavy Trash in elite Los Angeles neighborhoods to enable members of the public to ascend some steps and peer into the gardens of the wealthy (see Pool and Chong 2005).

Others interventions may be quite transparently self-serving or even anti-social in impact. Consider the well-known phenomenon of parking space 'dibs' or spot saving, which is common in snowy cities where car owners will place furniture or other objects in a shoveled out parking spot to save it for their return (and which happens even on warm sunny days in some places). More covert—if surprisingly common—is people painting curbs red themselves to prevent others from parking there, or painting red curbs grey to create a new spot. Residents of the hilltop Whitley Heights area of Hollywood have been known to do both, and more, in their local parking turf wars (see MacGregor 2014). Their wealthy compatriots in nearby Beachwood Canyon, frustrated with tourists trying to reach the famous Hollywood sign from neighborhood streets, have likewise painted curbs red and installed a variety of signage—some of it illegal, some of it mean spirited—to discourage parking or even driving through the area (Walker 2014). Such instances of curb painting abound in Los Angeles and other Californian cities, despite the fact that the act can be a felony and, in the case of making a red curb grey, may involve removing a fire access lane or other safety feature. People also post disingenuous No Trespassing, Private, or No Access signage in efforts to keep people off of public trails and beaches, and they remove official signage for similar reasons (Fig. 19.1).

These are extreme cases, and those for whom some assumption of positive intentionality is part of the very definition of bottom-up urbanism might

Fig. 19.1 An illegal sign posted by a beachfront homeowner in Malibu, California, 2009. Photo by Rosa Morrow. Courtesy of LA Urban Rangers

quibble with counting them even as negative examples. But they are similar in every other way, and they draw our attention to the fact that DIY urbanism in general is an inherently individualistic act, reflecting the values and priorities of one person or small group acted out upon urban space without permission or oversight. In this sense, we are reminded that these things may be no more democratic in terms of participation, or even outcomes, than the privatization of planning and development decision-making to any other private interest.

This is not necessarily problematic when the actor and their intervention are locally appropriate, or do not raise the ire of other community members. Indeed, the production of spontaneous, responsive, local improvement is the great promise of bottom-up urbanism, and is borne out in the many DIY projects that have been embraced by their surrounding communities or simply feel normalized there. In a twisted sense, perhaps even the anti-tourist restrictions of Beechwood Canyon, or the illegal beach enclosures in wealthy seaside communities, fit the shared values of the elite local residents (though certainly these latter examples do differ in their explicitly exclusionary intention). Plenty of other interventions are so subtle as to be innocuous (DIY signage or repair efforts, for instance), or so simple and generous in spirit as to be uncontroversial (community book exchanges are quite broadly beloved). But other interventions, as we have seen, do not meet this standard and are viewed as out of place and unwelcome by some residents, whether for enacting particularly unfitting or selfish priorities, or simply because their creators are *perceived* as relative newcomers who have no right to try to improve a neighborhood that they do not understand.

The potential for insensitive interventions is particularly concerning given a final factor: participation in the creation of DIY urbanism does not appear to be equal. I found that while informal improvement efforts occur in all sorts of communities, the reasons people engage in such practices, the conditions under which they do so, and the forms, impacts, and public perceptions of the interventions themselves vary considerably along lines of socio-economic inequality (Douglas 2018). Most fundamentally, because people of color tend to be viewed very differently in public space than white people and are more likely to be targeted and hassled by police (and often with more severe consequences), they have strong disincentives to risk flouting the law or attracting attention from authorities for the sake of a streetscape improvement.[5] As the sociologist Eli Anderson (1990: 182) noted in his study of street life among

---

[5]An enormous volume of research has explored these prejudices faced by people of color—and especially young men—from police and from society in general in public space, and how this shapes everything from walking down the street to civic engagement. Elijah Anderson (e.g., 1990, 1999) has been particularly influential in this regard. See also, for instance, Feagin and Sikes (1994), Michelson (2003), and Franklin et al. (2006) on the psychological stresses of living with prejudice, and on attitudes toward police and other authorities in particular, see Menjivar and Bejarano (2004), Brunson (2007), Dottolo and Steward (2008), and Correia (2010) among many others.

African Americans, "The public stigma is so powerful that black strangers are seldom allowed to be civil or even helpful without some suspicion of their motives." Additional factors of social privilege, including class, education, and life experiences, also inform people's willingness to take part in at least the most visible or provocative forms of DIY urbanism.

People in predominantly Black or Latinx communities who do make informal interventions described being hassled by police in their daily lives, and many sought some added degree of normative or organizational legitimacy at the local level before making any visible public space interventions. White, educated, middle-class do-it-yourselfers, on the other hand, tended to act with greater confidence, assuming that their efforts would be viewed as positive, and some described friendly interactions with police even while making entirely illegal alterations to public space. So, although many low-income and predominantly Black or Latinx communities have considerable needs and a demonstrated lack of voice in getting local improvements made (Schumaker and Getter 1977; Hajnal and Trounstine 2014), cultural norms and stark racial and political realities tend to more narrowly define the types of informal activities that are not viewed as too risky and impractical. Innovative DIY activities certainly occur in communities of color, often where they are desperately needed, but many of the most illegal, the most individualistic, and the most visible and provocative interventions in the data were created by do-it-yourselfers of privilege. So, while people in underprivileged communities tend to make needed local improvements close to home and in keeping with local norms, do-it-yourselfers of privilege are often recent arrivals in rapidly changing neighborhoods, where they feel confident making what they view as improvements but who may not have the local legitimacy to act. Again, therefore, the priorities and values represented may be less than universally shared.

At the same time, broader fundamental societal prejudices have just the opposite implications, with those unauthorized interventions created by white, educated, middle-class do-it-yourselfers more likely to be viewed by mainstream (white, elite) society as positive creative acts and signs of vibrant local character, rather than as signs of disorder. In other words, the difference between workaday making do (or even vandalism) and trendy, creative transgression may have a great deal to do with class habitus and racial stereotypes. And it is the case that a large number of do-it-yourselfers are white members of the so-called creative class, and many DIY urban design interventions are informed by some technical or scholarly knowledge, which only gives their creators more confidence and makes them more likely to elicit a positive response from authorities (Douglas 2016).

As such, many of the most prominent and celebrated DIY interventions—the sorts that get featured in design exhibitions, that become embraced by the planning community, and are sometimes formalized by officials—are those created by people of privilege, often with a background in design. They thus tend to reflect certain values and foster particular aesthetics, potentially at the expense of others. And they often fit right into the world of hip, edgy cultural production that defines many gentrifying urban neighborhoods. And as these things are associated with the creative class and celebrated in the press, they can become looked to not simply for the immediate value of the interventions themselves, but as signs of a neighborhood's cultural and economic vitality. To see this last point at work, I turn now to the second group of bottom-up urbanism interventions, the event trendier ones that work through formal channels or find their way to official approval.

PLACEMAKING OR PLACETAKING: WHO ARE NEW PUBLIC SPACES FOR?

The biggest impact—and some might say greatest promise—of bottom-up urbanism on American urban planning has surely been through its slow but steady embrace as an officially recognized placemaking strategy by cities, communities, and private developers. This has opened up an exciting avenue for public participation in the design or repurposing of local public spaces, and has inspired many well-respected architects, artists, and designers to produce "spontaneous" and innovative interventions for the common good (see Merker 2010; Ho and Douglas 2012). Major urban revitalization efforts, such as New York's High Line park and the restoration of the Los Angeles River, come from citizen initiated (and initially not entirely legal) beginnings. And smaller placemaking efforts in numerous cities that have embraced the tenets of tactical urbanism—and in some cases brought in its leading practitioners—also take cues from DIY efforts, their tactics, aesthetics, and even language, while promoting a looser, lighter, and ostensibly more populist local planning and economic development process.

Take one of the most prominent examples of a grassroots effort that has achieved the sort of official adoption and potential for "long term change" that tactical urbanism aims for (Merker 2010; Lydon and Garcia 2015): sidewalk- extending parklets and other pavement to parks initiatives that create small public spaces out of streets and parking spaces. Many have roots in informal and somewhat radical DIY efforts, including the Park(ing) Day phenomenon started by the group Rebar in San Francisco and the community street closures and 'reclaimed' intersections of the sorts especially attributed to Mark Lakeman's City Repair project in Portland, Seattle, and elsewhere.[6] Many cities, in almost every American state and beyond, now formally permit parklets—small public spaces, usually in the form of seating or dining areas,

[6]The first Park(ing) intervention by Rebar, in which they fed the meter for a San Francisco parking space and installed a temporary park rather than parking a car, and the rapid spread of this idea into the worldwide phenomenon it has become, is one of the great stories of twenty-first-century urbanism (see Merker 2010; Rebar Group 2012). And it was a direct inspiration for many cities' official parklet programs, including San Francisco's (Roth 2010).

built in the place of a former parking space—usually outside of a business that has applied to build and maintain one. New pedestrian plazas have also sprung up from coast to coast, as have simpler painted intersections and artistic crosswalks that can be found in hip neighborhoods from Baltimore to San José.

These trendy placemaking interventions can come about in different ways—many cities have online request and permitting processes for things like parklets—and they have all sorts of purposes, including reclaiming streets from cars and providing small amounts of public space; some include physical activities, cultural programming, services such as bicycle parking, or simply seating (see Fig. 19.2). But they are also used for economic development. Even if the opportunity to request or build new public spaces like these is technically open to anyone, the basic requirements in terms of application, funding, and ongoing maintenance make it unlikely to be worth the trouble of anyone besides a business interest, whether a restaurant, a neighborhood business improvement district (BID), or perhaps an adventurous architecture or design firm. As recently as 2015, all but one of San Francisco's 51 parklets was outside of a business (see Pilaar 2015). In some cities they exist only in this form, permitted directly to restaurants or other businesses to provide additional seating. Even Park(ing) Day, an annual event when people in cities around the world now temporarily occupy parking spaces with parks or art installations, is now embraced and promoted in many places by local BIDs, in the same way a street fair might be. New pedestrian plazas formed out of closed streets, such

Fig. 19.2 A parklet on a Manhattan street, 2016. Photo by the author

as those popularized in New York City, likewise rely on a model of financing and upkeep from local businesses. Some have claimed significant increases in sales "just by having the parklets outside" (Cho 2013). So, although these things are generally defined as public spaces, many of them do not quite appear or operate as such.

Many new public spaces appear to be designed primarily as sites for consumption, and through their aesthetics and visible uses may not seem welcoming to someone who is not willing to buy something. A parklet in an upscale neighborhood of San José, California, paid for by a local restaurant, spurred just this sort of concern, with confusion among community members and even staff at the restaurant about who was allowed to sit there (Baum 2016). Other residents complained about the removal of two parking spaces. And parklets, plazas, and other such spaces may also be more explicitly controlled than a typical street or sidewalk, with prohibitions on certain activities (from smoking and skateboarding to sleeping or loitering) that seem to favor particular uses and discourage unwanted populations. While they may be intended to reclaim space for positive uses and to promote street life, they can also implicitly claim them as spaces of elite consumption.

If some of these are extreme examples, questions about the intended uses and chosen aesthetics of these trendy public spaces apply to numerous other bottom-up urbanism interventions. Certainly, an orientation toward consumption and trendy consumers is common to many pop-up spaces, and such temporary uses, even when formed around cultural or educational themes, are typically about activating dormant spaces with economic development in mind. Other unquestionably innovative and beloved examples of grassroots placemaking— such as the Walk [Your City] pedestrian signage platform, or the Better Block neighborhood redevelopment model, or any number of street seating initiatives, brightly painted intersections, or artistic crosswalks—all designed as simple, community-engaged ways to enliven, rebrand, and pedestrianize unloved city streets, nonetheless tend to fit with particular ideals and aesthetics of trendy creative placemaking for economic development.

Part of this again stems from who is creating them in the first place. A majority of the DIY urban designers interviewed used sophisticated skills in the creation of their projects, with a significant number having professional or educational backgrounds of relevance to the interventions they make (Douglas 2016). For obvious reasons, even more of those in the exhibition dataset come from professional architecture, planning, or design contexts (a majority were submitted directly by studios). The results are often interventions with a contemporary design-forward aesthetic—reclaimed wood meets bright plastics and paints; innovative planters; stylized bike-friendly features; things that would not be out of place in a design showroom—speaking as much to that sort of trendy audience as to the everyday community members for whom they might ideally be intended.

In all of these ways, new creative public spaces and other examples of bottom-up urbanism can come across as intended for hip, white, affluent users.[7] In this sense, bottom-up urbanism is part of the broader mode of creative placemaking planning that combines progressive, green, arts-based, and New Urbanist ideals with economic development strategies. Because the phenomenon itself is increasingly visible and recognized as exciting and desirable, the simple presence of community-built or tactically initiated urban design interventions can often attract attention from urbanists or the local press and begin to mark an area as trendy. Intentionally or not, these projects can thus potentially have as much of an impact as part of the broad rebranding of an area as through the direct intended benefits of the project itself.[8] This can be especially damaging in already gentrifying neighborhoods, where they tend to be quite commonly employed as tools of economic development.[9] By creating spaces of (white, affluent) consumption and occupation, or simply highlighting the visible presence of design professionals and the creative class, they can work in congress with other features of the gentrification and displacement process and begin to change the character of whole areas.

This is not to say by any means that all participatory placemaking is about economic development. Many interventions have been used quite effectively to engage and legitimize underprivileged community members in the planning and placemaking process, and to provide valuable, needed, local improvements from community gardens, to public services, to the simple matter of something where there was very nearly nothing. Nor is it necessarily in any way problematic for improvements to have local development as part of their goals and impacts. Community economic development and more vibrant public spaces are noble goals, and they can be desperately needed in many places. We simply need to be aware that some of these seemingly progressive placemaking interventions can be exclusionary and may promote particular aesthetics and uses in public space, at the expense of others. As Catungal et al. (2009) argue, all manner of "creative city" planning and placemaking efforts can potentially be implicated in wider gentrification processes, almost by their nature. If communities and business

---

[7] Alkon and McCullen (2011) make such an argument about the cultural 'coding' of farmer's markets, to which I am indebted. Hoffmann (2016), and Hoffmann and Lugo (2014) have made similar observations about bicycle infrastructure. Both of these are spatial features of the progressively planned modern city with strong cultural associations to the priorities of bottom-up urbanism.

[8] Public art projects in general have long been employed, in the words of Sharp et al. (2005: 1013), to "reaestheticize areas within a city" as part of economic redevelopment schemes.

[9] Although there are certainly exceptions, my analysis of plaza locations in New York City and parklets in San Francisco, as well as anecdotal observations of similar new public spaces in Boston, Chicago, Los Angeles, Philadelphia, San José, and several other cities, found a predominance of these places in trendy or actively gentrifying areas of these cities, and almost always in front of businesses (see Douglas 2018).

owners are not mindful, the very tactics and aesthetics that make bottom-up urbanism so successful can further marginalize the voices and priorities of disadvantaged populations by making them feel as if these things are not for them.

Generous Urbanism and Authentic Placemaking

This chapter has offered a critical, even pessimistic take on some bottom-up urbanisms. But these sorts of interventions do not have to replicate the uneven priorities and development models of mainstream planning and development. Despite the individualistic nature of DIY urbanism and the prominence of elite aesthetics in much creative placemaking, bottom-up urbanisms are still uniquely positioned to serve as locally responsive and excitingly non-traditional models for improvements that emphasize community needs and priorities. Not only are many of the interventions I learned about certainly more positive (or at least harmless) than they are problematic, some set a new and inspiring standard for locally sensitive placemaking in places where it is most needed. Blane Merker (2010) refers to some of the quasi-informal creative placemaking efforts of his group, Rebar, as "generous urbanism." At their best, this is indeed what many acts of bottom-up urbanism can be.

For one thing, while perhaps not the norm, some exemplary projects go out of their way to prioritize local needs and avoid the sense that their creators are just outsiders parachuting in. In Queens, New York, the various collaborators engaged to redevelop a small public space called Corona Plaza made local input and representation from the highly diverse neighborhood a foundation of their design. The Baltimore Development Cooperative began their work on a community garden and gathering space in a disadvantaged area with months of community dialogue. And those working at a smaller scale can still focus on the least fortunate, as with the individual living spaces for those without homes designed by the Mad Housers of Atlanta and Tina Hovsepian in Los Angeles, or DIY projects like the Legal Waiting Zone, created by the group Ghana Think Tank, to empower immigrant laborers in the face of police harassment.

Of course, other interventions are simply more authentically of their communities in the first place. This is exemplified by a variety of what we might call ethnic urbanisms and other culturally relevant local placemaking interventions such as roadside shrines, neighborhood signs and murals, or what has been described as the Latino urbanism of California's Mexican American communities or the community garden *casitas* of Puerto Rican New York (e.g., Rios 2010; Rojas 2010). A little stone Buddha, simply placed at a crime- and trash-plagued corner in Oakland by a nearby resident was viewed as a gift by many in the surrounding Vietnamese community, who have since taken it on and built a large DIY shrine around it that people worship at daily. These things serve local needs in locally relevant ways and may even work to

Fig. 19.3 A DIY Buddhist shrine at a corner in Oakland, 2017. Photo by the author

resist gentrification and cultural displacement by stamping the visible presence of the long-term community on a public space in ways that code it as their own (Fig. 19.3).

On Chicago's South Side, Emmanuel Pratt and his Sweetwater Foundation have focused energy on what he calls the "regenerative placemaking" potential of urban agriculture in a sea of vacant lots and underemployed youth. Through a combination of formal and informal activities converting abandoned or underused sites into spaces for aquaponics, farming, gardening, carpentry, community events, and myriad skills training opportunities, Pratt's projects actively work to confront blight and to engage with issues ranging from crime and neglect to gentrification and economic development. His projects have won hearts in the local community, accolades in the planning and design profession, and city and foundation support. Other local efforts on the South Side, including the Chicago Resource Center, the Experimental Station, the building rehabilitations of Theaster Gates' Rebuild Foundation, and numerous community gardens, along with a host of traditional non-profits and community-based organizations, form a constellation of non-governmental forces bringing hope to troubled areas. There are many more hopeful examples like these that reflect local priorities and passions, from simple DIY safety improvements to powerful community-driven placemaking projects like the 78th Street Play Street in Queens, which went from temporary neighborhood street closure to beloved permanent public space. And these things often give us elegant design solutions that might never emerge from top-down bureaucracies.

Whatever the progressive intentions and positive impacts of many of these interventions, however, we must be wary that bottom-up urbanism is ultimately shaped by the same conditions that trouble mainstream planning and development processes. And, in many ways, it reflects these underlying social inequalities. Even as DIY urbanism efforts often aim to counter the ill effects of uneven development and market-driven planning, they can reinforce an individualistic and undemocratic paradigm in urban placemaking. Constraints on who can and does take part in this practice reveal the persistent incompleteness of democratic participatory citizenship in urban space and urban policy. Many do-it-yourselfers operate from a position of considerable privilege (in public space, in interaction with authorities, in everyday life) that emboldens their transgressive practices and in turn influences the sorts of projects that get built and legitimized. In all these ways, the trends described in this chapter may be less the panacea of participatory urbanism that some would have it, and are simply another inspiring form of urban space alteration with the same opportunities and pitfalls as any other. That is not a condemnation, but an invitation to think critically, and to build.

REFERENCES

Alkon, A. H., & McCullen, C. G. (2011). Whiteness and farmers markets: Performances, perpetuations... contestations? *Antipode, 43*(4), 937–959.

Anderson, E. (1990). *Streetwise: Race, class, and change in an urban community*. Chicago: University of Chicago Press.

Anderson, E. (1999). *Code of the streets*. New York: Knopf.

Baum, J. (2016, August 7). New Lincoln Avenue parklet sparks complaints, confusion. *Mercury News*. www.mercurynews.com/2016/08/02/san-jose-new-lincoln-avenue-parklet-sparks-complaints-confusion/. Accessed February 13, 2017.

Brunson, R. K. (2007). 'Police don't like Black people': African-American young men's accumulated police experiences. *Criminology & Public Policy, 6*, 71–102.

Catungal, J. P., Leslie, D., & Hii, Y. (2009). Geographies of displacement in the creative city: The case of liberty village, Toronto. *Urban Studies, 46*(5&6), 1095–1114.

CBS4 Denver. (2011, September 28). Slow down! Homeowners cited for making own speedbumps. *CBS4 Denver*. http://denver.cbslocal.com/2011/09/28/slow-down-homeowners-cited-for-making-own-speed-bumps/. Accessed February 16, 2017.

Cho, C. (2013). Parklet boosts local business sales. *KOMU 8 News website*, 3 October. http://www.komu.com/news/parklet-boosts-local-business-sales/. Accessed March 27, 2014.

Coreil-Allen, G. (2010). *The typology of new public sites: A field guide to invisible public space*. Baltimore: Self-Published.

Correia, M. (2010). Determinants of attitudes toward police of Latino immigrants and non-immigrants. *Journal of Criminal Justice, 38*(1), 99–107.

Dottolo, A. L., & Stewart, A. J. (2008). 'Don't ever forget now, you're a Black man in America': Intersections of race, class and gender in encounters with the police. *Sex Roles, 59*, 350–364.

Douglas, G. C. C. (2014). Do-it-yourself urban design: The social practice of informal 'improvement' through unauthorized alteration. *City & Community, 13*(1), 5–25.

Douglas, G. C. C. (2016). The formalities of informal improvement: Technical and scholarly knowledge at work in do-it-yourself urban design. *City & Community, 9*(2), 117–134.

Douglas, G. C. C. (2018). *The help-yourself city: Legitimacy and inequality in DIY urbanism*. New York: Oxford University Press.

Feagin, J., & Sikes, M. (1994). *Living with racism: The black middle class experience*. Boston, MA: Beacon.

Finn, D. (2014). DIY urbanism: Opportunities and challenges for planners. *The Journal of Urbanism, 7*(4), 381–398.

Franklin, A., Boyd-Franklin, N., & Kelly, S. (2006). Racism and invisibility: Race-related stress, emotional abuse, and psychological trauma for people of color. *American Journal of Drug and Alcohol Abuse, 6*, 9–30.

Guiney, A., & Crain, B. (Eds.). (2011). *By the city/for the city: An atlas of possibility for the future of New York*. New York: The Institute for Urban Design/Multi-Story Books.

Hajnal, Z., & Trounstine, J. (2014). Identifying and understanding perceived inequities in local politics. *Political Research Quarterly, 67*(1), 56–70.

Ho, C. L. (Commissioner). (2012, August–November). *Spontaneous interventions: Design actions for the common good exhibition*. United States Pavilion, 13th Venice International Biennale of Architecture, Venice.

Ho, C. L., & Douglas, G. (Eds.). (2012). *Spontaneous interventions: Design actions for the common good*. Catalogue for the exhibition of the same name. Washington, DC: Architect/Hanley Wood.

Hoffmann, M. L. (2016). *Bike lanes are white lanes: Bicycle advocacy and urban planning*. Lincoln and London: University of Nebraska Press.

Hoffmann, M. L., & Lugo, A. (2014). Who is 'world class'? Transportation justice and bicycle policy. *Urbanites, 4*(1), 45–61.

Idov, M. (2010, April 11). Clash of the bearded ones: Hipsters, Hasids, and the Williamsburg street. *New York Magazine*. http://nymag.com/realestate/neighborhoods/2010/65356/. Accessed February 14, 2017.

Iveson, K. (2013). Cities within the city: Do-it-yourself urbanism and the right to the city. *International Journal of Urban and Regional Research, 37*(3), 941–956.

Lydon, M., & Garcia, A. (2015). *Tactical urbanism: Short-term action for long-term change*. Washington, D.C.: Island.

MacGregor, H. (2014, February 12). My neighborhood parking wars: Midnight vigilantes and Phantom Valets battle over the curbs of one LA street. *Zocalo Public Square*. http://www.zocalopublicsquare.org/2014/02/12/my-neighborhood-parking-wars/ideas/nexus/. Accessed February 16, 2017.

Markusen, A., & Gadwa, A. (2010). *Creative placemaking. White paper for the Mayors' Institute on City Design*. Washington, DC: The National Endowment for the Arts.

Menjivar, C., & Bejarano, C. (2004). Latino immigrants' perceptions of crime and police authorities in the United States: A case study from the Phoenix metropolitan area. *Ethnic and Racial Studies, 27*, 120–418.

Merker, B. (2010). Taking place: Rebar's Absurd Tactics in generous urbanism. In J. Hou (Ed.), *Insurgent public space: Guerrilla urbanism and the remaking of contemporary cities*. New York: Routledge.

Michelson, M. R. (2003). The corrosive effect of acculturation: How Mexican Americans lose political trust. *Social Science Quarterly, 84,* 918–933.

Pilaar, D. (2015, February 27). Mapping all 51 awesome San Francisco Parklets. *Curbed San Francisco.* http://sf.curbed.com/maps/mapping-all-51-awesome-san-francisco-public-parklets. Accessed April 8, 2016.

Pool, B., & Chong, J.-R. (2005, April 26). Contrarians at the gates go to some heights to make a point. *Los Angeles Times.* http://articles.latimes.com/2005/apr/26/local/me-guerrilla26. Accessed February 16, 2017.

Project for Public Spaces. (2015, June 13). Creative communities and arts-based placemaking. *Project for Public Spaces.* https://www.pps.org/reference/creative-communities-and-arts-based-placemaking/. Accessed February 28, 2017.

Rebar Group. (2012). About Park(ing) Day. *Park(ing)Day.org.* http://parkingday.org/about-parking-day/. Accessed February 27, 2017.

Rios, M. (2010). Claiming Latino space: Cultural insurgency in the public realm. In J. Hou (Ed.), *Insurgent public space: Guerrilla urbanism and the remaking of contemporary cities.* New York: Routledge.

Rojas, J. (2010). Latino urbanism in Los Angeles: A model for urban improvisation and reinvention. In J. Hou (Ed.), *Insurgent public space: Guerrilla urbanism and the remaking of contemporary cities.* New York: Routledge.

Roth, M. (2010, November 2). From Park(ing) Day to permit: San Francisco's Parklets redefine public space. *Streetsblog SF.* http://sf.streetsblog.org/2010/11/02/from-parking-day-to-permit-san-franciscos-parklets-redefine-public-space/. Accessed February 27, 2017.

Schumaker, Paul, D., & Getter, R. W. (1977). Responsiveness bias in 51 American communities. *American Journal of Political Science, 21*(2), 247–281.

Sharp, J., Pollock, V., & Paddison, R. (2005). Just art for a just city: Public art and social inclusion in urban regeneration. *Urban Studies, 42*(5–6), 1001–1023.

Visconti, L. M., Sherry, J. F., Jr., Borghini, S., & Anderson, L. (2010). Street art, sweet art? Reclaiming the 'public' in public space. *Journal of Consumer Research, 37*(3), 511–529.

Walker, A. (2014). Why people keep trying to erase the Hollywood Sign from Google Maps. *Gizmodo,* 21 November 2014. http://gizmodo.com/why-people-keep-trying-to-erase-the-hollywood-sign-from-1658084644. Accessed November 22, 2014.

Weitzer, R., & Tuch, S. A. (2002). Perceptions of racial profiling: Race, class, and personal experience. *Criminology, 40*(2), 435–456.

CHAPTER 20

Pop-ups and Public Interests: Agile Public Space in the Neoliberal City

Quentin Stevens and Kim Dovey

INTRODUCTION

Spontaneous, or bottom-up, urbanism is often praised for bringing innovation and agility to urban design practices that are typically constrained by
context, convention, regulation, high cost, long timelines and complex stakeholder roles and needs. Such urbanism typically involves the rapid mobilization of new constellations of urban actors, resources, and spaces to quickly
address a range of previously unmet desires. It is clear that such ostensibly
spontaneous public space projects often involve complex intersections
between the agency of the state and that of citizens; there is rarely a simple
binary between bottom-up and top-down, informal and formal, temporary
and permanent, tactical and strategic. Temporary projects are often strongly
linked to the long-term transformation of urban property values and consumption patterns. Tactical urbanism has ironically become a new form of
top-down strategic planning—by both the state and private interests. Temporary urban interventions can serve as vectors of gentrification and neoliberal
planning, or they can be reactions to it.

This chapter explores the various claimed public benefits of a more agile
bottom-up urbanism. It also explores the paradoxes, entanglements, and
potential duplicity of such urbanism in the context of neoliberal urban planning regimes, through a critical examination of the range of interests that

Q. Stevens (✉)
RMIT University, Melbourne, Australia

K. Dovey
Melbourne School of Design, University of Melbourne,
Melbourne, Australia

© The Author(s) 2019
M. Arefi and C. Kickert (eds.), *The Palgrave Handbook of Bottom-Up
Urbanism*, https://doi.org/10.1007/978-3-319-90131-2_20

such projects might serve and the positive and negative impacts they can have. The broad field of urban design practices that is our focus here emerges under many titles, including DIY, informal, guerrilla, insurgent, pop-up, lean, and austerity urbanism. We will refer to it collectively as temporary/tactical, or T/T, urbanism. The phrase temporary urbanism is mostly prevalent in Britain and Europe and involves a focus on time horizons and rhythms of change; on physical transformations that are not intended to last more than a few years (Bishop and Williams 2012). Tactical urbanism is the more common term in North America and involves a focus on self-organized spatial practices and new forms of social agency in public spaces (Lydon and Garcia 2015). This is not a binary division since a good deal of the tactical (guerilla, DIY, insurgent) also characterizes British and European projects; likewise, the temporal dimension is fundamental to the Tactical Urbanism of Lydon and Garcia, (subtitled 'Short-term action for long-term change'). Our use of the term T/T is an attempt to collapse this cluster of practices into a twofold concept, where it is the intersection of the temporary and the tactical that comprises the core of this field.

The diverse projects so encompassed are typically short-term and low-cost adaptations of the urban fabric that also involve a circumvention of conventional urban planning rules and approaches. Such agile open space projects embrace a wide variety of forms, functions, scales, and durations. They range from relatively regulated artificial beaches, floating swimming pools, pop-up buildings, outdoor theaters, and container villages, to more informal projects such as instant plazas, unsanctioned guerrilla gardens, and parklets. Their functions include consumption, recreation, public art, performance, community engagement, and creative production (SfS Berlin 2007).

The focus here is on such projects in relatively formal cities of the Global North. T/T urbanism is, of course, widespread in the less formal cities of the Global South. In some ways it involves forms of learning from the high levels of informal urbanism embodied in those cities—the spontaneous DIY urbanism of informal settlements. However, the T/T urbanism focused on in this chapter is responding to conditions in more developed cities that are subject to rapid change: increases and decreases in urban density; rising social diversity; and economic volatility. As the neoliberal state cuts taxes and deregulates, the pressure increases to deliver better public spaces with fewer resources, and to adapt more rapidly. At the same time the vitality of urban public space is seen in connection with economic development and the production and exchange of new ideas that help drive a knowledge economy. Conventional urban development forms and planning tools are often considered inadequate to address these challenges (Oswalt et al. 2013). Agile open space projects are generally incremental and interstitial, emerging between large-scale and long-term strategies, and filling interim periods of time and underutilized urban space, particularly during economic downturns. Tactical urbanism blends into long-term strategies; some pop-ups never pop down.

The benefits that are argued to flow from T/T urbanism are many and inter-linked—summarized here in terms of five values: urban intensity; community engagement; innovation; resilience; and place identity. In each case there is also a downside; a danger that public interests cannot be easily presumed. T/T urbanism mobilizes and empowers particular urban actors; it unleashes particular flows of desire in public space with some clear public benefits. But there are also crucial questions about power, agency, and public interests that are much less explored. What follows first argues the case for each of these five benefits, followed by what are seen as their potential dangers. What are the prospects that temporary and tactical transformations of public space might reproduce or even exacerbate the urban problems they seek to address? Our interest in exploring such contradictions lies in developing more critical forms of T/T practice and a more rigorous critique of such projects.

Urban Intensity

Temporary and tactical urbanism is argued to increase the intensity of open space use. This concept of intensity can be difficult to define (Dovey and Pafka 2014), but here means an increase in both the volume and the variety of uses and users. T/T urbanism is opportunistic in seeking out underutilized public spaces and redesigning them for a greater volume and variety of peo-ple, activities, and experiences. T/T urbanism introduces new uses into exist-ing public spaces, and it produces new forms of public life in spaces that were not previously accessible. A good example is the replacement of low-intensity car-parking with high intensity parklets. By accommodating more uses and users within existing public spaces, temporary urbanism embodies a more effi-cient use of limited space, infrastructure, and other resources. This addresses the challenge of rising densities, demands for economic efficiency, and envi-ronmental sustainability (Oswalt et al. 2013; Nemeth and Longhorst 2014). T/T urbanism also explores and harnesses the underutilized capacities of the city by expanding the physical scope of public spaces, to bring people and activities to vacant and under-utilized spaces that were closed off or privately owned. These include empty parking lots, rooftops, infrastructure easements, vacated and ruined buildings, waste sites, and wild landscapes.

Intensification means a diversification of public space designs, as well as of their uses and users. The most expansive survey of temporary urban uses, in Berlin, classified over 100 open space projects into four main activity categories: culture, gardening, food, and playing sports (SfS Berlin 2007). Other studies have noted a range of uses including consumption, open space, active recre-ation, art and performance, community engagement, and creative production (Lydon and Garcia 2015; Hou 2001; Haydn and Temel 2006). While good public space design has always catered to a multiplicity of functions, tactical urbanism has a tendency to address underrecognized desires by spontaneously providing new amenities, such as street furniture, cycle lanes, and pedestrian crossings (Fabian and Samson 2014). Vacant, derelict sites often offer the potential to accommodate exploratory, experimental, and divergent activities that may be specifically excluded from formal public spaces. By expanding the range of functions of public space, T/T urbanism encompasses the desires of a greater variety of citizens and provides greater social equity. It cuts across the constraints of any 'one right way' of designing public space and creates a field of expression for differences of social class, culture, and ethnicity.

Finally, tactical open space interventions can increase intensity by enabling richer mixes of activity in both time and space. They can overcome the physical boundaries and legal controls that typically delimit and separate different uses and user groups in public spaces, often in the name of protecting amenity and preventing conflict (Valverde 2005). Agile urbanism can overcome the monotony and inflexibility of more conventional, top-down, long-term, urban development. It can intensify urban life by enabling synergies both among short-term, informal uses, and between them and the formal city (Loukaitou-Sideris and Mukhija 2014; Oswalt et al. 2013).

The other side of this argument for a more intensified city is to ask: why is it that so much urban space is underutilized? Why is the state not engaged in such intensification and diversification? To what degree do T/T projects fill the gaps created by the failures of a more permanent and strategic urban planning process? The boom/bust cycles of capitalist urban development produce uneven development pockmarked with sites that remain vacant for long periods. The neoliberal economic consensus that has prevailed globally since the 1970s involves a focus on economic growth as the goal of urban planning. Neoliberalism embodies regimes of inter-city competition, tax cutting, and deregulation, where the state retreats from investment in cities; it seeks instead to negotiate the private production of quasi-public space that couples public access to private control. Tactics for temporarily filling urban voids can be seen as well-adapted to the cycles of creative destruction that characterize neoliberal urban development. To what degree then might T/T urbanism be understood as austerity urbanism—band-aid urban planning for cities in decline (Tonkiss 2013)?

There are further questions that might be raised about T/T urbanism's claims to appeal to a greater diversity of users. To what degree is this simply a switch to a different range of uses and users? T/T urbanism can be integrated with practices of privatization, gentrification, and displacement, whereby some categories of users (consumers, creative actors, potential investors) are seen to add value to underutilized land, while others are subtly marginalized. Temporary commercial activities that are enabled by relaxed regulations can undermine the sustainability of local businesses as they privatize public space. New functions that emerge under the umbrella of creative innovation can disrupt the amenity of existing users and initiate a subtle displacement of previous functions. T/T urbanism has been criticized for encroaching on vacant, wild, and derelict urban spaces that have unrecognized uses and values—the city's "terrain vague" (Kamvasinou and Roberts 2014; Solà-Morales 1994). Such sites have long been recognized as sites of homeless refuge, urban memory, childhood imagination, and wild flora and fauna (Barron 2014; Carr and Lynch 1968, 1981). T/T interventions have been intentionally deployed in such spaces to displace such socially marginal and illicit activities (Tonkiss 2013; Douglas 2014).

COMMUNITY ENGAGEMENT

Engagement is a very broad term that embraces a range of roles people have in leading, facilitating, shaping, and experiencing T/T projects. Many studies have pointed to the wide range of actors involved, from community groups, artists and activist designers to landowners, entrepreneurs and varied branches of government (Lydon and Garcia 2015; Hou 2001; SfS Berlin 2007; Haydn and Temel 2006). Such engagement ranges from self-organized to state-led, from politically neutral citizen groups to politically-charged guerrilla urbanism by anonymous agents, from commercial enterprise to community volunteering (Hou 2001; Finn 2014). Three specific aspects of engagement are brought to the fore in the rhetoric and practice of T/T urbanism.

First is the engagement of new actors. A range of individuals are drawn into these production processes who have little or no prior experience with the development and management of urban open space. DIY urbanism emphasizes the empowerment of ordinary citizens in creating public space (Ferguson 2014; Finn 2014; Lydon and Garcia 2015). It has been suggested that T/T urbanism establishes niche sites and opportunities for those who are excluded from mainstream society, such as refugees and counter-cultural drop-outs (Urban Catalyst 2001). T/T urbanism also provides engagement opportunities for migrants and others who are moving between social roles and for whom there is no established niche. T/T urbanism can also serve as a professional incubator for start-up enterprises and those seeking to introduce new spaces, ideas, and activities into the wider society. This bottom-up engagement is often facilitated by initiatives from local governments to provide advice and liaison services, such as one-stop shops and fast-tracking for necessary permits (Bishop and Williams 2012; SfS Berlin 2007).

A second feature of engagement is that these varied actors bring with them new ideas, entrepreneurial skills, social networks, and resources to the processes of social engagement (Stevens and Ambler 2010; Stevens 2015; Colomb 2012). Non-government agencies emerge to facilitate such actors and projects, which then produce new working relations, site users, developers, artists, and designers—all of which increases the breadth and depth of citizen involvement (SfS Berlin 2007; Oswalt et al. 2013). Community benefits gain broad community support when T/T urbanism is successful, and this can lead in turn to official sanctioning and long-term changes to regulatory codes (Pagano 2013; Lydon and Garcia 2015).

A third distinctive feature of community engagement with T/T projects is that it changes relations between production and consumption. At varying levels, T/T projects encourage ordinary citizens who visit them to become 'prosumers,' who are actively involved in the co-production of the space, often through their performances as they make use of the space (Toffler 1980; Richards and Wilson 2006). In principle, such co-creation should enhance environmental fit (Lynch 1981).

However, there is a case for critiquing the ways that such opportunities and obligations for community engagement fit with the agendas of both the neoliberal state and predatory markets geared to privatization and exploitation. To what degree has a user-engagement model become a user-pays model in which the state retreats from the production of public space while communities fill the gap in a makeshift manner that we might call 'sweat without equity'? To what extent does community engagement become the end for which the design of public space is the means—where the design quality, durability, and sustainability of public space is sacrificed for the ideals of citizen participation? To what degree are private interests camouflaged as public interests? Expansions of community engagement are not always transparent or socially equitable. The opening up of public space to individual entrepreneurial initiative can be a form of uneven development that facilitates some desires better than others. Some new actors become engaged while others become disengaged.

Innovation

Large-scale, permanent urban design is time-consuming and expensive, which means that innovation is both risky and very slow (Dotson 2016). T/T urbanism accelerates creativity and innovation within this field by reducing both risk and cost (Bishop and Williams 2012). T/T urbanism is incremental and fast. Creative open space ideas can be developed, tested, and revised in practice as the city becomes a laboratory for urban experimentation. Design creativity is stimulated through a focus on underutilized, degraded, and problematic urban sites where there is little danger of damage. It is also stimulated through the new activities that are introduced into those sites and through the range of creative practitioners involved (Angst et al. 2009).

T/T urbanism enables an iterative testing of how creative solutions can best be implemented, managed, and made financially viable. Much tactical urbanism focuses on processes rather than outcomes; on ways that citizens can bypass local government bureaucracy and conventional project delivery processes, and on how such practices might be brought into the mainstream (Lydon and Garcia 2015). The rules and relationships that frame T/T projects are negotiated across a broad scope of conditions, including lease arrangements, uses, permits, rents, guarantees, insurances, and the provision of utilities and services (SfS Berlin 2007). Much of this experimentation involves T/T urbanism testing formal codes; testing the boundaries between permitted and proscribed practices and forms. T/T projects often transgress existing codes in a manner that provokes a response from the state, which then enforces, or adapts by tolerating non-conformity or turning a blind eye. Increasingly, local governments respond by engaging creatively with such processes of enabling innovative open space outcomes (Bishop and Williams 2012; SfS Berlin 2007; Kamel 2014). The innovation is both bottom up and top down.

T/T urbanism enables experimentation with a wider range of forms and functions than permanent urban design projects that require larger budgets and face larger risks. They can cater to smaller and more specialized audiences. The public has a higher tolerance of unconventional and controversial projects when they occupy marginal spaces that were previously unused and when they are of limited duration (Bishop and Williams 2012; Kamvasinou 2011).

The idea of the city as a laboratory in which we can test ideas within a temporary framework is compelling, but also has some limits. To what degree does this surrender public space to a cycle of fashion, novelty, and gimmickry? There is a case to be made that a minimalist design of public space is the framework that creates the greatest openness of access and appropriation for the widest gamut of citizens—a design that neither prescribes nor proscribes particular uses or users. Such an argument suggests that the role of the state is to create the minimum formal framework for a robust and convivial public space; protecting rights of access and use without overdetermining forms or functions. What is the danger that T/T urbanism is engaged in creative ways of privatizing and appropriating public space through a production of novelty rather than enduring public interest? Is T/T urbanism a backdoor way of getting approval for something that would not otherwise be approved— where no case needs to be made for public interests beyond the fact that it is innovative?

RESILIENCE

The idea of the resilient city is often invoked in terms of the capacity of cities to adapt to shocks and stresses in a manner that preserves and protects the core vitality and identity of the place. Resilience thinking involves a focus on complex adaptive systems, where the parts adapt to each other in unpredictable ways—outcomes cannot be determined in advance but rather emerge from practices of adaptation and self-organization (Walker and Salt 2006; Johnson 2001). Key properties of resilient systems include the diversity and redundancy of parts, such that each performs a multiplicity of functions— no single part is crucial to success, so the system can adapt by moving around forms, functions, and flows. Rigid hierarchical systems by contrast are fragile, in that any single dysfunction can collapse the entire system.

The diversification of actors, place types, locations, and outcomes outlined above all demonstrate the potential of T/T urbanism to rapidly and flexibly adapt spaces to unforeseen changes and localized opportunities and challenges. These factors enhance urban resilience in the face of uncertain futures (Lydon and Garcia 2015; Greco 2012).

Temporary schemes do not require the major 'sunk' investments in large, fixed projects that generate inertia. Temporary and flexible forms allow easy reconfiguration and redeployment as conditions change. Their technologies and designs allow ready replication and adaptation to new sites and to changing social, economic, and environmental contexts. Resilient systems are distributed through networks rather than being strictly hierarchical. In contrast to the inflexible hierarchies of top-down planning and management, T/T urbanism involves a broader distribution of know-how, social connections, and resources for urban placemaking (Radywyl and Biggs 2013). In contrast to conventional public and private-sector real-estate development, T/T urbanism can be sustained by a broader and more flexible mix of resources, most importantly the 'sweat equity,' discretionary spending power, and political support of local communities. Resilience thinking suggests strong two-way interconnections between scales—from individuals to institutions, and from sidewalks to neighborhoods— where micro- and macro-scale activities and forms are each informed by the other. T/T urbanism is fundamentally an activity of connecting urban processes both vertically and horizontally, forging new kinds of partnerships, and finding new ways of engaging with external influences (Dobson and Jorgensen 2014; Greco 2012; SfS Berlin 2007).

The idea of a resilient city is not necessarily positive. There are parts of cities that are seriously dysfunctional, where a spiral into poverty, dereliction, or crime requires transformational change. Such neighborhoods can be deeply resistant and resilient to such change. New public infrastructure may be vandalized, crime may be displaced from one public space to the next, and poor communities may resent any attempt at top-down change. The key task may be to overcome resilience. Here we also see an echo of the problems outlined earlier in relation to neoliberal planning. To what degree does T/T urbanism become the means of filling the gaps left by the withdrawal of the state from urban planning and investment? To what degree is the boom/bust cycle of capitalism and the exodus of manufacturing industry from rich Western cities the very disaster to which the city needs to adapt? In some cases T/T urbanism is produced by policies of austerity that force cities in decline to become more resilient (Tonkiss 2013; Färber 2014). One illustration here is adaptation to climate change. Many examples of T/T urbanism—from cycle lanes to urban orchards—are driven by green imperatives. Yet such T/T projects alone cannot effect the necessary transformation to low-carbon cities that also needs to be driven on a much larger political and urban scale.

PLACE IDENTITY

The final dimension we will discuss concerns the contribution of T/T urbanism to place identity and urban character. T/T urbanism has a metamorphic capacity to renew and transform the image of derelict public spaces and neighborhoods. Public space produced by the state often becomes identified with top-down control, regularity, and uniformity—a law-and-order image of the city, lacking in vitality and character (Nemeth and Longhorst 2014). T/T urbanism by contrast brings informality and irregularity; it signifies a multiplicity of differences that emerge unpredictably from bottom-up appropriations of the "right to the city" (Lefebvre 1996). T/T urbanism is a form of creative placemaking (Krauzick 2007). The production of place identity through T/T urbanism is generally emergent rather than imposed. It often signifies the dynamism of insurgency, resistance, and change that is introduced through social and political difference (Sandercock 2003).

Studies of place identity and urban character show that conceptions of place range from a relatively closed, stable, and deeply-rooted sense of place to a more open and dynamic sense of place, where a diverse mix becomes essential to its character (Dovey 2013). T/T urbanism shifts the balance toward more open, multiple, and dynamic conceptions of place. T/T urbanism often challenges the grounding of identity in a single authorized history and its preservation. Temporary projects can question, reevaluate, and enrich local identity, and inform future development by engaging critically with the historical legacy of places. Even where heritage controls ensure a fixity of built form, the provisional and ephemeral nature of temporary projects enables the city to embody more than one story.

T/T urbanism thus enables the city to encompass both preservation and change, to embody the tensions between history and progress, singularity and multiplicity. It enables more transparent forms of placemaking in the sense that one can read the values and concerns of citizens from the form of the city. Because T/T interventions often occur in the most marginalized spaces, they engage with the contradictions and uneven power relations that generate social and spatial marginality. The city becomes twofold—both formal and informal, top-down and bottom-up, permanent and temporary, strategic and tactical. T/T projects are spatial performances that can transform the image and the social interpretation of both formal and informal urbanism (Rios 2014). Constructing new communities of interest in this way underpins the extension and resilience of place identity across time and space.

While T/T urbanism can have a powerful impact on urban character and place identity, this is not always positive. A key concern lies in questions of design quality. While a large proportion of T/T urbanism emerges under conditions where the state has abrogated responsibility for quality, there are many parts of great cities where a consistently high-quality design of architecture, landscape, street furniture, signage, paving, and the shaping of public space—often minimalist and underdetermined—makes a crucial contribution to the vitality, accessibility, and multiplicity of public life, and to the construction of urban character. T/T urbanism has the potential to lower the bar in this regard—shoddy public seating, dead plants in planter boxes, rusty shipping containers, and bad public artworks can be too readily accepted if they are free and temporary.

A second concern is that the powerful impact that T/T urbanism has on place identity can become subsumed into place marketing. T/T urbanism often creates a buzz and adds a value to urban sites that is relatively easily capitalized. Pop-up urbanism can be a form of place marketing that enhances private property values. Such projects often facilitate the privatization of public space through advertising and commerce (Biddulph 2011; Rios 2014). Transformations of place identity under new regimes of T/T urbanism can become a cheap and superficial form of rebranding for derelict public spaces; a replacement for the investment and design thinking that might integrate sites of disadvantage with the larger city, and a means of colonizing disinvested land for an elite clientele. While T/T urbanism can add significant depth of urban character and place identity, these new values also become available for reappropriation, reduced to a brand, and privatized.

Researching T/T Urbanism

Having explored some of the critiques that might be applied to existing T/T projects, we now suggest both a research agenda and a conceptual rethinking of this field. We begin with three key dimensions for future research into T/T urbanism. The first area for further enquiry would be to carefully study its varied physical forms, sites, and morphological contexts. Where does T/T urbanism emerge and why? What particular forms of urban space are seen as ripe for intervention, and what are the forms of intervention that take place? What is the materiality of T/T urbanism, and how is it geared to technologies of instant transformation, such as shipping containers and synthetic turf? To what degree is it formally innovative? How are T/T designs adapted to local morphology and history?

The second focus is a more rigorous study of the processes of developing, regulating, constructing, and managing T/T projects. Who are their agents? To what degree are these transformations authorized by the state, or are transgressive? How are design and development decisions shaped by existing urban design and planning codes, or by financing and leasing arrangements? How are planning regulations relaxed, circumvented, broken, or revised? What kinds of informal and tacit codes emerge, and with what effects? Which project development pathways lead to more efficient, equitable, innovative, resilient, and engaging outcomes? Public interests lie as much in knowing 'how' as in knowing 'what'.

Third, such research needs to investigate the outcomes of T/T projects. Who benefits and who is displaced from such interventions? What specific activities and meanings do these projects give rise to? Public interests need to be evaluated rather than deduced or presumed. What is the impact of T/T projects on everyday urban life? This requires examining the new users, uses, perceptions, and interpretations that temporary projects introduce, as well as recognizing the invisibility of prior uses and users that they displace. What are the political impacts? How do T/T projects change the state's approach to the permanent and strategic? To what degree does T/T urbanism become the harbinger of gentrification?

These three approaches to understanding T/T urbanism in terms of form, process, and outcome need to be applied across a broad range of cases to rigorously compare and contrast international practice in this field. While a comparative approach has been attempted within Western Europe (Urban Catalyst 2001; Oswalt et al. 2013), the spread of T/T urbanism through the UK, North America, and Australasia, and a growing awareness of parallels to practices in the former Soviet Bloc and the developing world, make global comparisons possible. The prospect here is to identify the global scope of such practices, and to better understand the forms, processes, and impacts of T/T urbanism.

On Planning for Uncertainty

Earlier this chapter explored various theoretical critiques that can be applied to T/T urbanism, to unpack the different interests and impacts, revealing both public benefits and dangers. It now addresses the deeper research questions that are opened up: how to plan for the unplanned; how to design for the undesignated; how to expand and protect public interests while enabling self-organization and spontaneity. How are we to understand the affinities between a more agile public space and neoliberal economic regimes? The challenge lies in a fundamental re-think of urban planning and design in both research and practice. The ideal of rational urban planning working systematically toward master plans with fixed outcomes that drove the urban planning profession from its inception has little scope for the high levels of adaptation, informality, and uncertainty that prevail in T/T urbanism. Yet, if the structures and certainties of comprehensive rational planning are weakened, how are public interests to be protected?

In unpacking the relationships between T/T urbanism, social needs, uncertainty, and neoliberal capitalism, it is useful to note the important distinction between capitalism and markets. This distinction derives originally from the economic historian Braudel (1981–1984), for whom the economy can be understood as an interconnected triad of capitalism, markets, and everyday material life. Markets and capitalism both involve entrepreneurial activity. But while markets emerge from the bottom up to fill the material desires of everyday life, capitalism is a top-down system involving private ownership of the means of production and a global division of labor. Capitalism requires an alliance with the state to ensure a free flow of capital, a deregulated economy, and a hands-off approach to urban development. In the current era, it is the multinational neoliberal regime that keeps this system intact (Harvey 2007), ensuring that economic growth dominates other public interests in urban development. Neoliberalism involves a privatization of the production of public space. Indeed, public space becomes a zone for profit seeking, and much of the space thereby produced becomes 'quasi-public'—publicly accessible but privately controlled. Profit-making need not be excluded from T/T projects, but the difficult tensions between markets and capitalism need to be recognized and effectively managed. The privilege of using public space for private profit needs to be grounded in a critical understanding of public interests. The distinction between markets and capitalism is crucial here because the challenge of T/T urbanism—of planning for innovation and uncertainty—is one of enabling markets while resisting capitalism.

Assemblage

One useful framework for researching T/T urbanism is what might be called an assemblage approach, based in the work of Deleuze and Guattari (1987). Assemblage involves a rethinking of cities in ways that prioritize the connections between buildings, places, projects, and people over things-in-themselves; differences over identities; co-functioning over particular functions (Dovey 2010, Chapter 2). Assemblage thinking involves understanding the morphogenetic processes through which places emerge, based on a philosophy of transformation rather than of fixed forms and identities. There are several ways in which assemblage thinking can contribute to the critique of T/T urbanism.

First, assemblage engages with the concept of public interests in a direct and creative manner through a conception of urban life as an assemblage of flows of desire. Desires—for shelter, territory, security, aesthetic pleasures, views, privacy, and profit—are productive forces that produce the city as they also become a product of it. Collective desires congeal into shared interests, either formally or spontaneously. Public interests do not preexist waiting to be satisfied; rather, they emerge through urban life and collective action. Thus T/T urbanism is a means of directly engaging with and expanding public interests in an experimental manner.

T/T urbanism cuts across authorized narratives and practices of everyday urban life; it expresses and practices the right to the city in creative ways, expressing a claim over public space that does not wait for the state to determine where the public interest lies. By constructing dialectic images and spatial practices where authorized and unauthorized codes intersect, T/T urbanism changes the way in which public interests are conceived and constructed.

Second, assemblage thinking allows us to contextualize urban design within the larger frame of neoliberal capitalism without reducing the city to economic conditions. Neoliberalism is the embodiment of contradictory ideals; an open and deregulated economy within a notional framework of democracy and social justice. The neoliberal city embodies a reduction of urban design values to consumption and economic growth. While T/T urbanism has synergies with the neoliberal economy, it can also be a potent antidote. In filling the gaps and papering over the cracks produced by the creative destruction of capitalism, T/T legitimates this larger order. Yet T/T also violates the urban order, expanding capacities for the use and meaning of public space. These two actions operate at different scales; assemblage thinking connects them.

Third, assemblage thinking is a form of critical urbanism that valorizes both top-down and bottom-up practices of power. It provides a useful framework through which to rethink questions of urban informality. T/T practices are relatively informal, yet they emerge within the formal city with its existing morphology and regulatory codes. T/T urbanism is experimental with regard to these formal codes of governance; it pushes the boundaries of what is possible in public spaces in terms of both built form and spatial practice. The *raison d'etre* of urban codes lies in some notion of public interests such as safety—codes are congealed and formalized public interests—yet they often overdetermine outcomes. T/T urbanism unleashes desires in a temporary manner that enables a reassessment of the relations between formal codes and informal practices.

Finally, assemblage opens up questions of urban capacity and possibility—how can we better understand the range of possible futures—environmental, social, economic, aesthetic—that are embodied in an existing city but not yet actualized? We often speak about the development capacity of a site or a neighborhood as if it were fixed and waiting to be filled like a cup. Yet cities embody a much more complex set of capacities for transformations of form, function, and vitality. Assemblage thinking opens up the city as a space of possibility, not just a site for design speculation, but for a rigorous testing of what works and what does not. The capacity for change is discovered and produced through T/T experimentation in the city as a living laboratory.

At its best, T/T urbanism represents a relatively free market in the production of public space, where the drivers of innovation lie in a mix of market-based competition (some T/T designs are simply far better than others) and local creative networks and industries. T/T urbanism produces and embodies the emergence of new and more agile forms of urban governance; less rigid structures that allow room for creative experimentation and unpredictable outcomes. We need to learn more about what these practices are: how they work to enable and produce urban intensity, engagement, resilience, innovation, and place identity; how they work to expand and protect public interests; how and why they fail. In the end, the value of temporary and tactical urbanism is that it exposes and forces a creative engagement with one of the great dilemmas of urban design: How to organize the city while also enabling and enhancing its self-organizing capacity, how to plan for the unplanned, and how to govern spontaneity.

REFERENCES

Angst, M., Klaus, P., Michaelis, T., Müller, R., & Wolff, R. (2009). *Zone*imaginaire: Zwischennutzungen in Industriearealen*. Zürich: Vdf Hochschulverlag.

Barron, P. (2014). Introduction. In P. Barron & M. Mariani (Eds.), *Terrain vague: Interstices at the edge of the pale*. New York: Routledge.

Biddulph, M. (2011). Urban design, regeneration and the entrepreneurial city. *Progress in Planning, 76*, 63–103.

Bishop, P., & Williams, L. (2012). *The temporary city*. New York: Routledge.

Braudel, F. (1981–1984). *Civilization and capitalism, 15th–18th century* (3 vols.). New York: Harper and Row.

Carr, S., & Lynch, K. (1981). Open space: Freedom and control. In L. Taylor (Ed.), *Urban open spaces*. New York: Rizzoli.

Carr, S., & Lynch, K. (1968). Where learning happens. *Daedalus, 97*(4), 1277–1291.

Colomb, C. (2012). Pushing the urban frontier: Temporary uses of space, city marketing, and the creative city discourse in 2000s Berlin. *Journal of Urban Affairs, 34*(2), 131–152.

Deleuze, G., & Guattari, F. (1987). *A thousand plateaus*. London: Athlone Press.

Dobson, S., & Jorgensen, A. (2014). Increasing the resilience and adaptive capacity of cities through entrepreneurial urbanism. *International Journal of Globalisation and Small Business, 6*(3/4), 149–162.

Dotson, T. (2016). Trial-and-error urbanism: Addressing obduracy, uncertainty and complexity in urban planning and design. *Journal of Urbanism, 9*(2), 148–165.

Douglas, G. (2014). Do-it-yourself urban design. *City and Community, 13*(1), 5–25.

Dovey, K. (2013). Planning and place identity. In G. Young et al. (Eds.), *The Ashgate research companion to planning and culture*. London: Ashgate.

Dovey, K. (2010). *Becoming places*. London: Routledge.

Dovey, K., & Pafka, E. (2014). The urban density assemblage: Modelling multiple measures. *Urban Design International, 19*, 66–76.

Fabian, L., & Samson, K. (2014). DIY urban design: Between ludic tactics and strategic planning. In B. Knudsen, D. Christensen, & P. Blenker (Eds.), *Enterprising initiatives in the experience economy: Transforming social worlds*. New York: Routledge.

Färber, A. (2014). Low-budget Berlin: Towards an understanding of low-budget urbanity as assemblage. *Journal of Regions, Economy and Society, 7*, 119–136.

Ferguson, F. (2014). *Make_shift city: Renegotiating the urban commons*. Berlin: Jovis.

Finn, D. (2014). DIY urbanism: Implications for cities. *Journal of Urbanism, 7*(4), 381–398.

Greco, J. (2012). From pop-up to permanent. *Planning, 78*(9), 15–16.

Harvey, D. (2007). *A brief history of neoliberalism*. Oxford: Oxford University Press.

Haydn, F., & Temel, R. (Eds.). (2006). *Temporary urban spaces: Concepts for the use of city spaces*. Basel: Birkhäuser.

Hou, J. (Ed.). (2001). *Insurgent public space: Guerrilla urbanism and the remaking of contemporary cities*. New York: Routledge.

Johnson, S. (2001). *Emergence: The connected lives of ants, brains, cities and software*. London: Penguin.

Kamel, N. (2014). Learning from the margin: Placemaking tactics. In V. Mukhija & A. Loukaitou-Sideris (Eds.), *The informal American city: Beyond taco trucks and day labor*. Cambridge: MIT Press.

Kamvasinou, K. (2011). The public value of vacant urban land. *Municipal Engineer, 164*(3), 157–166.

Kamvasinou, K., & Roberts, M. (2014). Interim spaces. In P. Barron & M. Mariani (Eds.), *Terrain vague: Interstices at the edge of the pale*. New York: Routledge.

Krauzick, M. (2007). *Zwischennutzung als Initiator einer neuen Berliner Identität?* Berlin: Universitätsverlag der TU Berlin.

Lefebvre, H. (1996). *Writings on cities*. Oxford: Blackwell.

Loukaitou-Sideris, L., & Mukhija, V. (2014). Conclusion: Deepening the understanding of informal urbanism. In V. Mukhija & A. Loukaitou-Sideris (Eds.), *The informal American city: Beyond taco trucks and day labor*. Cambridge: MIT Press.

Lydon, M., & Garcia, A. (2015). *Tactical urbanism: Short-term action for long-term change*. Washington, DC: Island Press.

Lynch, K. (1981). *Good city form*. Cambridge: MIT Press.

Nemeth, J., & Longhorst, J. (2014). Rethinking urban transformation: Temporary uses for vacant land. *Cities, 40*, 143–150.

Oswalt, P., Overmeyer, K., & Misselwitz, P. (2013). *Urban catalyst: The power of temporary use*. Berlin: DOM publishers.

Pagano, C. (2013). DIY urbanism: Property and process in grassroots city building. *Marquette Law Review, 97*(2), 335–389.

Radywyl, N., & Biggs, C. (2013). Reclaiming the commons for urban transformation. *Journal of Cleaner Production, 50*, 159–170.

Richards, G., & Wilson, J. (2006). Developing creativity in tourist experiences: A solution to the serial reproduction of culture? *Tourism Management, 27*, 1209–1223.

Rios, M. (2014). Learning from informal practices: Implications for urban design. In V. Mukhija & A. Loukaitou-Sideris (Eds.), *The informal American city: Beyond taco trucks and day labor*. Cambridge: MIT Press.

Sandercock, L. (2003). *Cosmopolis II: Mongrel cities in the 21st century*. London: Continuum.

SfS Berlin (Ed.). (2007). *Urban pioneers: Temporary use and urban development in Berlin*. Berlin: Jovis.

Solà-Morales, I. (1994). Terrain vague. In C. Davidson (Ed.), *Anyplace*. Cambridge: MIT Press.

Stevens, Q. (2015). Sandpit urbanism. In B. Knudsen, D. Christensen, & P. Blenker (Eds.), *Enterprising initiatives in the experience economy: Transforming social worlds*. New York: Routledge.

Stevens, Q., & Ambler, M. (2010). Europe's city beaches as post-fordist placemaking. *Journal of Urban Design, 15*, 515–537.

Toffler, A. (1980). *The third wave*. New York: Bantam Books.

Tonkiss, F. (2013). Austerity urbanism and the makeshift city. *City, 17*(3), 313–324.

Urban Catalyst. (2001). *Analysis report Berlin study draft*. Berlin: Technische Universität Berlin.

Valverde, M. (2005). Taking land use seriously: Toward an ontology of municipal law. *Law, Text, Culture, 9*, 34–59.

Walker, B., & Salt, D. (2006). *Resilience thinking: Sustaining ecosystems and people in a changing world*. Washington, DC: Island Press.

Final Words

Mahyar Arefi and Conrad Kickert

In closing, we would like to reflect on and summarize the goals of this anthology. Its 19 chapters on the nature, locations, and functionalities of bottom-up urbanism offer far more than the problem-solving title of any handbook typically suggests. The chapters demonstrate that bottom-up urbanism cannot be summarized into best practices or simple formulas, but instead captures far more diverse practices than most urbanisms of the twentieth century. This book offers a plethora of examples that demonstrate both the relevance and the difficulty of defining the interplay between public institutions and local agents. The conventional center-periphery model in which institutions muster legitimacy by regulating physical expansion and development, and citizens represent the peripheral forces that influence spatial decisions in the face of traditional forms of authority, has faded.

Instead, relations between urban stakeholders have become far more complex. Agents do not just follow a center of authority and a periphery of insurgency, but act along a kaleidoscope of DIY urbanism classifications, including temporary and permanent, public and private, authored and anonymous, collective and individual, legal and illegal, old and new, or unmediated and mediated (Iveson 2013). That is why we believe that bottom-up urbanism means different things to different people, and that treating it as a one-size-fits-all approach with preconceived solutions to preordained problems is certainly misleading and erroneous. Thus, in addition to the how to aspect of a handbook in general, with richness both in terms of substance and style, the contributions in this edited volume address four other fundamental questions surrounding bottom-up urbanism: the what, the who, the why, and the when. We believe that addressing these fundamental questions do justice to the variegated and multi-faceted nature of bottom-up movements, providing policymakers, researchers, planners, and designers with a better understanding of their potentials and pitfalls.

© The Editor(s) (if applicable) and The Author(s) 2019
M. Arefi and C. Kickert (eds.), *The Palgrave Handbook of Bottom-Up Urbanism*, https://doi.org/10.1007/978-3-319-90131-2

These related but different questions reflect the planned versus the unplanned dichotomy where the conventional forces of land, labor, and capital adhere to formal planning strategies and singular narratives, rather than the bottom-up forces of informality and insurgency. The chapters in this book define a wide range of bottom-up approaches that move in between, on top, below, and beyond this dichotomy—some authors even argue for a reverse order. Closely related to what constitutes bottom-up urbanism, is the question of identity, or of who gets involved in its formation. Even though the unplanned or informal typically purports the disenfranchised, the contributors in this volume demonstrate bottom-up forces from a wide range of actors, from across a wide social and cultural landscape. Policymakers, planners, and politicians often face questions of social or cultural identity as a quintessential dilemma in dealing with various forms of bottom-up urbanism. The question is, do these types of urbanism have their own unique identities, or does each type of conventional urbanism share similarities with an informal counterpart—anything that defies formality, order, or planning?

Regardless of the identities of those involved in their formation, the manifestations of informality and unplanned-ness, are quite diverse, messy, and unique in terms of their inspirational sources, types of resiliency, adaptive capabilities, and future transformations, and therefore deserve acknowledgment. These aspects typically revolve around the questions that have to do with their modus operandi focusing on how these settlements are not only spatially formed, operated, or managed, but also how they survive.

This brings us to another relevant question: Why do these urbanisms occur in the first place? In the Global North, bottom-up urbanism has often sprouted from a frustration with the urban status quo. Early iterations of bottom-up urbanism were often acts of insurgency against an undemocratic or incompetent system of institutionalized planning. From this perspective, self-initiated interventions can help citizens adapt their urban environment to cope with physical and social conditions that do not fit their needs or lifestyles. Much of the discourse with this perspective has focused on the role of traditionally marginalized urban agents, although Crawford et al's concept of "familiarization" or Hou's Guerilla Urbanism can easily extend to different contexts (Crawford et al. 1999; Hou 2010). Indeed, more recent iterations of bottom-up urbanism have expanded its instigators to the creative class and have increasingly institutional roots. Tacticaland Lean Urbanism both sprouted from the established nest of New Urbanism, and they reflect a fascinating combination of conceptual similarities and paradigmatic frustration with their shared parent. If anything, the power behind these new bottom-up urbanisms runs the risk of delegitimizing institutional planning without a clear democratic alternative, furthering the demise of the role of national and local governments in urban development. Douglas, Stevens, and Dovey rightfully question the motives behind the contemporary surge in bottom-up initiatives in the Global North.

In the Global South, bottom-up urbanism represents less a frustration with a lethargic system, and more a continuing system of survival. Scott Shall provides us with a stark reminder that most urban growth is still informal, and is likely to continue to be so in the foreseeable future. Many citizens in the Global South simply do not have the time, agency, or resources to follow the traditional model of professional construction, and governments do not have the capacity to intervene either. He argues that since Mike Davis' 2006 publication of *Planet of Slums* (2006), the growth of informal settlements is still driven by "the reproduction of poverty." Yet informal urbanism is also rapidly changing shape, as it reflects changing goals. Just because globalization, economic liberalization, and immigration, or even gentrification and the emergence of the creative class have intensified the plight of the disadvantaged and the poor, they do not necessarily come to peace with their fates and have to give in. In many ways, new bottom-up forms of urbanizationincluding DIYand Guerrilla Urbanism, insurgent urbanism in many cases, reflect more than complacency with inequality and instead celebrate innovative ways to combat poverty and segregation, as several chapters in this book have portrayed.

Clearly, the role, power, and forces behind bottom-up urbanism strongly depend on where initiatives take place. This book demonstrates that informal interventions take vastly different shapes in different parts of the world. Within cities, bottom-up initiatives also strongly vary in nature, based on their location, from repurposing abandoned and underutilized spaces in downtowns to suburbs, or from sidewalks and street intersections to street furniture for parkour-like activities, or parklets, and/or markets around local mosques. The biggest difference between continents and urban locations is reflected in who is involved in these bottom-up movements. The stakeholders—including the people, facilitators, planners, designers, NGOs, and so on—involved in bottom-up interventions directly influences their form, reception, and hence their longevity. While informality may have traditionally been perceived as a coping strategy of marginalized citizens, AlSayyad and Eom nuance the misconception that all bottom-up initiatives align with desires of the urban poor, and that top-down planning only reflects the elite. Furthermore, Brain warns against romanticizing bottom-up urbanism as an individual act of self-expression or improvement, calling for a middle ground between citizens and institutional structures that balances customization and the common good. He advocates for design charrettes as a democratic tool to facilitate a mutually beneficial dialogue between institutions, professionals, and citizens, although Shall expressly warns of the power and information imbalances that may persist in this seemingly horizontal arrangement. In the end, most authors agree to a balance between planned and lived city, finding a middle ground between the false but easy binary of top-down versus bottom-up urbanism.

Against this conceptual backdrop, several chapters in this volume aptly address one or more of the questions on the nature and dimensions of bottom-up urbanism this conclusion poses. In an atmosphere of growing mistrust and overwhelming destabilizing international political and economic pressures, Khirfan, for example, discusses four types of reactions to urbanism (the what) in the context of Greater Amman, Jordan. These four reactions (coerced apathy, revolt, subversion, and innovative negotiation) represent the juxtaposition of planning and unplanning processes against the backdrop of a top-down, bottom-up urbanism discourse.

Dittmar and Kelbaugh explore the typologies of lean urbanism that target "small scale and incremental interventions" in contemporary urbanism. As a prime example of these enterprises, pink zones focus on infill development in marginal neighborhoods, where relaxing or lightening local regulations help to redevelop vacant lots, to repurpose dilapidated buildings, or to activate incubator startups—addressing both how and where this kind of bottom-up urbanism can succeed. Unlike their focus on North America, lean urbanism in Africa represents modest means of building and livelihood in order to accommodate to steady population and urbanization growth rates. Coyle's chapter explores how, when, and in what sequence these, so-called, lean strategies can help build resilient places for people that require fewer resources to initiate, incubate, and mature. For example, playfields also serve as water management systems that ensure people's open-source and meaningful participation in the development process.

Danenberg and Haas examine the complementary rather than the competitive dimension of the top-down and bottom-up processes. They provide their definition of bottom-up urbanism, and how it works. Drawing from two case studies on Stockholm and Istanbul, they identify the missing links in this mutual relationship. The discrepancy between bottom-up initiatives and institutional structures leads to surprisingly similar frictions of distrust, inflexibility, and institutional incompatibility in both cities. Arefi and Mohsenian-Rad's chapter revisits another aspect of the bottom-up, top-down urbanism debate. Dwindling public resources, along with growing rates of urbanization, poverty, and natural disasters, have increased the mushrooming of informal/squatter settlements in the Global South. Chapter 14 examines how the disbursement of a physical upgrading World Bank loan to Iran kickstarted the enablement process of informal settlements. Adaptation, formalization, and integration conceptualized the ways in which these, largely bottom-up, efforts were facilitated by top-down policies. This tripartite continuum shows promise in terms of scope and expectations of engaging people with future placemaking strategies.

As Iranian institutions struggle to connect their plans with the lived reality of informal settlement residents, Mehta describes the structure behind the seeming disorder of Indian streets, reminiscent of Jacobs' "organized complexity" (1961). As a complementary element of the urban economy, informal activities on Indian streets present inextricable linkages with the formal economy where, regardless of their social, political, or cultural status, a number of different players and stakeholders negotiate how and where they use and live on the street and how they exercise their rights to the city. This outlook is a far cry from considering the informal economy as a parasite and a burden on the formal economy. The art of negotiation among these actors provides rich answers to the complex relationship between people, place, and politics. Negotiation and engagement also play major roles in the European context. Exploring the nexus between engagement, negotiation, and place-making, Kuppinger addresses not only who is involved in this process, but also where and how such incremental spatial transformations unfold. Faith-based, bottom-up efforts in and around the Salam Mosque complex in the inner city of Stuttgart potently represent how successful spatialities promote vernacular creativity that end up revitalizing a deteriorating part of the city. These efforts include negotiations with the city officials to incrementally energize the mosque's local shops and eateries.

Elsewhere in Germany, families are also successfully finding their place in cities, as described by Ring's Chapter 9 on recent bottom-up housing efforts in Berlin, in which groups of likeminded residents pursue more than temporary real-estate profits from niche markets, and instead, seek to serve other, more long-term purposes. Unlike conventional developments with limited or even negative impacts on urban neighborhoods, these more recent types of grassroots self-made cities offer alternative participatory development models with more efficient ecological and architectural prospects. Ring outlines how this development can take place as a *Baugruppe*, a shared approach that relies heavily on dense, mixed-use forms of neighborhood living, along with customized cost-effective solutions for accommodating change and adaptability over time. Citizens joining forces to adapt the city to their wishes is also the central topic of Serraos' and Asprogerakas' Chapter 11 on green bottom-up urbanism in the new age of austerity. The chapter combines the what and the who of bottom-up urbanism by presenting a typology of organizational models that ranges from the insurgent to the planned, instigated citizens ranging from the marginalized to the multibillionaire. The chapters demonstrate how citizen-led efforts can flourish and complement the European tradition of relatively strong public planning institutions.

The relationship between institutions and citizens is quite different in China, however, where Stefan Al describes the urban village as an endangered expression of bottom-up urbanism in light of the world's largest urbanization drive. Adding 450 million people to its urban population in only 25 years in a very homogeneous manner, these villages are under threat of extinction, suffering from social-cultural stigmas and a lack of political interest. Al demonstrates that these villages have the potential to accommodate growth, while maintaining adaptability to China's rapidly changing urban conditions.

AlSayyad and Eom offer a deeper insight into the dynamic relation between institutions and citizens, arguing that instead of the typical top-down, what they call the top-up characterizes the concentration of capital and power in the development process in many countries. Also, informal interventions are not always instigated by marginalized groups, and can also reflect unauthorized (but usually tolerated) actions of urban elites and developers. Conversely, the bottom-down rather than the bottom-up represents the downward spiral of the plight of the poor. Gated communities and informal settlements epitomize these two discourses where, even in the best cases, improvements for the latter group are far from significant, and hence, downward—especially in the Global South.

Similar to AlSayyad and Eom, several other authors specifically focus on who instigates bottom-up urban interventions, nuancing the overly romantic bottom-up perspectives that reflect democracy, emancipation, and longevity. Hess, on the other hand, directly connects bottom-up actions to top-down planning through a coping perspective. He views the changes made to the pathways surrounding the apartments in suburban Toronto "the art of the weak" in convincing or, even better, imposing their "politics of expropriation." Apartment dwellers who do not own cars adapt to drivable suburban form by cutting through or across fence lines, using private driveway systems as major walking routes, and crossing major arterial roadways where there are no traffic signals or crosswalks. These transgressions are tolerated, if not fully legalized, by the planning authorities and, once done, they are rarely changed.

Like Hess, Shall describes bottom-up urbanism as an inherently insurgent, even extra-legal, type of urban practice that forms the cornerstone of future urbanization. Addressing how bottom-up urbanisms change the profession, Shall argues that cities of the future will materialize with less input from traditional experts such as architects, planners, engineers, or politicians. To respond to the shifting landscape of urbanization, professionals must shift role from author to instigator, or act less like an engineer and more like a guerrilla. Rather than aligning with existing power structures, professionals should aim to improve and intensify the dialogue with and between residents, allowing them to construct and inhabit their own environments, and arming rather than restricting them with expert knowledge.

Hess' and Shall's examples of bottom-up urbanism as acts of insurgency by marginalized citizens are countered by the critical perspectives of Douglas, Stevens and Dovey that question the complacency that many interventions have in furthering gentrification and neoliberal governance. In many ways, bottom-up initiatives in the Global North may inadvertently intensify the inequality and segregation that underlies the informality in the Global South. The absence of adequate government provision of housing, infrastructure, and public amenities in many developing countries may contrast with the incremental, small-scale improvements in Western cities, but they often respond to similar institutional shortcomings.

Within an atmosphere of uncertainty and unpredictability, Stevens and Dovey zoom in on forms, processes, and outcomes of bottom-up interventions from the perspective of market capitalism and neoliberal orthodoxy. Using the concept of assemblage, the authors address the challenges facing the role of T/T urbanism in creating bottom-up placemaking practices that are often analogous or even detrimental to mainstream top-down planning. Furthermore, Douglas demonstrates that many bottom-up interventions are instigated by the urban elite, adapting urban environments to suit their tastes while ignoring questions of equality and democracy.

Similarly, Talen contends that DIY neighborhood planning can play into the hands of nimbyism and receding governance, under the guise of democracy. Instead, she argues for a combination of "incremental, resident-controlled, planner-supported, and politically-enabled" strategies to support bottom-up neighborhoods. The rise of populism, especially in the Global North, and the diminished voice of constituency in local governance and increasing loss of identity and purpose at the local level, have prompted new ways of engaging DIY neighborhoods.

To emerge from this stalemate of self-interest and lack of accountability, Brain calls for a middle ground between the technical language of the top-down and the informal but inclusive practices of the bottom-up. He sees hope in actualizing urban design charrettes as a self-conscious practice of urbanism that can bring the best of both worlds together. In addressing the how question, he explores legitimizing "engagement" rather than seeking the persuasive power of sheer "participation" in bottom-up urbanism. By combining citizen and expert knowledge in frequent feedback loops, charrettes can reflect a symbiosis of social practice and institutional support, a process that is better suited to the established planning systems in the Global North than the uncertainty of urban authority in developing countries.

Mukhija and Loukaitou-Sideris also offer hope for bottom-up urbanism. First, they define the location and nature of bottom-up interventions by tracing the concepts and geographies of informal urbanism in the American context. Informal urbanism manifests in the myriad landscapes of everyday life in American cities, from "sidewalks and street corners to lawns, garage apartments, parking lots, and community gardens." But, in addition to describing where bottom-up interventions take place, they explore how urban designers can engage in these activities that often happen under the radar of formal developments. The economics and political contexts of marginal-neighborhoods promoting "more informal participatory processes" and creative practices (i.e., food carts facilitate formalizing pods for food vendors, or thinking outside the box in creating soft edges of negotiating between two types of spatial transitions) exemplify some of the ways in which urban design can engage in the bottom-up urbanism discourse. This integration between expert knowledge and citizen action can inform a more democratic, equitable, and lasting path forward for bottom-up urbanism.

Many authors have also described the temporal and temporary nature of bottom-up urbanism. While its name mostly refers to its instigators, its nature is strongly linked to short-term interventions which may or may not link to long-term improvements. The when dimension of bottom-up urbanism is paramount, as it embeds and distinguishes this movement from the canon of urban trends that have preceded it. Jeffrey Hou aptly explains that "rather than an overnight sensation," bottom-up movements have formed a constant counter-narrative in American urbanism throughout its history, arguably even starting formal urbanism itself in the nineteenth century. The everyday lifeworld and desires of citizens have consistently informed American urban theory and interventions, from the civic improvement committees that counterbalanced the commercial city in the late nineteenth century, the human ecology of the exploding metropolis described by the Chicago School in the early twentieth century, the ballet of Hudson Street in 1961, New York plazas and Copenhagen's public spaces in the decades after, to the informality of diverse cities that fueled the Everyday Urbanism movement from the late twentieth century onwards. Hou's description of a counter-narrative begs the question of what exactly is countered, especially as the traditional narrative of urban planning has been so diverse and self-contradictory.

In many ways, bottom-up initiatives are merely the frontrunner of institutionalized planning trends—indicators from the general populace or specific groups that their voices need to be heard, and that current urban environments or trends are ripe for change. Some of these initiatives did not align with the current perception of bottom-up urbanism as a piecemeal strategy to cope with rigid urban environments. The civic improvements advocated by local activists in the late nineteenth century were an effort to bring order to the escalating commercial American metropolis, and tied in with the call for hygienic and social improvements that pervaded throughout Europe. While originating from a narrow slice of urban elite, these movements spawned the birth of American city planning, and propelled the profession in Europe from merely structuring urban speculation to organizing urban society (Hall 2002; Wagenaar 2011). Similarly, Jacobs and Alexander's counter-Modern plea for acknowledgment of the city as a system of organized complexity have shifted the profession's notion of agency (Alexander et al. 1977; Christopher 1965; Jacobs 1961), and many of their lessons have been institutionalized and codified in the language of Structuralism in Europe (Habraken and Teicher 1998) and of New Urbanism in North America. David Brain describes in this volume how New Urbanism has informed policies in many large American cities, redefining the profession to focus on "local experience, observation and common sense," tenets that also define many European policies. Simultaneously with Jacobs' plea for observing citizens, Paul Davidoff taught planners to listen to them in his model of advocacy planning (1965), which had a similarly profound impact on American and European planning models to include more diverse citizen feedback.

The city as a social construct has entered the urban planning mainstream over the past decades, as governments understood that aligning their policies to lived experience could increase their efficacy, efficiency, and ultimate lon-gevity. The democracy behind William Whyte and Jan Gehl's studies of the everyday use of Western public spaces, and to Kaplan's studies of natural environments, has been mainly in their acceptance by governments in evi-dence-based policies to improve the urban and natural environment. The evo-lution of technology enables increasingly detailed observations and models of human behavior in cities, informing increasingly tailored urban policies (Ratti and Claudel 2016; Spek et al. 2013). While environmental observation and modeling has provided evidence for more effective urban interventions, its view of urban agency has been mostly passive, with citizens not customizing their environments much beyond moving a public chair (Hillier and Leaman 1973; Whyte1980, 1988). The most recent wave of bottom-up urbanism fea-tured in this volume acknowledges citizens as far more active participants in shaping their environments.

As with previous bottom-up movements, this most recent, and arguably most active, movement is also showing clear signs of preempting a paradigm shift in institutional urban policy. As Hou, Douglas, Stevens, and Dovey describe in this volume, the initial insurgency of informal urban interven-tions in North America in the past decades has become professionalized and institutionalized into the Lean and Tactical Urbanism movements, embraced by an increasing number of local governments and developers, many as con-scious of their wallets, as of appealing to a trend-conscious audience. As described, this maturation falls in line with previous evolutions and is not necessarily undesirable, but many authors in this volume warn that the trans-fer of agency that accompanies this evolution neatly plays into the hands of neoliberal governance and development. As this latest iteration of bottom-up urbanism institutionalizes in the Global North, we may be witnessing a new global esthetic of austerity, topped with a thin veneer of creative class verbi-age. What new bottom-up movement will rebel against this trend, and when?

We are delighted to have a group of unique and world-renowned scholars from various fields on board with this project, who provided a comprehensive discourse surrounding the epistemology and ontology of bottom-up urban-ism. We hope that this project has instigated even more excitement and inter-est for future research. We also hope that the contributions to this anthology will depict a promising trajectory, both in conducting more scholarly research on the role bottom-up urbanism plays in complementing if not shaping our future cities, and in helping policymakers to make better decisions.

References

Alexander, C., Ishikawa, S., & Silverstein, M. (1977). *A pattern language: Towns, buildings, construction*. New York: Oxford University Press.

Christopher, A. (1965). *A city is not a tree*. Paper presented at the Architectural Forum.

Crawford, M., Chase, J., & Kaliski, J. (1999). *Everyday urbanism*. New York: Monacelli Press.

Davidoff, P. (1965). Advocacy and pluralism in planning. *Journal of the American Institute of planners, 31*(4), 331–338.

Davis, M. (2006). *Planet of slums*. London: Verso.

Habraken, N. J., & Teicher, J. (1998). *The structure of the ordinary: Form and control in the built environment*. Cambridge, MA: MIT Press.

Hall, P. (2002). *Cities of tomorrow: An intellectual history of urban planning and design in the 20th century*. Oxford: Blackwell.

Hillier, B., & Leaman, A. (1973). The man-environment paradigm and its paradoxes. *Architectural Design, 78*(8), 507–511.

Hou, J. (2010). *Insurgent public space: Guerrilla urbanism and the remaking of contemporary cities*. New York: Routledge.

Iveson, K. (2013). Cities within the city: Do-it-yourself urbanism and the right to the city. *International Journal of Urban and Regional Research, 37* (3), 941–956.

Jacobs, J. (1961). *The death and life of great American cities*. New York: Random House.

Ratti, C., & Claudel, M. (2016). *The city of tomorrow: Sensors, networks, hackers, and the future of urban life*. New Haven and London, CT: Yale University Press.

Spek, S. C. V. d., Langelaar, C. M. V., & Kickert, C. C. (2013). Evidence-based design: Satellite positioning studies of city centre user groups. *Proceedings of the ICE-Urban Design and Planning, 166*(4), 206–216.

Wagenaar, C. (2011). *Town planning in the Netherlands since 1800: Responses to enlightenment ideas and geopolitical realities*. Rotterdam: 010 Publishers.

Whyte, W. H. (1980). *The social life of small urban spaces*. Washington, DC: Conservation Foundation.

Whyte, W. H. (1988). *City: Rediscovering the center* (1st ed.). New York: Doubleday.

INDEX

CPSIA information can be obtained
at www.ICGtesting.com
Printed in the USA
BVHW04*2137080718
521040BV00022B/83/P